THEORY OF LIFT

Aerospace Series List

Sense and Avoid in UAS: Research and Applications	Angelov	April 2012
Morphing Aerospace Vehicles and Structures	Valasek	April 2012
Gas Turbine Propulsion Systems	MacIsaac and Langton	July 2011
Basic Helicopter Aerodynamics, 3rd Edition	Seddon and Newman	July 2011
Advanced Control of Aircraft, Spacecraft and Rockets	Tewari	July 2011
Cooperative Path Planning of Unmanned Aerial Vehicles	Tsourdos *et al.*	November 2010
Principles of Flight for Pilots	Swatton	October 2010
Air Travel and Health: A Systems Perspective	Seabridge *et al.*	September 2010
Design and Analysis of Composite Structures: With applications to aerospace Structures	Kassapoglou	September 2010
Unmanned Aircraft Systems: UAVS Design, Development and Deployment	Austin	April 2010
Introduction to Antenna Placement & Installations	Macnamara	April 2010
Principles of Flight Simulation	Allerton	October 2009
Aircraft Fuel Systems	Langton *et al.*	May 2009
The Global Airline Industry	Belobaba	April 2009
Computational Modelling and Simulation of Aircraft and the Environment: Volume 1 - Platform Kinematics and Synthetic Environment	Diston	April 2009
Handbook of Space Technology	Ley, Wittmann and Hallmann	April 2009
Aircraft Performance Theory and Practice for Pilots	Swatton	August 2008
Surrogate Modelling in Engineering Design: A Practical Guide	Forrester, Sobester and Keane	August 2008
Aircraft Systems, 3rd Edition	Moir and Seabridge	March 2008
Introduction to Aircraft Aeroelasticity And Loads	Wright and Cooper	December 2007
Stability and Control of Aircraft Systems	Langton	September 2006
Military Avionics Systems	Moir and Seabridge	February 2006
Design and Development of Aircraft Systems	Moir and Seabridge	June 2004
Aircraft Loading and Structural Layout	Howe	May 2004
Aircraft Display Systems	Jukes	December 2003
Civil Avionics Systems	Moir and Seabridge	December 2002

THEORY OF LIFT
INTRODUCTORY COMPUTATIONAL AERODYNAMICS IN MATLAB®/OCTAVE

G. D. McBain,

School of Aerospace, Mechanical, & Mechatronic Engineering
The University of Sydney, Australia

A John Wiley & Sons, Ltd., Publication

Registered office
John Wiley & Sons Ltd, The Atrium, Southern Gate, Chichester, West Sussex, PO19 8SQ, United Kingdom

For details of our global editorial offices, for customer services and for information about how to apply for permission to reuse the copyright material in this book please see our website at www.wiley.com.

Library of Congress Cataloging-in-Publication Data

McBain, G. D. (Geordie G.)
    Theory of lift : introductory computational aerodynamics with MATLAB®/
OCTAVE / G.D. McBain.
        p. cm.
    Includes bibliographical references and index.
    ISBN 978-1-119-95228-2 (hardback)
    1. Lift (Aerodynamics)–Mathematical models.   2. Aerodynamics–Data
processing.   3. MATLAB.   I. Title.
    TL574.L5M33 2012
    629.132′33028553–dc23   2012004735

A catalogue record for this book is available from the British Library.

Print ISBN: 9781119952282

Set in 10/12pt, Times Roman by Thomson Digital, Noida, India
Printed in Malaysia by Ho Printing (M) Sdn Bhd

In memory of Dr Jonathan Harris

Contents

Preface

I found while teaching the course in introductory aerodynamics at The University of Sydney to third-year undergraduate aeronautical engineering students that I was frequently referring to my shelf of classic texts (Glauert 1926; Lamb 1932; Prandtl and Tietjens 1957; Abbott and von Doenhoff 1959; Batchelor 1967; Milne-Thomson 1973; Ashley and Landahl 1985; Katz and Plotkin 2001; Moran 2003), to the extent and effect that the university's Engineering Library's several copies of each of these were constantly out in use by some of the students and therefore unavailable to the rest. Thus it was that I began to collect together a précis of the relevant passages. This immediately necessitated a standardization of nomenclature, which unconsciously evolved into an organization of the elements into something that began to resemble a supporting theoretical framework for the course, something which had begun to cohere. The lectures began to rely on this background, by relegating to it much of the rigour and drudgery of detail in derivation so that they were more freed up to range over appeals to intuition, evocative illustration, and imprecise image and analogy. This combined approach was found to work well. Naturally some students took quickly to the printed notes and only attended the lectures to ask questions, which was a good course for them, as they were usually rapid learners; but the majority used the printed notes more as had been originally intended: these students continued to take the lectures as their main access to the subject, availing themselves of the carefully typeset equations to avoid having to squint at and transcribe so many squiggles and Greek letters on the blackboard.

As the years went by, the course increasingly made use of the proliferation of excellent computing facilities. Since all students had ready access to what were, despite really being only off-the-shelf desktop machines, more powerful digital computers than Abbott and von Doenhoff (1959) could have dreamed of, and since very good interactive software environments for matrix computation had become freely available, it was natural to exploit this. Other introductory textbooks had appeared discussing computational methods (generally in FORTRAN) as successors to the classical approaches (in some cases at the expense of the latter), but the initial approach at Sydney was more first to use computers to speed up those more laborious numerical parts of aerodynamics as already practised and only then, with the time thus saved, to begin to assay such extensions of the old methods as had only become feasible with automatic matrix computation. Then, following this path, it became apparent that the new and old methods had more in common than had been thought. To take a single example, the complex velocity, despite the ultimate simplicity of plane ideal flow in this form, had been dropped from the textbooks of the 1980s, but modern computer languages work as easily with complex variables as real. The concision of the complex-variable panel method presented in Chapter 8 is testament to this. It should be noted that the neglect of the one-to-one analogy between the Cauchy–Riemann equations and the equations of continuity and irrotationality to make way for monolithic FORTRAN programs was not universal, having retained its rightful central place in the French language textbooks (Darrozes and François 1982; Bousquet 1990; Paraschivoiu 1998).

This history is getting very close to giving an idea of this book: an introductory synthesis of theoretical aerodynamics, in a classical spirit, but done as perhaps classical aerodynamics might have been had modern matrix computation techniques and systems been available.

Thus, a few words on what this book is and isn't: it is not a compendium of results, but an introduction to methods and methodologies; and it is not the script to a course of lectures, but an accompanying almost self-contained reference.

It does make many references to the literature, but this is rarely to delegate detail or derivation but more partly out of respect to the authors that went before and moreover out of the firm belief that no twenty-first-century aerodynamicist reading Glauert (1926), Milne-Thomson (1973), the nine chapters of Abbott and von Doenhoff (1959) or anything written by Ludwig Prandtl will be wasting their time—if not amidst the flurry of the third year of an undergraduate degree, subsequently. That is, there is rarely any need to follow any of these pointers, rather the reader is invited to at their later leisure.

An exception to this policy of referencing applies to experimental evidence adduced either in validation of the theoretical models presented or to circumscribe their limits. These references do invoke authoritative results from outside. This book is avowedly theoretical, but aerodynamics is neither primarily theoretical nor primarily experimental, it is engineering and requires both theoretical and experimental wings to bear it aloft. At Sydney, a coextensive course on practical experimental aerodynamical measurement in wind tunnels was given simultaneously with the theoretical course from which this text arose.

While numerous quantitative results for lift, pitching moment, and skin-friction coefficients and similar quantities will be found throughout, these are only incidental: a numerical method necessarily results in numerical output, and comparison with these forms one of the easiest ways of verifying the correctness of an implementation or subsequent modification. Some of the results might have indicative value, but all are derived from models simple enough to fit into an introductory course; judgement on when models are applicable can only follow validation of these results, necessarily involving rigorous comparison with experiment, and very often depending on the details of the particular physical application. Where remarks of some generality on the limits of applicability of the methods discussed can be made, they are, but validation remains unfortunately and notoriously difficult to teach from a textbook; though an exhortation as to its paramount importance in computational aerodynamics is not out of place here—a modeller's every second thought should be of validation.

And, as noted above, while it is as discursive as an introduction should be and a compendium should not be, it is very far from being the transcript of a lecture course. It was and is envisaged as accompanying a lecture course, either as delivered simultaneously, or perhaps following, even years after, a remembered course, when a trained and practising engineer realizes that there are aspects of the theoretical underpinnings that aren't as well understood as they might be, or that they are curious to see how to set about implementing some simple aerodynamical models in the now so readily available interactive matrix computation systems.

Neither does it pretend to teach the design of lifting surfaces. Kuethe and Chow (1998) gave their book *Foundations of Aerodynamics* the subtitle 'Bases of Aerodynamic Design', which was perhaps reasonable, but more and more design involves conflicting requirements and becomes multidisciplinary and specific to domains that are difficult to foresee at this remove.

Given the number of constraints almost any practical project is subject to, it is almost always the case that the designer is pushed onto the corners of limits rather than having the luxury of optimizing within an open subset of the parameter space. This often means operating further from ideal conditions and involving secondary effects, which cannot be adequately treated in an introduction. Another aspect to the application of the theory of lift to design problems is that the ideas originally conceived for wings of aircraft have applications far beyond them, from the blades of wind turbines to Dyson's Air Multiplier^{TM}fans, and it seems unfair to favour one area of application with coverage at the expense of the rest, including of course those not yet thought of.

One of the organizing principles of this book is the combination, for each of the fundamental problems addressed, of

1. the historical development of the subject, including the first successful model,
2. the classical formulation,
3. the simplest possible model that can reproduce the essential physics, and
4. a modern (interactive, scripted, matrix-based) computational model.

Actually in most instances this is a recombination, since the subsequent approaches grew up amidst a knowledge of the history and are by no means such independent alternatives as one might think from some presentations of the subject. We would argue that even if no-one were ever to use Glauert's expansion for the lifting line theory again, for example (though we expect this eventuality remains a good few years off yet), learning it is of more use than merely as an aid to understanding the aerodynamical literature of the twentieth century; on contemporary reinspection, it also turns out to be an example of an orthogonal collocation technique for solving the integral equation, and so in fact when we increase the accuracy of our modern panel methods not merely by spatial refinement but also increase of order, we see that the old and the new are related after all. Our hope is that this uncovering of forgotten relationships will serve to demythologize both the sometimes seemingly arcane old ways and more importantly the contemporary computational methods which can too easily be, and which too often are, trusted blindly as black boxes while a knowledge of the classical foundations of the theory is deemed a luxury.

However, such an organizing principle, which might be suitable for a comprehensive treatise of the type that our subject is perhaps overdue in its current state of development—which one hesitates to call maturity but possibly emergence from infancy—is here always subjugated to the pedagogical imperative. This is a textbook and a broad introduction, and in many cases is much more selective than comprehensive.

It does differ from other introductions though. An endeavour has been made to take each element of the theory that has been presented far enough to enable the student to actually use it to compute some quantity of practical engineering interest, even if the methods thus arrived at are either not exactly in the same form as would be used in practice or if they are only expected to be applied here to somewhat simplified cases. This has been preferred to passive descriptions of, say finite-volume Euler or Reynolds-averaged Navier–Stokes solvers, which while undeniably important in contemporary practical aerodynamics certainly cannot be implemented in an undergraduate course. We firmly believe that engineers understand best by doing, and are persuaded therefore that better users of the state-of-the-art codes will ultimately be bred from students conversant with the mechanics of some small programs than from those whose first introduction to computational aerodynamics is by way of a black box.

As to the interdependence of the sections of the book, the core consists of the plane ideal theory of lift in Chapters 2–6, optionally extended by Chapters 7 and 8 on discrete singularity methods. This core is then modified to take into account three factors: three-dimensionality (Chapters 9–14), viscosity (Chapters 15–17), and compressibility (Chapters 18–19). Although in practice these departures from ideality do interact, here in this introduction only first-order corrections are attempted and so these any of these three modules can be taken in any order after the core. The chapters of each should be taken in sequence but within each, the later chapters may be omitted. Chapters 13 and 14 on three-dimensional discrete vortex methods presuppose mastery of the two-dimensional methods in Chapters 7 and 8.

A note on the included computer programs

A note should be made here on the programs developed and presented in the text. These are not intended to be production programs! They stand in relation to such as the chalkings on a classroom blackboard stand to the real analysis or design calculations of a working engineer: introduced intermediate quantities are not always defined, physical units of measurement are omitted, side-cases are ignored, and checks are entirely absent. These snippets are purely offered as educational illustrations of some of the key methods of computational aerodynamics. To that end, they are necessarily clear and brief rather than robust, general, or even efficient. No effort has been put into such essential aspects of software engineering as modularity, data encapsulation and abstraction, input validation, exception handling, unit testing, integration testing, or even documentation beyond the surrounding discussion in the text which they serve. They do do what they are supposed to, with the inputs given, but very close behind that primary goal of basic functionality have been brevity and clarity, since it is believed that these virtues will best facilitate the conveying of the essential ideas embodied.

To write real aerodynamical computer programs— i.e. those to be saved and used at a later date, perhaps by other than their author, perhaps by users who will not read let alone analyse the source code, and perhaps even for some definite practical end—one requires not just an understanding of the physics, which is what this book was written to provide an introduction to, but also good grounding in numerical analysis and software engineering. The former is touched on here, as there is often some true analogy between the refinement of a numerical approximation and its closer approach to the underlying physics, but the latter is completely neglected as sadly out of scope. Those who hope to write computer programs which will be relied upon (this being perhaps the most significant criterion of merit) must not necessarily themselves master *The Art of Computer Programming* (Knuth 1968, 1969, 1973, 2011) but at least gain an appreciation of software engineering and most often also engage and collaborate with those with more skills in that area. Real programs today, and this will be even more the case in the future, are, as aircraft have been for many decades, composite machines of many subassemblies and subsystems created and put together by many people at different times and separate locations; even the little programs here provide a simple example of this last point in that whereas those listed by Kuethe and Chow (1998), Katz and Plotkin (2001), and Moran (2003) all contained the complete code for solving systems of linear equations, here we are able to avail ourselves of the standard operations provided for this. Whereas William Henson and John Stringfellow might have constructed just about all of their Model of 1848 by hand or at least in their own workshops (Davy 1931), only hobbyists would consider doing this for any aircraft today.

Foremost, we trust that the little snippets served here will be more easily digested than the slabs of FORTRAN to be found in the previous generation of textbooks on aerodynamics. They might serve as points of departure or suggestions, or, without harm, left as mere illustrations of discussions of physics. Those continuing into research into computational aerodynamics will not be served far past their first steps with these training programs, but the hope is that having learned to read, use, and modify these, it will be found easier to use and adapt real programs and to craft the next generation of them.

Acknowledgements

My first debt is to my teachers, in particular the late Dr Jonathan Harris. I am also very grateful to Professors Douglass Auld and Steven Armfield for first inviting me to give the introductory aerodynamical course at Sydney, and then allowing me such scope in adapting the curriculum to make use of contemporary computational systems.

This book was developed by the author entirely using free software, that is, not so much software distributed free of charge, but software distributed along with its source code under licences such as the Free Software Foundation's General Public Licence. The programs used, besides GNU Octave, include LaTeX, GNU Emacs, GNU Make, Asymptote, and matplotlib. The author is deeply appreciative of the skill and time that their several authors, distributed around the world, have put into them, and equally appreciative of the power and elegance of these programs, which now form a most useful part of the common intellectual patrimony of the aeronautical engineering profession.

My last debt is to my readers, beyond the students of aerodynamics at Sydney such as Sujee Mampitiyarachchi, David Wilson, Christopher Chapman, and Thomas Chubb. Here I can only thank the earliest of these, João Henriques, Andreas Puhl, and Pierre-Yves Lagrée, for bringing errors, inconsistencies, ambiguities, and obfuscations to my attention—I have endeavoured to fix these and more, but apologize for the multitude that undoubtedly remain.

G. D. McBain
Dulwich Hill, New South Wales

References

Abbott, I.H. and von Doenhoff, A.E. (1959) *Theory of Wing Sections*. New York: Dover.

Ashley, H. and Landahl, M. (1985) *Aerodynamics of Wings and Bodies*. New York: Dover.

Batchelor, G.K. (1967) *An Introduction to Fluid Dynamics*. Cambridge: Cambridge University Press.

Bousquet, J. (1990) *Aérodynamique: Méthode des singularités*. Toulouse: Cépaduès.

Darrozes, J.S. and François, C. (1982) *Mécanique des Fluides Incompressibles, Lecture Notes in Physics*, vol. 163, New York: Springer.

Davy, M.J.B. (1931) *Henson and Stringfellow, Their Work in Aeronautics*. Board of Education Science Museum. London: His Majesty's Stationery Office.

Glauert, H. (1926) *The Elements of Aerofoil and Airscrew Theory*. Cambridge: Cambridge University Press.

Katz, J. and Plotkin, A. (2001) *Low-Speed Aerodynamics*, 2nd edn. Cambridge: Cambridge University Press.

Knuth, D.E. (1968) *Fundamental Algorithms. The Art of Computer Programming*, vol. 1. Boston: Addison-Wesley.

Knuth, D.E. (1969) *Seminumerical Algorithms, The Art of Computer Programming*, vol. 2. Boston: Addison-Wesley.

Knuth, D.E. (1973) *Searching and Sorting. The Art of Computer Programming*, vol. 3. Boston: Addison-Wesley.

Knuth, D.E. (2011) *Combinatorial Algorithms. The Art of Computer Programming*, vol. 4A. Boston: Addison-Wesley.

Kuethe, A.M. and Chow, C.Y. (1998) *Foundations of Aerodynamics*, 5th edn. Chichester: John Wiley & Sons, Ltd.

Lamb, H. (1932) *Hydrodynamics*, 6th edn. Cambridge: Cambridge University Press.

Milne-Thomson, L.M. (1973) *Theoretical Aerodynamics*, 4th edn. New York: Dover.

Moran, J. (2003) *An Introduction to Theoretical and Computational Aerodynamics*. New York: Dover.

Paraschivoiu, I. (1998) *Aérodynamique Subsonique*. Éditions de l'École de Montréal.

Prandtl, L. and Tietjens, O.G. (1957) *Applied Hydro- and Aeromechanics*. New York: Dover.

Series Preface

The field of aerospace is wide ranging and multi-disciplinary, covering a large variety of products, disciplines and domains, not merely in engineering but in many related supporting activities. These combine to enable the aerospace industry to produce exciting and technologically advanced vehicles. The wealth of knowledge and experience that has been gained by expert practitioners in the various aerospace fields needs to be passed onto others working in the industry, including those just entering from University.

The *Aerospace Series* aims to be a practical and topical series of books aimed at engineering professionals, operators, users and allied professions such as commercial and legal executives in the aerospace industry, and also engineers in academia. The range of topics is intended to be wide ranging, covering design and development, manufacture, operation and support of aircraft as well as topics such as infrastructure operations and developments in research and technology. The intention is to provide a source of relevant information that will be of interest and benefit to all those people working in aerospace.

Aerodynamics is the key science that enables the aerospace industry world-wide – without the ability to generate lift from airflow passing over wings, helicopter rotors and other lifting surfaces, it would not be possible to fly heavier-than-air vehicles, or use wind turbines to generate electricity. Much of the development of today's highly efficient aircraft is due to the ability to accurately model aerodynamic flows and thus design high-performance wings. Although there are many readily available sophisticated CFD codes, a thorough understanding of the fundamental aerodynamics that they are based upon is vital for engineers to comprehend and also verify the results that are obtained from them.

This book, *Theory of Lift: Introductory Computational Aerodynamics in MATLAB®/Octave*, provides an important addition to the Wiley Aerospace Series. Aimed at undergraduates and engineers new to the field, it provides a comprehensive grounding to the fundamentals of theoretical aerodynamics, illustrated using modern matrix based computational techniques. A notable feature of the book is how each element of theory is taken far enough so that students can use it to compute quantities of practical interest using the accompanying computer codes.

Peter Belobaba, Jonathan Cooper, Roy Langton and Allan Seabridge

Part One

Plane Ideal Aerodynamics

1

Preliminary Notions

1.1 Aerodynamic Force and Moment

An aircraft in flight is subject to several forces: gravity causes the weight force; the propulsion provides a thrust, and the air the *aerodynamic force* A.[1]

The central problem of aerodynamics is the prediction of the aerodynamic force; as important is its line of action, or equivalently its moment.

The motion of the aircraft through the air forces the air to move, setting up aerodynamic stresses. In turn, by Newton's Third Law of Motion, the stress in the air is transmitted back across the surface of the aircraft. The stresses include pressure stresses and viscous stresses. The aggregates of the stresses on the surface are the aerodynamic force and moment.

1.1.1 Motion of the Frame of Reference

Newton's equations of motion are unchanged if the frame of reference is replaced with one moving at a constant relative velocity; that is, the aerodynamic force can be computed or measured equally well by an observer in the aircraft in steady flight as by an observer on the ground. In the aeroplane's frame of reference, it's stationary and the air moves at a velocity $-V$.

The equivalence is very useful in aerodynamics; e.g. instead of mounting models of wings on force-measuring apparatus atop an express train as John Stringfellow did in the first half of the nineteenth century, the aerodynamic force can be

- measured in a wind tunnel where a model of the aircraft is held fixed in an air-stream; or
- computed by a numerical solution of the governing equations on a grid fixed to the aircraft surface.

Of course, other factors are involved in the interpretation of wind tunnel data (e.g. the effect of the walls), or computational fluid dynamics (e.g. grid dependence).

---

[1] Herein **bold** denotes a vector, *italic* a variable, and so ***bold italic*** a variable vector.

Theory of Lift: Introductory Computational Aerodynamics in MATLAB®/Octave, First Edition. G. D. McBain.
© 2012 John Wiley & Sons, Ltd. Published 2012 by John Wiley & Sons, Ltd.

1.1.2   Orientation of the System of Coordinates

Various choices of coordinate system in the frame of reference are possible; e.g.

- a geocentric system with coordinates latitude, longitude, and altitude;
- a Cartesian system defined by the instantaneous velocity, curvature, and torsion of the flight; or
- a Cartesian system fixed to the aircraft, moving and rotating with it.

Although all physical results obtained must be independent of the choice of orientation, in aerodynamics we almost always use the last of these, after carefully and explicitly defining it in each case.

Most aircraft have an approximate plane of symmetry naturally dividing the craft into left and right halves. (This symmetry is deliberately broken during independent deployment of the left and right control surfaces.) The reference line is always chosen within this plane, and generally approximately coincides with the long direction of the craft and the usual direction of travel.

Our system of coordinates then consists of:

- the aircraft reference line (x, *longitudinal* or *axial*, positive 'downstream' or 'behind' for the usual direction of flight);
- one axis at right angles to the reference line but still in the plane of symmetry (y, positive in the direction considered up by seated pilots and passengers); and
- a third perpendicular to the plane of symmetry (z, *spanwise*, positive left).

These definitions of 'backwards', 'upwards', and 'leftwards' for increasing x, y, and z give a *right-handed* coordinate system.

1.1.3   Components of the Aerodynamic Force

In simple cases, symmetry considerations imply that the aerodynamic force acts parallel to the plane of symmetry. In such cases, the aerodynamic force A can be resolved into two perpendicular component forces in two different ways.

First, $A = L + D$. The *drag* D is opposed to the direction of motion V of the aircraft (or parallel to the direction of the airstream relative to the aircraft, $-V$). The *lift* L is directed at right-angles to the direction of motion.

Second, we have the simple Cartesian components: $A = A_x \mathbf{i} + A_y \mathbf{j}$; here A_x and A_y are the longitudinal and normal components. The two decompositions are illustrated in Figure 1.1. Remember that in both diagrams, the horizontal x-coordinate corresponds to the (backward) aircraft reference line and is not necessarily the same as horizontal with respect to the ground.

1.1.4   Formulation of the Aerodynamic Problem

The basic task of aerodynamics is to predict the aerodynamic force A, or the lift and drag forces.

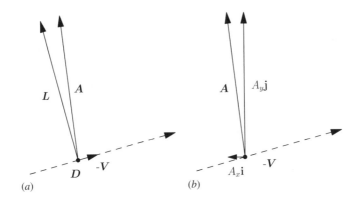

Figure 1.1   Resolving the aerodynamic force into (*a*) lift and drag and (*b*) Cartesian components

Let's see what we can say about this problem without knowing the details of the air flow around the aircraft.

We assume:

- steady flight in a straight line;
- that the air extends to infinity in all directions around the aircraft, and that the air far from the aircraft is otherwise undisturbed.

What then constitutes a specification of the problem?

- We need to know the geometry of the aircraft.
- We need to know the direction and magnitude of the aircraft velocity (or airstream velocity relative to the aircraft).
- We need to know the properties of the air.

These three items are considered in Sections 1.2–1.4. In Section 1.5, we see how *dimensional analysis* can be used to reduce the size of the problem, still without recourse to any detailed fluid mechanics. Finally, in Section 1.6, we look at a real aerodynamical study and see how it incorporates the preliminary notions presented in this chapter.

1.2   Aircraft Geometry

Aircraft have complicated shapes and require a lot of numbers for their complete specification. A number of common terms are in use, e.g. the *span* and *chord* of the wings, but these don't always have standard definitions. For example, Milne-Thomson (1973) lists several definitions of the chord of a two-dimensional wing section:

As a general definition the *chord* of any profile is an arbitrary fixed line drawn in the plane of the profile.

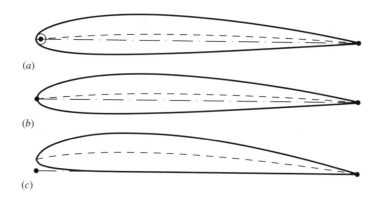

(a)

(b)

(c)

Figure 1.2 Definitions of wing section chord: the line joining the leading edge centre of curvature to the trailing edge (a); the line joining the ends of the camber line (b); and the double-tangent to the lower surface (c). The NACA 2412 aerofoil (a, b) has no double-tangent to the lower surface, so (c) is illustrated for a Clark Y profile. Dashed curve is the camber line and dot-dashed the chord

The chord has direction, position, and length. The main requisite is that in each case the chord should be precisely defined, since the chord enters into the constants which describe the aerodynamic properties of the profile.

The official definition is the line which joins the centres of the circles of curvature of minimum radius at the nose and tail.

Another definition is the longest line which can be drawn to join two points of the profile.

A third definition which is sometimes convenient is the projection of the profile on the double tangent to its lower surface (i.e. the tangent which touches the profile at two distinct points).

This definition fails if there is no such point.

A fourth definition was used by NACA: NACA defined wing sections relative to a given *camber line* (actually a curve) and then defined the chord as the straight line joining its ends. For many wing sections, this will coincide at least approximately with the second of Milne-Thomson's definitions.

The definitions are illustrated in Figure 1.2. Notice that, apart from the double-tangent, the other definitions give similar results; this is true for most typical aerofoils.

All of these are or have been in use, and we can do no better than warn of this and stress that while how the chord is defined is relatively unimportant, it is essential that in each case it be defined clearly and precisely. A similar state of affairs persists with regard to the other dimensions of wings. Here we provide a list of common terms with rough definitions; see also the Glossary and Section 1.8, Further Reading.

1.2.1 Wing Section Geometry

Of particular importance in the study of aerodynamics, as we shall see in Chapters 2–8, are the two-dimensional sections of the wing parallel to the aircraft's plane of symmetry; these are called *wing sections* or *aerofoils*. Wing sections are parameterized by:

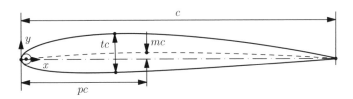

Figure 1.3 NACA 2412 wing section, showing: the camber line (dashed); chord defined as its se-cant (dot-dashed); leading edge circle of curvature (solid); chord length c; maximum thickness ratio t; maximum camber ratio m; and chordwise fractional position of maximum camber p

chord As above (Section 1.2); see Figure 1.2.

maximum camber The *camber line* is the curve lying halfway between the upper and lower surfaces. The maximum camber is its greatest distance from the chord. The 'halfway' can be measured perpendicular to either the chord or the camber line.

chordwise location of maximum camber

maximum thickness measured perpendicular to either the chord or camber line

location of maximum thickness

leading edge radius of curvature

trailing edge radius of curvature or angle if the radius is zero

These concepts are illustrated in Figures 1.2 and 1.3.

1.2.2 Wing Geometry

The wings are parameterized by:

span b, the length of the line joining the tips of the wings, being the points on the wings furthest from the plane of symmetry

planform area the projection of the wings on the spanwise–longitudinal (z–x) plane, including or excluding the area where the wings and fuselage coincide or meet

root chord c_r, the chord where the wings meet the fuselage or $c(0)$, the chord of the wing section in the plane of symmetry (projected, if required)

tip chord $c_t = c(\pm b/2)$ the chord of the wing sections furthest from the plane of symmetry; zero for a delta-wing

average chord $\bar{c}$, the ratio of the planform area to the span

aspect ratio $\mathcal{R}$, the ratio of the span to the average chord

taper ratio the ratio of the tip and root chords; zero for a delta-wing, and typically ≤ 1

sweep angle of the wing backward from the spanwise (z) axis, which is perpendicular to the root chord, can be measured at leading edge, trailing edge, or one-quarter of the way along the chord

which are all properties of the wing's planform; i.e. its projection on the z–x plane. See Figure 1.4.

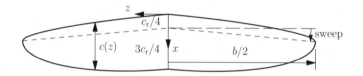

Figure 1.4 Planform of a wing, showing a chord $c(z)$, the root chord c_r, the span b, and the angle of sweep. The plan projection of all chords are parallel. The dashed line joins the quarter-chord points. In this wing, the tip chord is zero

Some additional parameters describe three-dimensional properties:

dihedral angle the upward tilt of the wings relative to the longitudinal–spanwise plane (called *anhedral* if negative)
twist variation of the angle of the chord to the root chord along the span, called *wash in* or *out* as the angle increases or decreases towards the wing tip.

1.3 Velocity

As noted in Section 1.1.2, a reference line for the aircraft must be defined to specify the coordinate system. The usual choice for a whole wing or aircraft is that of the root chord; for a two-dimensional study of a wing section, the section's chord may be used.

For two-dimensional motion (parallel to the plane of symmetry), the angle α between the velocity and the aircraft reference direction is called the *geometric angle of incidence* or *geometric angle of attack*. Note that it depends on the definition of the reference direction. The sign of the angle is that of the y-component of the velocity of the air relative to the aircraft.

In terms of the incidence, and referring to Figure 1.1, the axial and normal components of the aerodynamic force are related to the lift and drag by

$$L = A_y \cos \alpha - A_x \sin \alpha \qquad (1.1a)$$
$$D = A_x \cos \alpha + A_y \sin \alpha. \qquad (1.1b)$$

1.4 Properties of Air

After density, the two most important properties of air in aerodynamics are its compressibility and its viscosity.

1.4.1 Equation of State: Compressibility and the Speed of Sound

The *equation of state* of a material relates its pressure and density. A material resists changes in density by changing its pressure. The relevant material property is the *compressibility*. Using a

Taylor series for the pressure as a function of density for small changes in density about some reference level ρ_0,

$$p(\rho) \approx p(\rho_0) + (\rho - \rho_0)p'(\rho_0)$$
$$\equiv p(\rho_0) + \frac{\rho - \rho_0}{\rho_0}\frac{1}{\kappa} \tag{1.2}$$

where $\kappa \equiv 1/\rho_0 p'(\rho_0)$ is the compressibility. (The Taylor series approximation to a function matches at a given point first the value, then the derative, then the second derivative, and so on.) Sound waves travel through the material with a speed $a = 1/\sqrt{\rho_0\kappa}$. It is often more convenient to use the speed of sound a as the parameter rather than κ, and we will do so here. In terms of a, the first-order Taylor series approximation to pressure–density Equation (1.2) is

$$p(\rho) \approx p(\rho_0) + (\rho - \rho_0)a^2.$$

The Adiabatic Speed of Sound in an Ideal Gas

The compressibility of a gas depends on how quickly the compression occurs; specifically, how the temperature varies, or is allowed to vary, during the compression. If the compression is very fast, there may be insufficient time for heat transfer between the parcel of gas of interest and its surroundings; this *adiabatic* compression applies to the passage of sound waves.

The work done during expansion by a unit mass is $p\delta(1/\rho)$, and the increase in internal energy is $c_v\delta T$. Without heat transfer, the conservation of energy implies that these sum to zero and so $\mathrm{d}T = p\,\mathrm{d}\rho/c_v\rho^2$.

This can be applied to the differential of the thermal equation of state of an ideal gas

$$p = \rho RT, \tag{1.3}$$

which besides the thermodynamic properties of pressure p, temperature T, and density ρ, involves the gas constant R which for air is about 287.0 J/kg K.

$$\mathrm{d}p = R(T\,\mathrm{d}\rho + \rho\,\mathrm{d}T)$$
$$= R(T\,\mathrm{d}\rho + p\,\mathrm{d}\rho/c_v\rho)$$
$$= R(T + p/c_v\rho)\,\mathrm{d}\rho$$
$$\rho\,\mathrm{d}p = pR(1/R + 1/c_v)\,\mathrm{d}\rho$$
$$= p(1 + R/c_v)\,\mathrm{d}\rho$$
$$= p\gamma\,\mathrm{d}\rho$$
$$\kappa = p\gamma$$

where we have defined the dimensionless number $\gamma \equiv 1 + R/c_v$, a property of the gas, which is independent of temperature and pressure insofar as the gas constant R and specific heat c_v are. When we return to this topic in more detail in Section 18.3.5 and Equation (18.9) we will see that γ is the ratio of the isobaric and isochoric specific heat coefficients.

Thus,

$$a = \frac{1}{\sqrt{\rho\kappa}} = \sqrt{\frac{\gamma p}{\rho}} = \sqrt{\gamma RT}, \tag{1.4}$$

Listing 1.1 `speed_of_sound`: compute the speed of sound in air in m/s at a temperature (or array of temperatures) in Kelvin.

```
function a = speed_of_sound (T), a = sqrt (1.4 * T * 287.0);
```

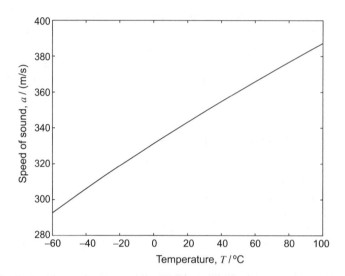

Figure 1.5 Variation of the speed of sound in air with temperature at atmospheric pressures, according to Equation (1.4)

The theoretical value of γ for diatomic gases is $\frac{7}{5}$, and this is a good approximation for air, which, as far as its mechanical properties are concerned, largely consists of diatomic nitrogen and oxygen. Thus the speed of sound depends only on temperature, and at $T = 15°C$ $= 288.15$ K is $a \doteq 340$ m/s. Values at other (absolute) temperatures are conveniently calculated with a one-line Octave function (Listing 1.1), which was used to generate Figure 1.5.

1.4.2 Rheology: Viscosity

The *constitutive law* of a material relates the stress and stain. A material resists deformation by changing its stress. A fluid has no preferred shape, but resists the rate of change of shape.

  Rheology in general is very complicated, but a good model for air is the *linearly viscous fluid*, for which the stress is the sum of an *isotropic* part (the pressure) and a part proportional to the rate of strain. Even for the linearly viscous fluid the constitutive law is complicated, as we shall see in Section 15.2.2. The relevant point here is that the constitutive law introduces another parameter: the coefficient of dynamic viscosity, μ, which has SI units of Pa s (or kg m^{-1} s^{-1}).

Listing 1.2 `viscosity.m`.

```
function m = viscosity (T)
  m = 1.495e-6 * sqrt (T) ./ (1 + 120 ./ T);
```

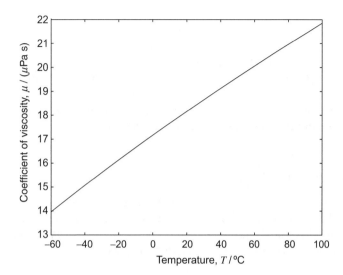

Figure 1.6 Variation of the coefficient of dynamic viscosity of air with temperature at atmospheric pressures, according to Equation (1.5)

Correlating the Viscosity of Air with Sutherland's Law

For sub-hypersonic aerodynamics within the lower parts of the atmosphere, it is usually adequate to take the viscosity of air as depending only on the temperature. A simple correlation is the *Sutherland law*

$$\mu = \frac{S\sqrt{T}}{1 + \frac{C}{T}} \tag{1.5}$$

where T is the absolute temperature and S and C are coefficients determined from correlation with experiments. Reasonable values are $S = 1.495 \ \mu\text{Pa s/K}^{-1/2}$ and $C = 120$ K. This is coded in Octave as shown in Listing 1.2 and used to plot Figure 1.6.

Kinematic Viscosity

It is often more convenient to work with the coefficient of kinematic viscosity defined by

$$\nu = \frac{\mu}{\rho} \tag{1.6}$$

which evidently has SI units m^2/s.

1.4.3   The International Standard Atmosphere

For convenience in comparing data gathered at different times and places, an *International Standard Atmosphere* has been defined. This specifies, among other properties, the pressure p, density ρ, speed of sound a, and coefficients of dynamic μ and kinematic ν viscosity as functions of altitude. These values are widely tabulated, but also easily computed.

The lowest layer of the International Standard Atmosphere is the *troposphere*. It begins at sea level at 15°C and 101.325 kPa and extends upwards to the base of the *tropopause* at 11 km with the temperature decreasing by 6.5 K/km. The next lowest layer, up to 20 km, is the lower part of the *stratosphere*, which is isothermal.

The variation of pressure and density with height is then computed from:

- the ideal gas Equation (1.3), relating pressure p, temperature T, and density ρ; and
- hydrostatic equilibrium, relating pressure, density, and altitude y.

1.4.4   Computing Air Properties

The speed of sound and viscosity can be computed directly from Equations (1.4) and (1.5), respectively, and the temperature, so they're easy. Computing the pressure and density requires the solution of the differential equation of hydrostatic equilibrium – which is essentially what remains of the Euler Equations (2.9) in the absence of velocity:

$$\mathrm{d}p = -\rho g\, \mathrm{d}y. \tag{1.7}$$

Troposphere

The temperature in the standard troposphere is defined to be

$$T(y) = T(0) - \Gamma y, \tag{1.8}$$

where $T(0) = 15°C = 288.15$ K is the standard temperature at sea-level, and $\Gamma = 6.5$ K/km is the standard constant *lapse rate*.

Expressing the density in terms of the ideal gas Equation (1.3) in the hydrostatic balance with $\mathrm{d}y = -\mathrm{d}T/\Gamma$

$$\mathrm{d}p = \frac{pg}{RT\Gamma}\, \mathrm{d}T$$

$$\mathrm{d}\ln p = \frac{g}{\Gamma R}\, \mathrm{d}\ln T$$

$$\frac{p}{p(0)} = \left\{ \frac{T}{T(0)} \right\}^{g/\Gamma R},$$

where $p(0) = 101325$ Pa $= 1$ atm is the standard atmospheric pressure at sea level.

Listing 1.3 `atmosphere.m`.

```
function [p, T, rho, a, mu] = atmosphere (y)

    g = 9.80665; R = 287.0; cpcv = 7/5;
    T0 = 15 + 273.15; p0 = 101325e0;
    L = 6.5e-3; yt = 11e3; top = 20e3;

    troposphere = y <= yt;
    strat = ~troposphere & (y <= top);
    T = NaN (size (y)); p = T;

    T(troposphere) = T0 - L * y(troposphere);
    Ts = T0 - L * yt;
    T(strat) = Ts;

    p(troposphere) = p0 * (T(troposphere) / T0) .^ (g/L/R);
    pt = p0 * (Ts / T0) ^ (g/L/R);
    p(strat) = pt * exp (g/R/Ts * (yt - y(strat)));

    rho = p ./ (R * T);
    a = speed_of_sound (T);
    mu = viscosity (T);
```

Stratosphere

The lower part of the standard stratosphere is isothermal, so the hydrostatic balance reduces to

$$dp = -\rho g \, dy$$

$$= \frac{-p}{RT} g \, dy$$

$$d \ln p = \frac{-g}{RT} dy$$

$$\frac{p}{p_t} = \exp \left\{ \frac{g}{RT} (y_t - y) \right\}.$$

These properties for these two lowest layers of the atmosphere can be computed with an Octave function (`atmosphere.m`, in Listing 1.3), as shown in Figure 1.7.

1.5 Dimensional Theory

Following the considerations in Section 1.1.4, we expect that the aerodynamic force A for steady symmetrical flight in a straight line is a function of the airspeed and incidence, the aircraft geometry, and the air properties. This involves a lot of parameters.

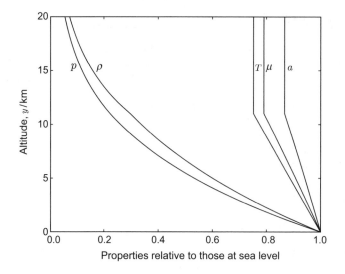

Figure 1.7   Relative variation of the lower International Standard Atmosphere with altitude

First, consider just the two-dimensional flow over a wing section of given shape but variable size parameterized, say, by its chord length c. Then the lift (per unit span) is

$$\ell = f(V, \alpha, c, \rho, \mu, a).$$

This is still a lot of parameters to have to vary in an experimental program. Some further reduction can by achieved using *dimensional theory*.

An equation like 'two metres plus three seconds equals five kilograms' doesn't make any sense. We can only add or compare for equality quantities that have like units. However, we can multiply different physical quantities; e.g.

$$1 \text{ kg} \times 9.8 \text{ m/s} = 9.8 \text{ N}$$

does make sense, since $1 \text{ N} \equiv 1 \text{ kg m/s}^2$.

Say we attempt to correlate the aerodynamic force by a functional form like

$$\ell = \sum k_{pqrstu} V^p \alpha^q c^r \rho^s \mu^t a^u$$

where $p, q, \ldots, u$ are undetermined powers and k_{pqrstu} is a dimensionless coefficient. Then every term in the summation must have the same units as ℓ:

$$\frac{\text{N}}{\text{m}} = (\text{m/s})^p (-)^q (\text{m})^r (\text{kg/m}^3)^s (\text{Pa s})^t (\text{m/s})^u.$$

Reduce all derived units—those except the kilogram, metre, and second—to the fundamental ones using $\text{N} = \text{kg m/s}^2$ and $\text{Pa} = \text{kg/m s}^2$, and equate powers of kg, m, and s. This leads to

$$\frac{\text{kg}}{\text{s}^2} = (\text{m/s})^p(-)^q(\text{m})^r(\text{kg/m}^3)^s(\text{kg/m.s})^t(\text{m/s})^u$$

$$= \frac{\text{m}^p}{\text{s}^p}(-)^q\text{m}^r\frac{\text{kg}^s}{\text{m}^{3s}}\frac{\text{kg}^t}{\text{m}^t\text{s}^t}\frac{\text{m}^u}{\text{s}^u}$$

$$= \text{m}^{p+r-3s-t+u}\text{kg}^{s+t}\text{s}^{-p-t-u}$$

so

$$\text{for kg:} \qquad\qquad\qquad s + t = 1 \qquad\qquad\qquad\qquad (1.9a)$$

$$\text{for m:} \quad p + r - 3s - t + u = 0 \qquad\qquad\qquad (1.9b)$$

$$\text{for s:} \qquad\qquad -u - p - t = -2. \qquad\qquad\qquad (1.9c)$$

This is three equations in five unknowns, so we have two free parameters; say t and u. In terms of them, the solution for the other three is $s = 1 - t$, $p = 2 - t - u$, and $r = 1 - t$. Then the formula for the force is

$$\ell = \sum k_{qtu} V^{2-t-u}\alpha^q c^{1-t}\rho^t a^u$$

$$= \sum k_{qtu} V^2\alpha^q c\rho \left(\frac{\mu}{\rho V c}\right)^t \left(\frac{a}{V}\right)^u$$

where the indices q, t, and u remain arbitrary. Divide through by $\rho V^2 c$ to get

$$\frac{\ell}{\rho V^2 c} = \sum k_{qtu}\alpha^q \left(\frac{\mu}{\rho V c}\right)^t \left(\frac{a}{V}\right)^u.$$

Define

$$C_\ell \equiv \frac{\ell}{\frac{1}{2}\rho V^2 c} \qquad \text{(coefficient of lift)} \qquad\qquad (1.10)$$

$$\text{Re} \equiv \frac{\rho V c}{\mu} \equiv \frac{V c}{\nu} \qquad \text{(Reynolds number)} \qquad\qquad (1.11)$$

$$\text{Ma} \equiv \frac{V}{a} \qquad\qquad \text{(Mach number)}. \qquad\qquad (1.12)$$

These three quantities have no units: i.e. are pure numbers or *dimensionless* quantities. Then

$$C_\ell = 2 \sum k_{qtu}\alpha^q \text{Re}^{-t}\text{Ma}^{-u},$$

so that the lift coefficient must depend only on the incidence and Reynolds and Mach numbers

$$C_\ell = C_\ell(\alpha, \text{Re}, \text{Ma}).$$

Similarly, for the *drag coefficient*

$$C_d = \frac{d}{\frac{1}{2}\rho V^2 c} = C_d(\alpha, \text{Re}, \text{Ma}),$$

where d is the drag per unit span.

In general, *Buckingham's Π theorem* states that a function of n parameters involving d fundamental dimensions (e.g. mass, length, time) can be reduced to a dimensionless function of $n - d$ dimensionless parameters. Here we reduced the lift correlation formula from $n = 6$ parameters to $n - d = 6 - 3 = 3$.

We see that the Reynolds number represents the influence of viscosity, and the Mach number the influence of compressibility.

The same result can also be obtained by *nondimensionalizing* the full equations governing the fluid motion: the *Navier–Stokes equations*.

To consider more general problems (e.g. other wing section shapes) we need more parameters. These should also be dimensionless, for example, quantities with dimensions of length like

- maximum camber
- position of maximum camber
- maximum thickness
- position of maximum thickness

are usually introduced using their ratio to the chord length, as in Figure 1.3.

Another advantage of working with dimensionless quantities is that they have the same value, regardless of whether the underlying dimensional quantities are measured in Imperial, SI, or whatever other system of units.

1.5.1 Alternative methods

The Π theorem can be exploited to derive simpler procedures for finding the dimensionless parameters. A popular one is the Hunsaker–Rightmire method; here we present another, taken from Bradshaw (1964).

For each of the relevant dimensions, e.g. mass, length, and time, choose a quantity involving that dimension; e.g. ρ for mass, V for time, and c for length. Then take each of the remaining quantities and multiply or divide by an appropriate power of each of these in turn to eliminate the dimensions. Thus, writing [=] to indicate dimensional equivalence,

$$\frac{\ell}{\rho}[=]m^3.s^{-2}$$

$$\frac{\ell}{\rho V^2}[=]m$$

$$\frac{\ell}{\rho V^2 c}[=]1,$$

which leads us to the lift coefficient. Operating similarly on μ and a would lead us to the Reynolds and Mach numbers, respectively.

1.5.2 Example: Using Octave to Solve a Linear System

Although the system of three linear Equations (1.9a) in five variables was easy enough to solve by hand, this also makes it ideal for demonstrating the use of Octave for these problems. Much larger systems will be encountered later in the lumped vortex (Chapter 7), panel (Chapter 8), and vortex lattice (Chapter 14) methods and they will really require automatic computation.

First, let us rewrite the three Equations (1.9a) as a single matrix equation.

$$
\begin{bmatrix} 0 & 0 & 1 & 1 & 0 \\ 1 & 1 & -3 & -1 & 1 \\ -1 & 0 & 0 & -1 & -1 \end{bmatrix} \begin{Bmatrix} p \\ r \\ s \\ t \\ u \end{Bmatrix} = \begin{Bmatrix} 1 \\ 0 \\ -2 \end{Bmatrix}
\tag{1.13}
$$

Then *partition* the five variables into those remaining free (*t* and *u*) and those that will be expressed in terms of them (*p*, *r*, and *s*).

$$
\begin{bmatrix} 0 & 0 & 1 \\ 1 & 1 & -3 \\ -1 & 0 & 0 \end{bmatrix} \begin{Bmatrix} p \\ r \\ s \end{Bmatrix} + \begin{bmatrix} 1 & 0 \\ -1 & 1 \\ -1 & -1 \end{bmatrix} \begin{Bmatrix} t \\ u \end{Bmatrix} = \begin{Bmatrix} 1 \\ 0 \\ -2 \end{Bmatrix}
$$

Now, hoping that the square matrix at the left is nonsingular, left-multiply through by its inverse. (If it were singular this step would fail, which would indicate that the partitioning should be reconsidered; this doesn't apply in this example.)

$$
\begin{Bmatrix} p \\ r \\ s \end{Bmatrix} = \begin{bmatrix} 0 & 0 & 1 \\ 1 & 1 & -3 \\ -1 & 0 & 0 \end{bmatrix}^{-1} \left(\begin{Bmatrix} 1 \\ 0 \\ -2 \end{Bmatrix} - \begin{bmatrix} 1 & 0 \\ -1 & 1 \\ -1 & -1 \end{bmatrix} \begin{Bmatrix} t \\ u \end{Bmatrix} \right)
\tag{1.14}
$$

Carrying out the matrix solutions (omitting the details here since it will shortly be redone automatically in Octave using its backslash operator)

$$
\begin{Bmatrix} p \\ r \\ s \end{Bmatrix} = \begin{Bmatrix} 2 \\ 1 \\ 1 \end{Bmatrix} - \begin{bmatrix} 1 & 1 \\ 1 & 0 \\ 1 & 0 \end{bmatrix} \begin{Bmatrix} t \\ u \end{Bmatrix} = \begin{Bmatrix} 2 - t - u \\ 1 - t \\ 1 - t \end{Bmatrix},
\tag{1.15}
$$

which is exactly the answer obtained previously.

The matrix from Equation (1.13) can be set up and then Equation (1.14) executed in Octave quite directly, as shown in Listing 1.4, with output in Listing 1.5.

Listing 1.4   Octave code for solving the 5 × 3 linear system, implementing Equation (1.14) for reducing the number of governing parameters from 5 to 3.

```
A = [ 0,   0,   1,   1,    0;
      1,   1,  -3,  -1,    1;
     -1,   0,   0,  -1,   -1];
bound = 1:size (A, 2) <= size (A, 1);
A(:,bound) \ [1; 0; -2]
A(:,bound) \ A(:,~bound)
```

Listing 1.5 Octave output for Listing 1.4, being the 3 constants and 3×2 coefficients in Equation (1.15).

```
ans =
   2
   1
   1

ans =
   1   1
   1   0
   1   0
```

1.6   Example: NACA Report No. 502

Here we look at an example of an actual wind-tunnel study of the aerodynamic force on a wing.

> Silverstein A 1935 Scale effect on Clark Y airfoil characteristics from NACA full-scale wind-tunnel tests. Report 502, NACA.

It exhibits many of the points noted above.

Noting discrepancies in previously published wind-tunnel data, Silverstein (1935) carried out a new set of force measurements on a wing in a wind-tunnel, using different airstream speeds V, wing sizes (varying chord length c and span b), and geometric angle of incidence α.

Silverstein (1935, table I) defines the geometry of the Clark Y wing section using the dimensionless ordinates of the upper y_U and lower y_L surfaces as functions of the dimensionless distance along the chord. Using this data, reproduced here in Table 1.1, we can reconstruct the profile and the camber line, as in Figure 1.8. We see that:

- The maximum camber is about 6% of the chord length, and occurs about 30% along the chord from leading to trailing edge.
- The third of Milne-Thomson's definitions of chord has been used: a line tangent to the lower surface at two points. Note that if, instead, the chord had been defined as the straight line joining the ends of the camber line, the chord would have a different angle, and so the geometric incidence would be modified for the same airstream.
- The coordinates are all normalized by the chord.

Although the airspeed, chord length, and wing span were all varied, Silverstein enables easy and meaningful comparison between the results of different runs by quoting and presenting them in terms of the dimensionless Reynolds number and coefficient of lift (see e.g. his figure 5). Also, although Silverstein measured airspeed in miles per hour and chord length in feet, the dimensionless numbers are exactly the same as if he had used SI units.

Table 1.1 Upper and lower surface ordinates for the Clark Y aerofoil.

$100x/c$	$100y_U/c$	$100y_L/c$
0	3.50	3.50
1.25	5.45	1.93
2.5	6.50	1.47
5	7.90	0.93
7.5	8.85	0.63
10	9.60	0.42
15	10.68	0.15
20	11.36	0.03
30	11.70	0.00
40	11.40	0.00
50	10.52	0.00
60	9.15	0.00
70	7.35	0.00
80	5.22	0.00
90	2.80	0.00
95	1.49	0.00
100	0.12	0.00

Source: After Silverstein (1935, table I).

Figure 1.8 The Clark Y profile, from Silverstein (1935, table I); the upper nodes are marked with plus signs

1.7 Exercises

1. A wing has span b and root chord c_r, compute the planform area, mean chord, and aspect ratio if the planform is
   (a) elliptic;
   (b) triangular, with zero tip chord;
   (c) trapezoidal, with taper ratio t.

2. Invert Equations (1.1a) and (1.1b) to obtain the Cartesian components.
   Ans.:

$$A_x = D\cos\alpha - L\sin\alpha$$
$$A_y = D\sin\alpha + L\cos\alpha.$$

3. In the Introduction to NACA Report No. 463, Stack (1933) wrote

> The advantages of model testing as an aid to the solution of full-scale problems are often neutralized by the inaccurate reproduction of the full-scale flow in the model test. The conditions which must be fulfilled in the model test so that the results may be directly applicable to the full-scale problem are twofold. First, the model must be geometrically similar to the full-scale object – a condition usually obtained – and second, the model flow pattern must be similar to the full-scale flow pattern – a condition generally not fulfilled. The principal factors that determine flow similarity and the Reynolds nuber $\rho Vl/\mu$ and the compressibility factor V/V_c where V_c is the velocity of sound in the gas.

To what quantity in Section 1.5 does the ratio V/V_c correspond?

4. Consider a wing at $12°$ incidence. The lift and drag coefficients are 1.2 and 0.1092, respectively. Calculate the axial and normal force coefficients. (These figures are from Silverstein 1935, table II.)

5. (a) If Re $= 1.12 \times 10^6$ for a wing section of chord $c = 4$ ft, what must the air-speed be, assuming sea-level air? (Again, these figures are from Silverstein 1935, table II.)
   (b) What would the free-stream Mach number have been?
   (c) If the chord and span were $c = 4$ ft and $b = 24$ ft, the wing-planform was rectangular, and the lift and drag coefficients were as given in Exercise 4, what must the lift and drag forces have been?
   (d) What would the Reynolds number be in water at the same speed?
   (e) What speed in water would give the same Reynolds number?

6. Nondimensionalize the two-dimensional incompressible Navier–Stokes equations for constant density and viscosity:

$$\frac{\partial u}{\partial x} + \frac{\partial v}{\partial y} = 0$$

$$\rho \left(\frac{\partial u}{\partial t} + u \frac{\partial u}{\partial x} + v \frac{\partial u}{\partial y} \right) = -\frac{\partial p}{\partial x} + \mu \left(\frac{\partial^2 u}{\partial x^2} + \frac{\partial^2 u}{\partial y^2} \right)$$

$$\rho \left(\frac{\partial v}{\partial t} + u \frac{\partial v}{\partial x} + v \frac{\partial v}{\partial y} \right) = -\frac{\partial p}{\partial y} + \mu \left(\frac{\partial^2 v}{\partial x^2} + \frac{\partial^2 v}{\partial y^2} \right).$$

If the velocity is normalized by q_∞ and the coordinates by c, how should the pressure be nondimensionalized if the pressure term is to remain finite as the Reynolds number Re $\equiv \rho q_\infty c/\mu$ tends to: (a) 0; (b) ∞.

7. Golf balls (which have a standard minimum diameter of 42.7 mm) can exceed 300 km/h when driven, although recreational players generally achieve less than half this. What are the Reynolds and Mach numbers of the ball under these conditions, near sea level in the standard atmosphere?

8. The normal altitude of Qantas's A380-800 passenger aircraft is 10 700 – 13 100 m. How much do the properties of air vary over this range, and how different are they from those at sea level.

What is the Mach number at the cruising speed of 920 km/h? What is the cruising Reynolds number, based on the mean wing chord length of 10.6 m?

9. How does the coefficient of kinematic viscosity vary with altitude in the International Standard Atmosphere? Graph and tabulate it at each kilometre from sea level up to the top of the isothermal part of the stratosphere, both in absolute terms and as a ratio to the value at sea level.

10. Between 20 and 32 km, the stratosphere has a lapse rate of $\Gamma = -1$ K/km. Extend Listing 1.3 and Figure 1.7 to this height.

11. Reaction Engines Ltd's A2 hypersonic transport is envisaged as cruising at Mach five at an altitude of 100 000 ft. The total length of the craft is 132 m. Using the results of Exercise 10, what would be the A2's air-speed? What would be its Reynolds number (based on the given overall length)?

12. For each of the properties defined by the International Standard Atmosphere, how high can one go above sea-level before it changes by 1%? By 10%? Given a steady cruising altitude of 5 km, what variations in altitude correspond to these same variations in atmospheric properties?

13. Does humidity affect the properties of air? Under atmospheric conditions, water vapour and dry air combine together ideally; i.e. the water vapour and dry air sharing a given volume of space have a common temperature and their ('partial') densities and pressures are what they would be if they were alone, with each behaving as an ideal gas. The total density and pressure is the sum of that for each component. The gas constant for water vapour is about 461.5 J/kg K.

The 'saturation pressure' of water is a function of temperature, roughly

$$\log_{10} \frac{p}{\text{mmHg}} = 8.07131 - \frac{1730.63}{T/\text{K} - 39.724},$$

where 760 mmHg = 101 325 Pa. The partial pressure of water at a given temperature and relative humidity is the saturation pressure times the relative humidity. The partial pressure of dry air, then, is the difference between the total pressure and the partial pressure of water vapour.

Consider humid air at 101.325 kPa, 15°C, and 50% relative humidity. What is the total density? How different is this from the density of dry air at the same temperature and (total) pressure?

14. If the partial pressure of water vapour in humid air exceeds the saturation pressure, the moisture will begin to condense out as droplets.

Say a given volume of humid air at the conditions considered in the last question is raised in altitude through the standard atmosphere, so that the temperature and total pressure fall, while the composition (absolute humidity) is conserved. How does the relative humidity of the parcel vary with height? At what height is saturation achieved?

15. What about the droplets of liquid water in a cloud, do they affect atmospheric properties? Consider a typical cloud at 10 km above sea level consisting of 50 000 000 droplets per cubic metre, each of diameter 10 μm. What mass do these droplets add to each unit volume of air?

1.8 Further Reading

The pioneering aeronautical experiments of Henson and Stringfellow were described by Davy (1931). For an introduction to the use of wind tunnels, see Glauert (1926, chapter 14), Prandtl and Tietjens (1957), Bradshaw (1964, chapter 2), and Liepmann and Roshko (1957).

For frames of reference and how to define the geometric angle of incidence, see Milne-Thomson (1973) or Anderson (2007).

For alternative quantitative descriptions of wings and their sections, see Dommasch *et al.* (1967), Kuethe and Chow (1998), Bertin (2002), Houghton and Carpenter (2003), or Anderson (2007). Details on the NACA wing sections can be found in Abbott *et al.* (1945) and Abbott and von Doenhoff (1959).

The Sutherland law is one of the most common forms of equations used to describe the dependence viscosity on temperature (Glasstone 1946; Montgomery 1947).

The International Standard Atmosphere (ISO 1975), its predecessors, and their importance in aerodynamics are discussed by Glauert (1926), Batchelor (1967), Milne-Thomson (1973), Hoerner and Borst (1985), and Kuethe and Chow (1998); tables may be found in Bertin (2002) and Anderson (2007).

The concept of dimensional analysis is fundamental and therefore difficult to reduce to simpler terms. Once grasped it is obvious but until then it can appear abstract. Such fundamental concepts are best acquired by reading different explanations until the darkness suddenly clears; for dimensional analysis, try Glauert (1926), Hunsaker and Rightmire (1947), Karamcheti (1966), Batchelor (1967), Milne-Thomson (1973), Streeter and Wylie (1983), Kuethe and Chow (1998), or Anderson (2007). All textbooks on aerodynamics or fluid mechanics will contain some explanation of dimensional analysis, not least because of the utility of dynamical similarity. Bertin (2002) demonstrates the approach mentioned of nondimensionalizing the Navier–Stokes equations. For the method of Hunsaker and Rightmire (1947), see Streeter and Wylie (1983) or Anderson (2007); the alternative method of Section 1.5.1 is taken from Bradshaw (1964).

References

Abbott, I.H. and von Doenhoff, A.E. (1959) *Theory of Wing Sections*. New York: Dover.

Abbott, I.H., von Doenhoff, A.E. and Stivers, L.S. (1945) Summary of airfoil data. Report 824, NACA.

Anderson, J.D. (2007) *Fundamentals of Aerodynamics*, 4th edn. New York: McGraw-Hill.

Batchelor, G.K. (1967) *An Introduction to Fluid Dynamics*. Cambridge: Cambridge University Press.

Bertin, J.J. (2002) *Aerodynamics for Engineers*, 4th edn. New York: Prentice Hall.

Bradshaw, P. (1964) *Experimental Fluid Mechanics*. Oxford: Pergamon.

Davy, M.J.B. (1931) *Henson and Stringfellow, their Work in Aeronautics*. Board of Education Science Museum. London: His Majesty's Stationery Office.

Dommasch, D.O., Sherby, S.S. and Connolly, T.F. (1967) *Airplane Aerodynamics*, 4th edn. London: Pitman.

Glasstone, S. (1946) *Textbook of Physical Chemistry*, 2nd edn. Van Nostrand.

Glauert, H. (1926) *The Elements of Aerofoil and Airscrew Theory*. Cambridge: Cambridge University Press.

Hoerner, S.F. and Borst, H.V. (1985) *Fluid-Dynamic Lift*, 2nd edn. Bakersfield, CA: Hoerner Fluid Dynamics.

Houghton, E.L. and Carpenter, P.W. (2003) *Aerodynamics for Engineering Students*, 5th edn. Oxford: Butterworth Heinemann.

Hunsaker, J.C. and Rightmire, B.G. (1947) *Engineering Applications of Fluid Mechanics*. New York: McGraw-Hill.

ISO (1975) 2533:1975 *Standard Atmosphere*. International Organization for Standardization.

Karamcheti, K. (1966) *Principles of Ideal-Fluid Aerodynamics*. Chichester: John Wiley & Sons, Ltd.

Kuethe, A.M. and Chow, C.Y. (1998) *Foundations of Aerodynamics*, 5th edn. Chichester: John Wiley & Sons, Ltd.

Liepmann, H.W. and Roshko, A. (1957) *Elements of Gasdynamics*. Chichester: John Wiley & Sons, Ltd.

Milne-Thomson, L.M. (1973) *Theoretical Aerodynamics*, 4th edn. New York: Dover.

Montgomery, R.B. (1947) Viscosity and thermal conductivity of air and diffusivity of water vapor in air. *Journal of the Atmospheric Sciences* **4**:193–196.

Prandtl, L. and Tietjens, O.G. (1957) *Applied Hydro- and Aeromechanics*. New York: Dover.

Silverstein, A. (1935) Scale effect on Clark Y airfoil characteristics from N.A.C.A. full-scale wind-tunnel tests. Report 502, NACA.

Stack, J. (1933) The N.A.C.A. high-speed wind tunnel and tests of six propeller sections. Report 463, NACA.

Streeter, V.L. and Wylie, B.E. (1983) *Fluid Mechanics*, 1st SI metric edn. New York: McGraw-Hill.

2

Plane Ideal Flow

In Chapter 2 we look at the equations governing two-dimensional flow in aerodynamics. Such a flow is defined by its components u and v with respect to a Cartesian x–y coordinate system in a plane.

The velocity of a flow can be described by giving the velocity of each particle of fluid, but these particles are many and difficult to label and keep track of. Instead of this *material description* it is more convenient to describe the velocity of a flow by the velocity at a given point in space; i.e. to give the velocity as a function of x and y:

$$u = u(x, y)$$
$$v = v(x, y).$$

The functions u and v specify a *velocity field*. This is called the *spatial description* and is used throughout hereafter; nevertheless, it should be remembered that physical laws like those of classical mechanics apply to particles of fluid and not to regions of space.

> By *classical* mechanics here is meant the mechanics embodied in Newton's three laws of motion, for example, rather than the later developments of special and general relativity which blur the distinction between space, time, and the 'bodies' which exist 'in' them, or quantum mechanics which blurs the idea of a 'particle'. Classical mechanics is adequate not merely for the introduction to aerodynamics provided by this book, but well beyond that too, for very nearly all aerodynamics practised today.

Although the velocity here has two components, it is really a single physical thing, and so it is often more apt to refer to it as a vector:

$$q(x, y) \equiv u(x, y)\mathbf{i} + v(x, y)\mathbf{j},$$

where $\mathbf{i}$ and $\mathbf{j}$ are the unit-vectors in the positive x- and y-directions.

2.1 Material Properties: The Perfect Fluid

A *perfect fluid* is a 'a continuous homogeneous medium within which no shearing stresses can exist' (Abbott and von Doenhoff 1959). This definition implies that

Theory of Lift: Introductory Computational Aerodynamics in MATLAB®/Octave, First Edition. G. D. McBain.
© 2012 John Wiley & Sons, Ltd. Published 2012 by John Wiley & Sons, Ltd.

- density is constant (since the material is homogeneous);
- viscosity is zero (since no shearing stresses can exist); and
- fluid stress only produces normal forces on surfaces (i.e. the only stresses are pressure stresses).

Many important problems in aerodynamics can be solved by treating air as a perfect fluid.

2.2   Conservation of Mass

2.2.1   Governing Equations: Conservation Laws

The two fundamental physical requirements that a flow must satisfy are:

- conservation of mass; and
- Newton's Second Law of Motion: force equals mass times acceleration.

2.3   The Continuity Equation

Consider a fixed closed loop $\mathcal{C}$ in the xy-plane. If the loop is filled with perfect fluid, the mass inside must remain constant, since it encloses a fixed area and the fluid density is constant. This means that the net rate of outflow must vanish; i.e. the integral around $\mathcal{C}$ of the component of velocity normal to $\mathcal{C}$ must be zero.

$$\oint_{\mathcal{C}} \boldsymbol{q} \cdot \hat{\boldsymbol{n}} \, \mathrm{d}s = 0. \tag{2.1}$$

A loop is shown in Figure 2.1 with a typical normal element; by convention, loop-integrals are always traced anticlockwise. From the figure, or as $\boldsymbol{q} = u\mathbf{i} + v\mathbf{j}$ and

$$\hat{\boldsymbol{n}} \, \mathrm{d}s = \mathbf{i} \, \mathrm{d}y - \mathbf{j} \, \mathrm{d}x \tag{2.2}$$
$$\boldsymbol{q} \cdot \hat{\boldsymbol{n}} \, \mathrm{d}s = u \, \mathrm{d}y - v \, \mathrm{d}x;$$

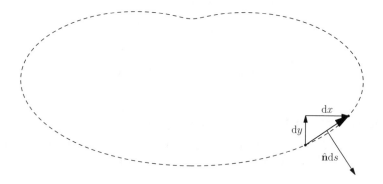

Figure 2.1   Relation between normal loop element and Cartesian coordinates: $\hat{\boldsymbol{n}} \, \mathrm{d}s = \mathbf{i} \, \mathrm{d}y - \mathbf{j} \, \mathrm{d}x$

i.e. the horizontal component of velocity u contributes to the outflow through the segment in proportion to its vertical height dy, and the vertical component in proportion to the horizontal width dx. The signs correspond to the segment of the loop being traced anticlockwise.

By the *divergence theorem*, Equation (2.1) can be converted to an integral over the region $\mathcal{R}$ enclosed by $\mathcal{C}$:

$$\oint_{\mathcal{C}} \boldsymbol{q} \cdot \hat{\boldsymbol{n}} \, ds = \iint_{\mathcal{R}} \nabla \cdot \boldsymbol{q} \, dA$$

$$\text{or} \quad \oint_{\mathcal{C}} (u \, dy - v \, dx) = \iint_{\mathcal{R}} \left(\frac{\partial u}{\partial x} + \frac{\partial v}{\partial y} \right) dA.$$

Since Equation (2.1) has to hold for all closed loops in the fluid, the integrands on the right-hand sides, the *divergence*, must vanish everywhere:

$$\nabla \cdot \boldsymbol{q} \equiv \Theta \equiv \frac{\partial u}{\partial x} + \frac{\partial v}{\partial y} = 0. \tag{2.3}$$

A two-dimensional flow satisfying Equation (2.3) is said to be *divergence-free*.

2.4   Mechanics: The Euler Equations

Newton's Second Law of Motion states that the rate of change of momentum of a particle is equal to the resultant of the forces acting on it.

2.4.1   Rate of Change of Momentum

Consider a particle of perfect fluid occupying a unit volume (and so possessing mass ρ) near (x, y) at t. It moves with the prevailing velocity (u, v) so that a short time δt later the particle has moved by $u\delta t$ in the x-direction and $v\delta t$ in the y-direction. At the new location and time, the prevailing velocity is $(u + \delta u, v + \delta v)$; see Figure 2.2.

Substantial Derivative

Say we have some field $T = T(x, y, t)$ defined in the two-dimensional space and changing in time (think of it as the temperature, say, which is why we've given it the symbol T). If a particle of fluid is at (x, y) at time t and has velocity components u and v there and then, it will

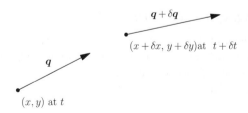

Figure 2.2   The change in position and velocity of a particle in time δt

move by $q\delta t = (u\mathbf{i} + v\mathbf{j})\delta t$; that is, the x-coordinate will change to $x + u\delta t$, the y-coordinate to $y + v\delta t$, and of course the time increases to $t + \delta t$. Therefore, the value of T the particle experiences at $t + \delta t$ is $T(x + u\delta t, y + v\delta t, t + \delta t)$. If δt is short, this is (using the first-order Taylor series in each independent variable)

$$T(x + u\delta t, y + v\delta t, t + \delta t) \sim T + u\frac{\partial T}{\partial x}\delta t + v\frac{\partial T}{\partial y}\delta t + \frac{\partial T}{\partial t}\delta t + O(\{\delta t\}^2)$$

$$= T + \left\{ u\frac{\partial T}{\partial x} + v\frac{\partial T}{\partial y} + \frac{\partial T}{\partial t} \right\}\delta t + O(\{\delta t\}^2)$$

$$\equiv T + \frac{\mathrm{D}T}{\mathrm{D}t}\delta t + O(\{\delta t\}^2), \tag{2.4}$$

where

$$\frac{\mathrm{D}}{\mathrm{D}t} \equiv \frac{\partial}{\partial t} + u\frac{\partial}{\partial x} + v\frac{\partial}{\partial y}$$

is the *substantial derivative*. It's the derivative following the particle, since from Equation (2.4)

$$\lim_{\delta t \to 0} \frac{T(x + u\delta t, y + v\delta t, t + \delta t) - T(x, y, t)}{\delta t} = \frac{\mathrm{D}T}{\mathrm{D}t}. \tag{2.5}$$

Substantial Derivative of the Momentum Field

The acceleration field then, being the acceleration of the fluid particles, at a point and instant, is given by

$$\frac{\mathrm{D}q}{\mathrm{D}t}$$

with components

$$\frac{\mathrm{D}q}{\mathrm{D}t} = \frac{\mathrm{D}u}{\mathrm{D}t}\mathbf{i} + \frac{\mathrm{D}v}{\mathrm{D}t}\mathbf{j}.$$

Since mass is conserved, the rate of change of momentum is just ρ times this; in vector form

$$\rho\frac{\mathrm{D}q}{\mathrm{D}t}.$$

2.4.2   *Forces Acting on a Fluid Particle*

In addition to any extraneous forces (f_x, f_y) per unit mass, a fluid particle experiences forces acting across its surface from the stresses in the neighbouring fluid. For a perfect fluid, these are due entirely to the pressure P, as in Figure 2.3.

The resultant (per unit volume) is

$$-\frac{\partial P}{\partial x}\mathbf{i} - \frac{\partial P}{\partial y}\mathbf{j} = -\nabla P.$$

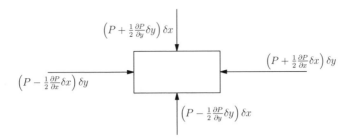

Figure 2.3   Pressure forces acting on an infinitesimal rectangle

2.4.3   The Euler Equations

Combining the results for the rate of change of momentum and the forces, Newton's Second Law of Motion for a unit volume of perfect fluid takes the form

$$\rho\frac{D\boldsymbol{q}}{Dt} = -\nabla P + \rho\boldsymbol{f} \qquad (2.6)$$

or

$$\rho\left(\frac{\partial u}{\partial t} + u\frac{\partial u}{\partial x} + v\frac{\partial u}{\partial y}\right) = -\frac{\partial P}{\partial x} + \rho f_x \qquad (2.7a)$$

$$\rho\left(\frac{\partial v}{\partial t} + u\frac{\partial v}{\partial x} + v\frac{\partial v}{\partial y}\right) = -\frac{\partial P}{\partial y} + \rho f_y. \qquad (2.7b)$$

These are known as either the *Euler equation* or the *Euler equations*, depending on whether it is the vector or component nature that is to be emphasized. The equation never stands alone but always with a corresponding equation for the conservation of mass.

2.4.4   Accounting for Conservative External Forces

If the external force per unit volume per unit span can be expressed as the gradient of a scalar *potential*:

$$\rho f_x = -\frac{\partial \Phi}{\partial x}$$

$$\rho f_y = -\frac{\partial \Phi}{\partial y},$$

the force terms of Euler's Equations (2.7), can be subsumed in the pressure terms:

$$-\frac{\partial P}{\partial x} + \rho f_x = -\frac{\partial P}{\partial x} - \frac{\partial \Phi}{\partial x} = -\frac{\partial}{\partial x}(P + \Phi) = -\frac{\partial p}{\partial x}$$

$$-\frac{\partial P}{\partial y} + \rho f_y = -\frac{\partial P}{\partial y} - \frac{\partial \Phi}{\partial y} = -\frac{\partial}{\partial y}(P + \Phi) = -\frac{\partial p}{\partial y},$$

introducing the *aerodynamic pressure*

$$p \equiv P + \Phi. \tag{2.8}$$

Thus, provided all external forces are conservative, Euler's equations can be expressed as if there were no external forces so long as the pressure is interpreted as the aerodynamic pressure:

$$\rho \left(\frac{\partial u}{\partial t} + u \frac{\partial u}{\partial x} + v \frac{\partial u}{\partial y} \right) = -\frac{\partial p}{\partial x} \tag{2.9a}$$

$$\rho \left(\frac{\partial v}{\partial t} + u \frac{\partial v}{\partial x} + v \frac{\partial v}{\partial y} \right) = -\frac{\partial p}{\partial y}. \tag{2.9b}$$

Gravity in a Perfect Fluid is Conservative

The most important application of the foregoing is the elimination of gravity from Euler's equations for a perfect fluid. If the density ρ is uniform, and say the y-axis is aligned vertically up, the weight on a unit mass of fluid has components

$$f_x = 0$$
$$f_y = -g$$

which can be expressed in terms of the *gravitational potential* $\Phi \equiv \rho g y$, as obtained by integrating the hydrostatic equilibrium Equation (1.7):

$$f_x = 0 = -\frac{1}{\rho} \frac{\partial (\rho g y)}{\partial x}$$

$$f_y = -g = -\frac{1}{\rho} \frac{\partial (\rho g y)}{\partial y},$$

or in vector form as

$$\boldsymbol{f} = -g\boldsymbol{j} = -\frac{1}{\rho} \nabla (\rho g y).$$

The true pressure is recovered from the aerodynamic pressure using Equation (2.8): $P = p - \rho g y$.

Hereafter, gravity will be ignored, as is usual in aerodynamics.

2.5   Consequences of the Governing Equations

2.5.1   The Aerodynamic Force

Given Euler's equations, the aerodynamic force on an object can be expressed as a surface-integral.

Say we have an impermeable two-dimensional obstacle occupying region $\mathcal{R}$ bounded by a closed curve $\mathcal{C}$ and held in place by an external force exactly balancing the aerodynamic force.

Since C is impermeable, the velocity throughout $\mathcal{R}$ can be taken as zero. Then integrate the vector form of Euler's momentum Equation (2.6) over $\mathcal{R}$:

$$\rho \iint_{\mathcal{R}} \frac{Dq}{Dt}\, dA = -\iint_{\mathcal{R}} \nabla p\, dA + \rho \iint_{\mathcal{R}} f\, dA$$

$$0 = -\oint_{C} p\hat{n}\, ds + \rho \iint_{\mathcal{R}} f\, dA.$$

Here we have used the fact that $q = 0$ throughout $\mathcal{R}$ to eliminate the left-hand side, and the formula for the area-integral of a gradient, a.k.a. the *gradient theorem*:

$$\iint_{\mathcal{R}} \nabla p\, dA = \oint_{C} p\hat{n}\, ds$$

to convert the pressure term to a loop-integral.

The external force must be the integral over $\mathcal{R}$ of ρf; thus the longitudinal and normal components of its opposite, the aerodynamic force (per unit span), are, recalling from Equation (2.2) that $\hat{n}\, ds = i\, dy - j\, dx$:

$$a = -\rho \iint_{\mathcal{R}} f\, dA = -\oint_{C} p\hat{n}\, ds.$$

or

$$a_x \equiv -\rho \iint_{R} f_x dA = -\oint_{C} p i \cdot \hat{n}\, ds \qquad = -\oint_{C} p\, dy \qquad (2.10a)$$

$$a_y \equiv -\rho \iint_{R} f_y dA = -\oint_{C} p j \cdot \hat{n}\, ds \qquad = +\oint_{C} p\, dx. \qquad (2.10b)$$

The Cartesian components, Equations (2.10), can then be combined according to Equations (1.1a) and (1.1b) to get the lift and drag.

Integrals along the Chord

These integrals in Equations (2.10) around the surface of the wing section can be expressed as integrals along the chord; i.e. with respect to x. Consider first the normal component of aerodynamic force per unit span Equation (2.10b). Beginning the anticlockwise loop at the trailing edge $x = c$, the integration passes over the upper surface as x decreases from c to 0, then the integration passes along the lower surface as x increases from 0 to c again:

$$a_y = +\oint_{C} p\, dx$$

$$= +\int_{c}^{0} p_U\, dx + \int_{0}^{c} p_L\, dx.$$

Then swap the limits of integration on the upper surface integral and change its sign:

$$a_y = -\int_0^c p_U \, dx + \int_0^c p_L \, dx$$

$$= \int_0^c (p_L - p_U) \, dx.$$

This has the physical interpretation that it's not so much either the lower pressure that pushes or the upper suction (negative pressure) that pulls the wing in the normal y direction, but their sum, or the excess of the lower pressure over the upper pressure.

Similarly, after first changing the integration variable in Equation (2.10a):

$$a_x = -\oint_C p \, dy = -\oint_C p \frac{dy}{dx} \, dx,$$

the axial component of force per unit span can be written as

$$a_x = \int_0^c \left\{ \left(p \frac{dy}{dx} \right)\Big|_U - \left(p \frac{dy}{dx} \right)\Big|_L \right\} \, dx.$$

Nondimensionalization and the Pressure Coefficient

Since the curve C is closed, the integral of any constant around it with respect to either coordinate vanishes:

$$\oint dy = 0.$$

Thus we can integrate instead of the pressure its excess over the value in the far-stream:

$$a_x = -\oint (p - p_\infty) \, dy$$

$$a_y = \oint (p - p_\infty) \, dx.$$

For a wing section of chord c and airspeed $q_\infty \equiv |-\mathbf{V}|$, this is conventionally nondimensionalized by dividing by the *dynamic pressure* $\rho q_\infty^2 c/2$ and introducing the dimensionless *pressure coefficient*

$$C_p \equiv \frac{p - p_\infty}{\frac{1}{2}\rho q_\infty^2} \tag{2.11}$$

to give

$$\frac{a_x}{\frac{1}{2}\rho q_\infty^2 c} = -\oint \frac{p - p_\infty}{\frac{1}{2}\rho q_\infty^2} \, d\left(\frac{y}{c}\right) = \int_0^1 \left[C_p \frac{d(y/c)}{d(x/c)} \right]_L^U \, d\left(\frac{x}{c}\right)$$

$$\frac{a_y}{\frac{1}{2}\rho q_\infty^2 c} = \oint \frac{p - p_\infty}{\frac{1}{2}\rho q_\infty^2} \, d\left(\frac{x}{c}\right) = \int_0^1 [C_p]_U^L \, d\left(\frac{x}{c}\right).$$

2.5.2   Bernoulli's Equation

Bernoulli's equation is derived from Euler's equations. It gives a very useful expression for the pressure.

As a fluid particle moves along its trajectory, $\delta x = u\delta t$ and $\delta y = v\delta t$. Therefore, eliminating δt along a trajectory,

$$v\delta x = u\delta y. \tag{2.12}$$

A velocity field independent of time is called *steady*. Multiplying the steady Euler Equations (2.9),

$$\rho\left(u\frac{\partial u}{\partial x} + v\frac{\partial u}{\partial y}\right) = -\frac{\partial p}{\partial x}$$

$$\rho\left(u\frac{\partial v}{\partial x} + v\frac{\partial v}{\partial y}\right) = -\frac{\partial p}{\partial y},$$

by δx and δy, adding, and using Equation (2.12), we obtain

$$\mathrm{d}p + \rho\,\mathrm{d}\left(\frac{1}{2}q^2\right) = 0, \tag{2.13}$$

where

$$q \equiv \sqrt{u^2 + v^2}$$

is the magnitude of velocity, or the speed. For a perfect fluid (using the constancy of density), this reduces to

$$p + \frac{\rho q^2}{2} = \text{const.} \tag{2.14}$$

Note that Equation (2.14)

- applies only to perfect fluids (constant density, no shearing stress);
- applies only along a fluid trajectory;
- applies only to steady flows (u and v independent of time); and
- applies only in the absence of external body forces, or when all external body forces are conservative.

2.5.3   Circulation, Vorticity, and Irrotational Flow

The *circulation* Γ around a circuit is defined as the negative of the integral of the anticlockwise tangential component of velocity around the circuit:

$$\Gamma \equiv -\oint_C \boldsymbol{q} \cdot \hat{\boldsymbol{\tau}}\,\mathrm{d}s \tag{2.15}$$

$$= -\oint u\,\mathrm{d}x - \oint v\,\mathrm{d}y.$$

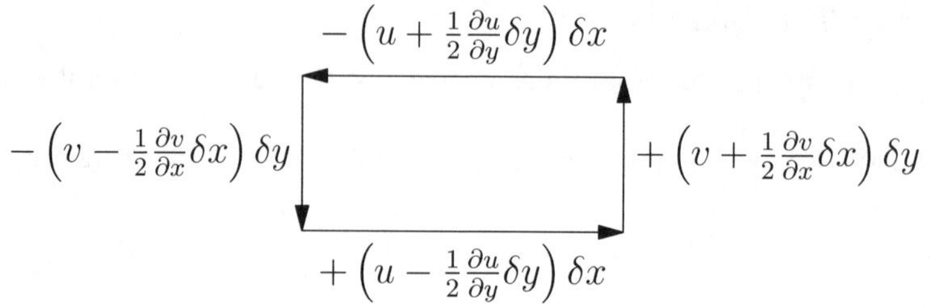

$$-\left(u + \tfrac{1}{2}\tfrac{\partial u}{\partial y}\delta y\right)\delta x$$

$$-\left(v - \tfrac{1}{2}\tfrac{\partial v}{\partial x}\delta x\right)\delta y \qquad\qquad +\left(v + \tfrac{1}{2}\tfrac{\partial v}{\partial x}\delta x\right)\delta y$$

$$+\left(u - \tfrac{1}{2}\tfrac{\partial u}{\partial y}\delta y\right)\delta x$$

Figure 2.4 Tangential components of flow around an infinitesimal anticlockwise loop; their sum is minus the circulation, and proportional to the local vorticity

Here $\hat{\boldsymbol{\tau}}$ is the unit vector tangent to the loop (taken anticlockwise), with (see Figure 2.1 again)

$$\hat{\boldsymbol{\tau}}\, ds = \mathbf{i}\, dx + \mathbf{j}\, dy.$$

Any region can be divided by a grid into rectangular cells. The circulation around the region is the sum of the circulation around each cell. Letting the grid size become infinitesimal, we find that the circulation in each cell is (Figure 2.4)

$$\delta\Gamma = -\zeta\,\delta x\,\delta y$$

where

$$\zeta \equiv \frac{\partial v}{\partial x} - \frac{\partial u}{\partial y} \tag{2.16}$$

is the *vorticity*. It follows that for an area $\mathcal{R}$ with boundary curve $\mathcal{C}$,

$$\Gamma_{\mathcal{C}} = -\iint_{\mathcal{R}} \zeta\, dx\, dy.$$

An equation governing the vorticity in the flow can be obtained by cross-differentiating and subtracting the vector-components of the Euler equations (without external body forces). Using the conservation of mass Equation (2.3), we obtain

$$\frac{\partial\zeta}{\partial t} + u\frac{\partial\zeta}{\partial x} + v\frac{\partial\zeta}{\partial y} = 0.$$

This is the same as

$$\frac{D\zeta}{Dt} = 0,$$

which says that the vorticity of a fluid particle doesn't change during a two-dimensional flow, and implies that if a particle ever has zero vorticity, it has zero vorticity for all time; i.e. is *irrotational*. This applies to the undisturbed air that an aircraft is flying into or the air in a uniform stream. Thus irrotational flow is of great relevance to aerodynamics.

2.5.4 Plane Ideal Flows

We are particularly interested in two-dimensional flows satisfying

$$\Theta \equiv \frac{\partial u}{\partial x} + \frac{\partial v}{\partial y} = 0 \tag{2.17a}$$

$$\zeta \equiv \frac{\partial v}{\partial x} - \frac{\partial u}{\partial y} = 0. \tag{2.17b}$$

Any flow satisfying these two equations satisfies Euler's equations (with no external body force), and is called an *ideal* flow. The associated pressure is

$$p = \text{const.} - \frac{1}{2}\rho q^2, \tag{2.18}$$

as may be easily verified by substitution.

Note that Equation (2.18) resembles Bernoulli's Equation (2.14) but does not suffer the restriction to particle trajectories; i.e. the constant in Equation (2.14) varies from streamline to streamline, but in Equation (2.18) is the same for the whole flow.

It is not true that all solutions of Euler's equations satisfying Equation (2.17a) satisfy Equation (2.17b), but those that do are of particular interest in aerodynamics as explained above.

2.6 The Complex Velocity

In Section 2.5.4 it was shown that plane ideal flows satisfying Equations (2.17a) and (2.17b) satisfy the continuity and Euler equations for a perfect fluid. In Section 2.6, we discover a connection with the theory of complex variables that provides many solutions to the flow equations.

2.6.1 Review of Complex Variables

A few useful properties of complex variables are collected here for convenience. These are all standard formulae.

- An imaginary number is formed by multiplying a real number y by the imaginary unit i, a constant with the special property

$$i^2 = -1. \tag{2.19}$$

- A complex number z is formed by adding a real number x to an imaginary number iy: $z = x + iy$. This gives a one-to-one correspondence between points of the plane and complex numbers; the complex number $x + iy$ is called the *complex coordinate* of the point (x, y).
- We define, for $z = x + iy$:

$$\Re z \equiv x \qquad \text{(real part), and}$$
$$\Im z \equiv y \qquad \text{(imaginary part).}$$

- Complex numbers can be added and subtracted by adding and subtracting the real and imaginary parts.

$$(x_1 + iy_1) \pm (x_2 + iy_2) = (x_1 \pm x_2) + i(y_1 \pm y_2).$$

- Complex numbers can be multiplied according to the usual rules of algebra, augmented by Equation (2.19):

$$(x_1 + iy_1)(x_2 + iy_2) = x_1 x_2 + i(x_1 y_2 + y_1 x_2) + i^2 y_1 y_2$$
$$= (x_1 x_2 - y_1 y_2) + i(x_1 y_2 + y_1 x_2).$$

- Complex numbers are equal if and only if their real and imaginary parts are both equal.
- The polar coordinates in the plane are related to the Cartesian by

$$x = r \cos \theta \qquad\qquad\qquad y = r \sin \theta \qquad\qquad (2.20a)$$

$$r = \sqrt{x^2 + y^2} \qquad\qquad\qquad \theta = \arctan \frac{y}{x}. \qquad\qquad (2.20b)$$

In terms of the polar coordinates,

$$z = x + iy = r \cos \theta + ir \sin \theta.$$

- De Moivre's Theorem:

$$e^{i\theta} \equiv \cos \theta + i \sin \theta. \qquad\qquad (2.21)$$

- Therefore a complex number can also be expressed in *polar form* as:

$$z = x + iy = re^{i\theta}.$$

- Multiplication is easier in polar form:

$$r_1 e^{i\theta_1} r_2 e^{i\theta_2} = r_1 r_2 e^{i(\theta_1 + \theta_2)}.$$

- The *modulus* (or *magnitude* or *absolute value*) of a complex number is defined by

$$\mathrm{mod}\, z \equiv |z| = |re^{i\theta}| = r.$$

- The *argument* (or *phase*) of a complex number is defined by

$$\arg z \equiv \arg(re^{i\theta}) = \theta.$$

The value of the argument is nonunique in that any multiple of 2π can be added, since the cosine and sine functions are periodic.
- The *complex conjugate* z^* of a complex number z has the same real part and opposite imaginary part, and same magnitude but opposite phase:

$$z^* \equiv \Re z - i\Im z = |z|e^{-i\,\arg z}.$$

- We have:

$$z + z^* = 2\Re z$$
$$z - z^* = i2\Im z$$
$$zz^* = |z|^2.$$

The last of these results is useful for rendering the denominator of a complex fraction real:

$$\frac{z_1}{z_2} = \frac{z_1 z_2^*}{|z_2|^2}.$$

Analytic Functions

A complex function of a complex variable is said to be *analytic* if it possesses a well-defined derivative. Say

$$f(z) \equiv f(x + iy) \equiv g(x, y) + ih(x, y),$$

where $g(x, y) \equiv \Re f(z)$ and $h(x, y) \equiv \Im f(z)$. Then the differential of $f(z)$ is

$$df = \frac{\partial g}{\partial x}dx + \frac{\partial g}{\partial y}dy + i\left(\frac{\partial h}{\partial x}dx + \frac{\partial h}{\partial y}dy\right),$$

the differential of z is

$$dz = dx + idy,$$

and the derivative of $f(z)$ is

$$\frac{df(z)}{dz} = \frac{\frac{\partial g}{\partial x}dx + \frac{\partial g}{\partial y}dy + i\left(\frac{\partial h}{\partial x}dx + \frac{\partial h}{\partial y}dy\right)}{dx + idy}.$$

When $dy = 0$, so that $dz = dx$, this becomes

$$\frac{df(z)}{dz} = \frac{\partial g}{\partial x} + i\frac{\partial h}{\partial x},$$

and when $dx = 0$, so that $dz = idy$, it becomes

$$\frac{df(z)}{dz} = \frac{\frac{\partial g}{\partial y} + i\frac{\partial h}{\partial y}}{i} = \frac{\partial h}{\partial y} - i\frac{\partial g}{\partial x}.$$

For the derivative of f to be well defined, these two expressions must be equal. Equating the real and imaginary parts then gives

$$\frac{\partial g}{\partial x} = \frac{\partial h}{\partial y} \tag{2.22a}$$

$$\frac{\partial h}{\partial x} = -\frac{\partial g}{\partial y}. \tag{2.22b}$$

These equations are the condition that a complex function be analytic; they are called the *Cauchy–Riemann* equations.

All the elementary functions of z (powers of z, trigonometric functions, exponential function, and hyperbolic functions) have the usual derivatives, e.g. $dz^k/dz = kz^{k-1}$, $de^z/dz = e^z$. These are all analytic, except at points where the derivatives are undefined; e.g. for $1/z$, the derivative is

$$\frac{d}{dz}\frac{1}{z} = \frac{-1}{z^2}$$

which is undefined at $z = 0$. Isolated points where an otherwise analytic function doesn't have a derivative are called *singularities*.

2.6.2 Analytic Functions and Plane Ideal Flow

Consider the analytic function

$$w(z) = u(z) - iv(z).$$

The Cauchy–Riemann Equations (2.22) require

$$\frac{\partial u}{\partial x} = -\frac{\partial v}{\partial y}$$

$$\frac{\partial v}{\partial x} = \frac{\partial u}{\partial y},$$

which are seen to be identical to the requirements that the plane velocity field with components u and v be divergence-free and irrotational; i.e. satisfy Equation (2.17a) and Equation (2.17b), respectively. Therefore, any analytic function defines a solution of the two-dimensional continuity and Euler equations. We call $w = u - iv$ the *complex velocity*.

The Polar Form of the Complex Velocity: Speed and Direction

For a complex velocity field $w(z) = u - iv$, the speed is given by

$$q(z) = |w| = \sqrt{u^2 + v^2}. \tag{2.23}$$

The direction of the velocity at a point, expressed as its slope, is

$$\frac{v}{u} = \tan(-\arg w).$$

Thus the complex velocity can also be expressed as

$$w(z) = u - iv = qe^{-i\arctan v/u} \equiv qe^{-i\beta},$$

where $\beta(z) \equiv -\arg w$ is the angle of the velocity at the point z to the positive x-axis.

General Component

For any angle χ, the vector $\mathbf{i}\cos\chi + \mathbf{j}\sin\chi$ has unit magnitude and is at an angle χ anticlockwise from the positive x-axis. Therefore, the component of a vector $\mathbf{q} = u\mathbf{i} + v\mathbf{j}$ in this direction is

$$(\mathbf{i}\cos\chi + \mathbf{j}\sin\chi)\cdot\mathbf{q} = (\mathbf{i}\cos\chi + \mathbf{j}\sin\chi)\cdot(u\mathbf{i} + v\mathbf{j})$$
$$= u\cos\chi + v\sin\chi.$$

Now consider the product

$$e^{i\chi}w = (\cos\chi + i\sin\chi)(u - iv)$$
$$= u\cos\chi + v\sin\chi + i(u\sin\chi - v\cos\chi).$$

Thus

$$\Re\left(e^{i\chi}w\right) = u\cos\chi + v\sin\chi$$

gives the component of velocity in the direction χ and

$$-\Im\left(e^{i\chi}w\right) = v\cos\chi - u\sin\chi \qquad (2.24)$$

the perpendicular component (reckoned positive in the direction rotated a quarter-turn anticlockwise, i.e. making an angle $\chi + \pi/2$ with the positive x-axis). These expressions are useful in aerodynamics for finding the components of velocity tangential and normal to a given surface.

In terms of the polar form:

$$e^{i\chi}w = qe^{-i(\beta-\chi)}$$
$$\Re e^{i\chi}w = q\cos(\beta - \chi)$$
$$-\Im e^{i\chi}w = q\sin(\beta - \chi).$$

Polar Components

In particular, the radial and polar components of a velocity field are related to the Cartesian components by

$$v_r = u\cos\theta + v\sin\theta$$
$$v_\theta = v\cos\theta - u\sin\theta.$$

This can be expressed as

$$v_r - iv_\theta = e^{i\theta}w = qe^{i(\theta-\beta)}. \qquad (2.25)$$

Note that $v_r - iv_\theta$, unlike $u - iv$, is not an analytic function. This is because the argument θ isn't and neither is the radial projection on the unit-circle $e^{i\theta}$; neither can be formed from the complex number z by analytical processes and it may be verified that neither satisfy the Cauchy–Riemann Equations (2.22).

2.6.3   Example: the Polar Angle Is Nowhere Analytic

1. Show that the argument θ of a complex number $z = re^{i\theta}$ is not analytic; i.e. does not satisfy the Cauchy–Riemann Equations (2.22). If $w \equiv u - iv = \theta$ were interpreted as a complex velocity field, show that it would either have nonzero divergence or vorticity or both.

Solution: First express the complex function in terms of x and y, here using Equation (2.20b): $\theta = \arctan y/x$. Then find its real and imaginary parts; here this is trivial as θ is real by definition, thus $\theta = u - iv = \arctan y/x - i0$, so $u = \arctan y/x$ and $v = 0$. The divergence is $\nabla \cdot q = \partial u/\partial x = -y/r^2$ and the vorticity is $\zeta = -\partial u/\partial y = -x/r^2$. Both of these are undefined at the origin and nonzero everywhere else in the plane.   □

Notes:
  (a) To arrive at the result more intuitively, consider any point z in the plane other than the origin. An infinitesimal step dz from the point along the ray from the origin doesn't change the polar angle θ, and so $d\theta = 0$, but a step dz along the circle centred on the origin and passing through the point (and so of radius $r = |z|$) does change θ according to $dz = r \, d\theta$.

2. Compute the outflow and circulation of that nonanalytic complex velocity for a loop, e.g. a quarter-circle in the first quadrant centred on the origin.

Solution: Split the loop-integral into three parts: outwards from the origin along the positive real axis, around the arc, and inwards along the positive imaginary axis.
On the arm of the quarter-circle along the positive real axis, $\theta = 0$ and so there is no contribution to either outflow or circulation: $Q_1 = \Gamma_1 = 0$.
   To deal with the circular arc centred on the origin, the polar components of velocity will be most convenient. Here these are $v_r - iv_\theta = \theta e^{i\theta}$ so $v_r = \theta \cos \theta$ and $v_\theta = -\theta \sin \theta$. On the circle $|z| = a$, $\hat{t} ds = a d\theta$ so (using integration by parts)

$$Q_2 = a \int_0^{\pi/2} v_r \, d\theta$$

$$= \int_0^{\pi/2} \theta \cos \theta \, d\theta$$

$$= \int_{\theta=0}^{\theta=\pi/2} \theta d \sin \theta$$

$$= [\theta \sin \theta]_0^{\pi/2} - \int_0^{\pi/2} \sin \theta \, d\theta$$

$$= \frac{\pi}{2} - [-\cos \theta]_0^{\pi/2}$$

$$= a \left(\frac{\pi}{2} - 1 \right).$$

Similarly

$$\Gamma_2 = a \int_0^{\pi/2} v_\theta \, d\theta = a \int_0^{\pi/2} \theta \sin \theta \, d\theta = -a.$$

Along the arm on the positive imaginary axis, $\theta = \pi/2$ so the velocity is constant. The flow is horizontal so there is no contribution to the circulation ($\Gamma_3 = 0$) while the outflow is $Q_3 = -a\pi/2$.

Thus, summing the contributions to the loop-integrals along the three parts of the loop, $Q = Q_1 + Q_2 + Q_3 = 0 + a(\pi/2 - 1) - a\pi/2 = -a \neq 0$ and $\Gamma = \Gamma_1 + \Gamma_2 + \Gamma_3 = 0 - a + 0 = -a \neq 0.$ □

3. Obtain these results for the outflow and circulation by integrating the divergence and vorticity over the interior of the loop.

Solution:

$$Q \equiv \int_0^a \int_0^{\pi/2} \frac{-\sin\theta}{r} r d\theta\, dr \qquad = -a \int_0^{\pi/2} \sin\theta d\theta = -a$$

$$\Gamma \equiv \int_0^a \int_0^{\pi/2} \frac{-\cos\theta}{r} r d\theta\, dr \qquad \equiv -a \int_0^{\pi/2} \cos\theta d\theta = -a. \qquad\qquad □$$

2.7   The Complex Potential

Define $W(z)$ by the differential equation

$$\frac{dW}{dz} = w \tag{2.26}$$

where $w(z)$ is the complex velocity $w = u - iv$. The function $W(z)$ is called the *complex potential*. It is evidently analytic, since it has a derivative: w. Adding any complex constant to the complex potential has no effect on the aerodynamics; it is sometimes convenient for the mathematics.

Let the real and imaginary parts of the complex potential W be ϕ and ψ:

$$W = \phi + i\psi. \tag{2.27}$$

Using the definition Equation (2.26),

$$dW = w\, dz$$
$$d(\phi + i\psi) = (u - iv)d(x + iy)$$

so

$$d\phi = u\, dx + v\, dy$$
$$d\psi = u\, dy - v\, dx$$

and the velocity components can be expressed in terms of either ϕ or ψ as

$$u = \frac{\partial\phi}{\partial x} = +\frac{\partial\psi}{\partial y} \tag{2.28a}$$

$$v = \frac{\partial\phi}{\partial y} = -\frac{\partial\psi}{\partial x}. \tag{2.28b}$$

Along a particle's trajectory, $dx = u\,dt$ and $dy = v\,dt$, and the infinitesimal change in the stream function during this time dt is

$$d\psi = u\,dy - v\,dx = uv\,dt - vu\,dt = 0$$

so that ψ is constant along streamlines. It is called, therefore, the *stream function*. This gives a convenient method of drawing flow fields: draw the curves along which ψ is constant.

The real part ϕ of the complex potential $W = \phi + i\psi$ is simply called the *potential*.

Note that the velocity can be obtained from either ϕ or ψ; this holds whenever the velocity is two-dimensional, divergence-free, and irrotational.

2.8   Exercises

1. Consider a uniform flow with velocity q_∞. Show that this flow is a physically possible flow for a fluid with constant density and that it is irrotational.

   Choose a coordinate system and then express the velocity field in complex form.

2. Show that the radial projection on the unit circle, $z/|z| \equiv e^{i\theta}$, of an arbitrary nonzero complex number $z = |z|e^{i\theta}$ is not analytic; i.e. does not satisfy the Cauchy–Riemann Equations (2.22).
   (a) If $w \equiv u - iv = z/|z|$ were interpreted as a complex velocity field, show that it would be either have nonzero divergence or vorticity or both.
   (b) Sketch the velocity field.
   (c) Take a loop in the plane (e.g. a quarter-circle centred on the origin with radius a in the first quadrant) and compute:

   i the net outflow across it and
   ii the circulation around it,

   both by loop and area integrals.

   Ans.:

   $$\Theta = -\cos 2\theta/r, \ \zeta = -\sin 2\theta/r, \ Q = 0, \ \Gamma = -a/2.$$

3. Consider the flow for $y > 0$ over the ground $y = 0$. If the velocity is unidirectional and everywhere parallel to the x-axis, say, what are the conditions that the distribution of the velocity must satisfy:
   (a) to have zero divergence,
   (b) to be irrotational?
   If the velocity is zero at the ground, show that it is either zero everywhere or rotational.

4. For the two-dimensional flow of incompressible air near the surface of a flat plate (in the plane $y = 0$), the component of velocity parallel to the plate may be approximated by the relation $u = a_1 yx^{-1/2} - a_2 y^3 x^{-3/2}$ where a_1 and a_2 are constants. Using the continuity equation, what is the velocity component v in the y-direction? Evaluate the constant of integration by noting that $v = 0$ at $y = 0$ if the plate is to be impermeable.

5. (a) Show that wherever Bernoulli's Equation (2.14) holds, the pressure coefficient as defined by Equation (2.11) can be expressed as

$$C_p = 1 - \left(\frac{q}{q_\infty}\right)^2.$$

(b) Show that the greatest increase in pressure relative to the free-stream value is $\frac{1}{2}\rho q_\infty$, and that the pressure coefficient is bounded above by unity.

(c) Show that there is no lower bound on the pressure or its coefficient within the scope of the theory developed thus far. (Actually increasing speed and decreasing pressure without bound eventually causes the density to drop if compressibility is admitted; see Chapter 18.)

(d) Say one wished to provide lift for a wing by arranging that the pressure beneath should increase over its value in the free-stream by half the maximum possible amount. By what fraction of its upstream speed should the air passing beneath be retarded?

(e) If also the pressure above is to be decreased by the same absolute amount, what fractional acceleration is required for the upper stream? [*Ans.:* $q_U = q_\infty\sqrt{3/2}$.]

6. Given the velocity field

$$q = (x^2 y - xy^2)i + \left(\frac{y^3}{3} - xy^2\right)j,$$

(a) is it divergence-free?
(b) is it irrotational?
(c) for the triangle with vertices $(0, 0)$, $(1, 0)$, and $(1, 1)$,

    i. evaluate the outflow and circulation across the perimeter; and
    ii. integrate the divergence and vorticity over the area.

7. Show that the dot product of a plane velocity and a plane vector $fi + gj$ can be expressed as the real part of the complex product of the complex velocity and the complex number $f + ig$.

8. Hence show that the square of the speed can be obtained as

$$q^2 = \Re(ww^*)$$

and therefore deduce Equation (2.23).

9. Also hence show that the component of velocity along an element of curve in Section 2.5.3 can be expressed as $\Re(w\,dz)$, and therefore the circulation along the curve as

$$\Gamma = \Re \int w\,dz.$$

10. Show that the imaginary part of the same complex product is the component of velocity normal to an element of the curve, as given by Equation (2.2). Deduce that the net outflow through a closed curve is

$$\Theta = \Im \oint w \, dz.$$

11. Derive Equation (2.25) from the preceding considerations and the complex product of the complex velocity and the unit vector directed away from the origin.

12. Assuming the standard integral

$$\int_0^1 \arctan \frac{t}{1-t} \, dt = \frac{\pi}{4},$$

show that if the quarter-circle in Section 2.6.3 were replaced by the right-angled isoceles triangle with the same three vertices, so that the integral along the arc is replaced by an integral along the hypoteneuse which would be $\pi a(1 - i)/4$ then the circulation and outflow remain equal to each other but both change from $-a$ to $-\pi a/4$.

13. Is $r(z) \equiv |z|$ analytic? Describe the velocity field given by $u(x, y) - iv(x, y) = |x + iy|$.

14. Is z^* analytic? Describe the velocity field given by $u(x, y) - iv(x, y) = (x + iy)^*$.

15. What are the conditions that a complex-valued function of x and y should have a well-defined derivative with respect to z^*?

2.9   Further Reading

The divergence and gradient theorems will be revisited in Equations (9.7) and (9.8) in the presentation of vector analysis required for three-dimensional flow; see also Bayin (2008) and Karamcheti (1966).

The substantial derivative can be approached in different ways (Abbott and von Doenhoff 1959; Moran 2003; Anderson 2007), as can Bernoulli's principle (Abbott and von Doenhoff 1959; Milne-Thomson 1973; Kuethe and Chow 1998; Bertin 2002; Anderson 2007).

The continuity Equation (2.3), can alternatively be derived from the mass balance on an infinitesimal rectangle (Abbott and von Doenhoff 1959); i.e. requiring zero net outflow from a rectangle with sides dx and dy. That approach is essentially a special rederivation the divergence theorem (Truesdell and Rajagopal 2000).

For the treatment of conservative forces such as gravity in aerodynamics, see Milne-Thomson (1973, section 2.11) and Karamcheti (1966, section 5.15).

Another brief summary of the parts of complex analysis required for aerodynamics was given by Pope (2009). The complex velocity was introduced very early in the introductions to the theory of wing sections and panel methods by Weissinger (1963) and Bousquet (1990), respectively. The complex treatment of polar components can also be found in Ashley and Landahl (1985). The complex potential is discussed by Abbott and von Doenhoff (1959), Milne-Thomson (1973), Paraschivoiu (1998), and Pope (2009).

References

Abbott, I.H. and von Doenhoff, A.E. (1959) *Theory of Wing Sections*. New York: Dover.

Anderson, J.D. (2007) *Fundamentals of Aerodynamics*, 4th edn. New York: McGraw-Hill.

Ashley, H. and Landahl, M. (1985) *Aerodynamics of Wings and Bodies*. New York: Dover.

Bayin, S.S. (2008) *Essentials of Mathematical Methods in Science and Engineering*. Chichester: John Wiley & Sons, Ltd.

Bertin, J.J. (2002) *Aerodynamics for Engineers*, 4th edn. New Jersey: Prentice Hall.

Bousquet, J. (1990) *Aérodynamique: Méthode des singularités*. Toulouse: Cépaduès.

Karamcheti, K. (1966) *Principles of Ideal-Fluid Aerodynamics*. Chichester: John Wiley & Sons, Ltd.

Kuethe, A.M. and Chow, C.Y. (1998) *Foundations of Aerodynamics*, 5th edn. Chichester: John Wiley & Sons, Ltd.

Milne-Thomson, L.M. (1973) *Theoretical Aerodynamics*, 4th edn. New York: Dover.

Moran, J. (2003) *An Introduction to Theoretical and Computational Aerodynamics*. New York: Dover.

Paraschivoiu, I. (1998) *Aérodynamique Subsonique*. Éditions de l'École de Montréal.

Pope, A. (2009) *Basic Wing and Airfoil Theory*. New York: Dover.

Truesdell, C. and Rajagopal, K.R. (2000) *An Introduction to the Mechanics of Fluids*. Birkhäuser.

Weissinger, J. (1963) Theorie des Tragflügels bei stationärer bewegung in reibungslosen, inkompressiblen medien In *Fluid Dynamics II* (ed. Flügge S) vol. VIII/2 of *Encyclopedia of Physics*. New York: Springer.

3

Circulation and Lift

In Chapter 2, the equations governing plane ideal flow were presented and it was shown that solutions to these equations representing physically possible flow could be obtained as analytic functions of a complex variable. In Chapter 3, this discovery will be exploited to build up a catalogue of divergence-free irrotational flows and then, with these available as definite examples, some general properties are developed along with computational and analytical tools, culminating in the remarkable short-cut to lift that is the Kutta–Joukowsky theorem.

3.1 Powers of z

As examples of flows generated by complex functions, consider the powers of z: $w(z) = z^k$.
From $w = z^k = r^k e^{ik\theta}$, we see that the speed and direction fields are:

$$q \equiv |z^k| = r^k \tag{3.1a}$$

$$\beta \equiv -\arg z^k = -k\theta. \tag{3.1b}$$

From De Moivre's Equation (2.21), we see that $w = z^k = r^k e^{ik\theta} = r^k \cos k\theta + i r^k \sin k\theta$ and the Cartesian components are

$$u = \Re z^k = r^k \cos k\theta$$

$$v = -\Im z^k = -r^k \sin k\theta.$$

From Equation (2.25), the polar components are given by $v_r - iv_\theta = e^{i\theta} z^k = r^k e^{i(k+1)\theta}$ so

$$v_r = r^k \cos(k+1)\theta \tag{3.2a}$$

$$v_\theta = -r^k \sin(k+1)\theta. \tag{3.2b}$$

Theory of Lift: Introductory Computational Aerodynamics in MATLAB®/Octave, First Edition. G. D. McBain.
© 2012 John Wiley & Sons, Ltd. Published 2012 by John Wiley & Sons, Ltd.

3.1.1 Divergence and Vorticity in Polar Coordinates

The continuity equation in polar coordinates is

$$\frac{1}{r}\frac{\partial(rv_r)}{\partial r} + \frac{1}{r}\frac{\partial v_\theta}{\partial \theta} = 0,$$

so that we can verify that these complex velocities satisfy conservation of mass:

$$\frac{1}{r}\frac{\partial}{\partial r}\left\{r^{k+1}\cos(k+1)\theta\right\} + \frac{1}{r}\frac{\partial}{\partial \theta}\left\{-r^k\sin(k+1)\theta\right\}$$
$$= (k+1)r^k\cos(k+1)\theta - (k+1)r^k\cos(k+1)\theta = 0.$$

The vorticity in polar coordinates is

$$\zeta = \frac{1}{r}\left\{\frac{\partial(rv_\theta)}{\partial r} - \frac{\partial v_r}{\partial \theta}\right\},$$

so for $w = z^k$,

$$\zeta = \frac{1}{r}\left\{\frac{\partial(-r^{k+1}\sin[k+1]\theta)}{\partial r} - \frac{\partial(r^k\cos[k+1]\theta)}{\partial \theta}\right\}$$
$$= \frac{1}{r}\left\{-(k+1)r^k\sin(k+1)\theta + (k+1)r^k\sin(k+1)\theta\right\} = 0.$$

3.1.2 Complex Potentials

The complex potential for the complex velocity $w(z) = z^k$ is

$$W(z) = \int z^k\,dz = \begin{cases} (k+1)^{-1}z^{k+1}, & (k \neq -1); \\ \ln z, & (k = -1). \end{cases}$$

Notice that the case $w = z^{-1}$ is special. The logarithm of a complex number z satisfies

$$\ln z = \ln\left(re^{i\theta}\right) = \ln r + \ln\left(e^{i\theta}\right) = \ln r + i\theta. \tag{3.3}$$

That is

$$\Re \ln z = \ln|z|$$
$$\Im \ln z = \arg z.$$

Thus for $w = z^{-1}$,

$$\phi = \ln r$$
$$\psi = \theta.$$

Listing 3.1 cmeshgrid construct an $n \times n$ rectangular grid in the complex plane between lower-left and upper-right corners ll and ur.

```
function Z = cmeshgrid (ll, ur, n)
  x = linspace (real (ll), real (ur), n);
  y = linspace (imag (ll), imag (ur), n);
  [X, Y] = meshgrid (x, y);
  Z = complex (X, Y);
```

3.1.3 Drawing Complex Velocity Fields with Octave

Octave is very useful for quickly drawing complex velocity fields.

First we need to define a grid of points. Octave has a **meshgrid** routine for this purpose, but it returns the two lots of coordinates as separate real arrays so here we define a simple utility function that converts these to complex form (Listing 3.1).

This will be used for drawing arrow plots and potential and streamlines for velocity fields defined by complex functions.

Arrow Plots of Velocity

The most obvious way to depict a two-dimensional vector field is to draw an arrow at each point of a grid, parallel to the field. This representation is similar to that which would be obtained experimentally by holding an array of thin light short threads of silk in the flow.

Octave provides the function **quiver** for this purpose, except it makes the length of the arrows proportional to the local magnitude of the field. This does provide more information, but in many important flows the speed varies widely through the flow-field which makes it difficult to choose a consistent scale for the arrows.

Here we define a function to generate a rectangular grid with given lower-left and upper-right corners and the number of points and then evaluate a given complex velocity function on the grid and pass it to Octave's **quiver**. To obtain a depiction closer to the experimental field of threads of fixed length, we normalize each arrow by dividing it by its modulus (Listing 3.2).

Listing 3.2 direction: plot the direction field on a 22×22 rectangular grid with defined lower-left and upper-right corners ll and ur of a complex velocity function wf.

```
function direction (ll, ur, wf)
  Z = cmeshgrid (ll, ur, 22);
  w = wf (Z);
  s = w ./ abs (w);
  quiver (real (Z), imag (Z), real (s), -imag (s))
  axis ([real(ll), real(ur), imag(ll), imag(ur)], ...
        'off', 'equal')
  box ('on')
```

Listing 3.3   `isopotentials`: plot the level sets in a rectangle with corners `ll` and `ur` of the real and imaginary parts of the complex potential function `Wf`.

```
function isopotentials (ll, ur, Wf)
  Z = cmeshgrid (ll, ur, 1e3);
  contour (real (Z), imag (Z), real (Wf (Z)), '--')
  hold ('on')
  contour (real (Z), imag (Z), imag (Wf (Z)), 'k-')
  axis ([real(ll), real(ur), imag(ll), imag(ur)], ...
        'off', 'equal')
  box ('on')
  hold ('off')
```

Isopotentials and Streamlines

Arrow plots are simple but can become cluttered. A useful alternative view is to draw the isopotentials and streamlines.

The underlying tool here is Octave's **contour** routine; here we define a function to generate a grid, evaluate the complex potential defined by the given function on it, and then draw the contours (Listing 3.3).

Note that Octave is not being asked here to integrate or differentiate the complex velocity or complex potential (this kind of symbolic manipulation is possible, but beyond the scope of the present discussion) so the complex velocity w passed to `direction` should be the derivative of the complex potential W passed to `isopotentials`.

3.1.4   Example: $k = 1$, Corner Flow

Set $w = z^1 = z$. The Cartesian components are $u = x$ and $v = -y$. This is illustrated in Figure 3.1 with an arrow plot.

Since the horizontal component $u = x$ vanishes on the vertical y-axis and the vertical component $v = -y$ vanishes on the horizontal x-axis, $w = z$ represents a flow with impermeable walls along the x and y axes.

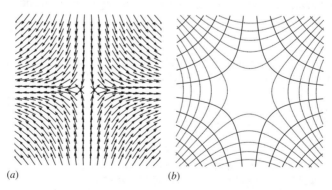

(a)   (b)

Figure 3.1   The complex velocity field $w = z$: arrow plot (a) and level curves of the imaginary (solid) and real (dashed) parts of the complex potential $W = z^2/2$ (b)

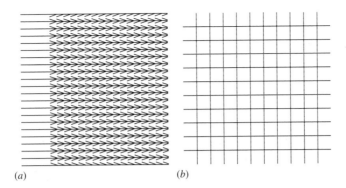

Figure 3.2 The complex velocity field $w = 1$: its arrow plot (a) and level curves of the real (dashed) and imaginary (solid) parts of its complex potential $W = z$ (b)

From Equation (3.1a), we see that the speed of the flow increases without bound with distance from the origin for positive k. We will therefore normally be more interested in nonpositive powers of z in aerodynamical applications.

3.1.5 Example: $k = 0$, Uniform Stream

Set $w = z^0 = 1$. Then the Cartesian components are $u = 1$ and $v = 0$; i.e. a uniform stream in the x-direction with unit speed. The field is plotted in Figure 3.2.

3.1.6 Example: $k = -1$, Source

Set $w = z^{-1}$. Then, referring to Equation (2.25), the polar components are given by

$$v_r - iv_\theta = e^{i\theta} w = e^{i\theta} z^{-1} = e^{i\theta} \left(r e^{i\theta} \right)^{-1} = r^{-1};$$

i.e.

$$v_r = 1/r$$
$$v_\theta = 0.$$

This could also have been obtained directly from Equations (3.2a) and (3.2b).

The velocity is purely radial out from the origin with a speed inversely proportional to the distance. Since the circumference of a circle grows in proportion to its radius, the net flow out of any circle centred at the origin is independent of the radius. This flow is called a *source* at the origin. It satisfies the continuity equation everywhere except at the origin, where the complex function z^{-1} is singular. The direction field is plotted in Figure 3.3a.

Since the magnitude of the velocity is unbounded at the origin, a simple **quiver** plot with shafts proportional to speed would be less useful.

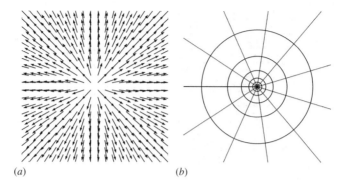

Figure 3.3 The complex velocity field $w = z^{-1}$ (a) and its complex potential $W = \log z$ (b) of the source (Section 3.1.6)

3.1.7   Example: $k = -2$, Doublet

Set $w = z^{-2}$. The Cartesian and polar components of the velocity are:

$$u = \frac{x^2 - y^2}{\left(x^2 + y^2\right)^2}$$

$$v = \frac{2xy}{\left(x^2 + y^2\right)^2}$$

$$v_r = \frac{\cos\theta}{r^2} \tag{3.4}$$

$$v_\theta = \frac{\sin\theta}{r^2}.$$

The flow, called a *doublet*, is plotted in Figure 3.4. Notice that unlike the isotropic source of Figure 3.3, the doublet has a preferred direction: along the positive real axis.

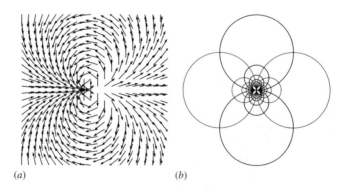

Figure 3.4 The complex velocity field $w = z^{-2}$ (a) and its complex potential $W = -1/z$ (b) of the doublet of Section 3.1.7

3.2 Multiplication by a Complex Constant

Say we have a complex velocity field $w(z)$ expressed in polar form $w = qe^{-i\beta}$, where q is the speed and β the angle to the positive x-axis. Multiplying an analytic function by a (complex) constant $Ae^{-i\alpha}$ gives another analytic function. The velocity field obtained is

$$w' = \left(Ae^{-i\alpha}\right)\left(qe^{-i\beta}\right) = (Aq)e^{-i(\beta+\alpha)},$$

so that the effect of multiplying by the complex constant $Ae^{-i\alpha}$ is to scale the speed to $q' = Aq$ and increment the direction to $\beta' = \beta + \alpha$.

Thus, for example, noting that $i = e^{i\pi/2}$, multiplication by the imaginary unit i keeps the speed the same but rotates the velocity vector at each point through a clockwise right-angle.

3.2.1 Example: $w = $ const., Uniform Stream with Arbitrary Direction

The uniform horizontal (rightward) stream $w(z) = 1$ becomes, on multiplication by $e^{i\pi/2}$, $w' = iw = i$, with components $u' = 0$ and $v' = -1$. This is a uniform vertical downward stream.

A uniform stream with speed q_∞ and angle α is therefore obtained by multiplying the complex velocity $w = 1$ by the complex constant $q_\infty e^{-i\alpha}$:

$$q_\infty e^{-i\alpha} = (q_\infty \cos\alpha) - i(q_\infty \sin\alpha). \tag{3.5}$$

This is plotted in Figure 3.5 for $\alpha = \frac{\pi}{6}$.

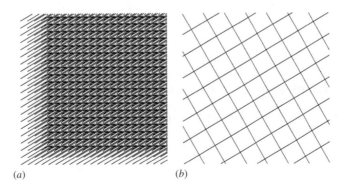

(a)   (b)

Figure 3.5 The complex velocity field given by Equation (3.5), $w = q_\infty e^{-i\alpha}$ (a) and its complex potential $W = q_\infty z e^{-i\alpha}$ (b) for $\alpha = \frac{\pi}{6}$

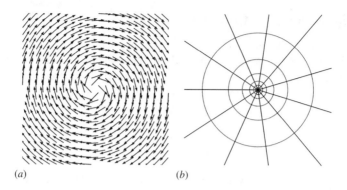

(a)   (b)

Figure 3.6   Arrow (a) and complex potential (b) fields for the complex velocity $w = i/z$

3.2.2   Example: $w = i/z$, Vortex

Since the velocity of a source is purely radial, multiplying its complex velocity z^{-1} by i rotates the velocity at each point through a right-angle to become purely azimuthal:

$$w = iz^{-1} = ir^{-1}e^{-i\theta}$$
$$v_r - iv_\theta = e^{i\theta}w = ir^{-1},$$

so $v_r = 0$ and $v_\theta = -1/r$. This flow is called a *vortex*; it is plotted in Figure 3.6.

3.2.3   Example: Polar Components

Another example of multiplying the complex velocity by a constant to rotate the velocity vectors at a point can be seen in the formula (2.25) for obtaining the polar components. This is because the polar components can be thought of as the components relative to a pair of perpendicular unit vectors that have been rotated so that one is directed away from the origin.

3.3   Linear Combinations of Complex Velocities

If $f(z)$ and $g(z)$ are analytic functions and α and β are complex constants, the linear combination $\alpha f(z) + \beta g(z)$ is also analytic.

3.3.1   Example: Circular Obstacle in a Stream

From Equations (3.2a) and (3.4) the radial components of velocity for the uniform stream $v_r = \cos\theta$ and the doublet $v_r = r^{-2}\cos\theta$ have the same dependence on the polar angle θ, and

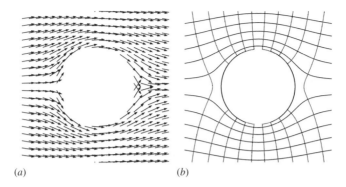

(a) (b)

Figure 3.7   The complex velocity field of Equation (3.6), $w = 1 - z^{-2}$, (a) and its complex potential (b); the field in the unit disk $|z| < 1$ is suppressed

are identical on the circle $r = 1$. Therefore, subtracting the complex velocity for a doublet from the complex velocity for a uniform stream

$$w = 1 - \frac{1}{z^2} \tag{3.6}$$

gives a velocity field

$$v_r = \left(1 - \frac{1}{r^2}\right)\cos\theta$$

$$v_\theta = -\left(1 + \frac{1}{r^2}\right)\sin\theta$$

that has zero radial component on the circle. This is as if the circle were an impermeable barrier, and so $w = 1 - z^{-2}$ for $|z| > 1$ is the complex velocity for a stream with a circular obstacle. The field is plotted in Figure 3.7 using Listings 3.4 and 3.5.

Notice the trick we have used to zero the velocity inside the circle: 'true' logical values count as unity in arithmetic computations in Octave and 'false' as zero; therefore multiplying pointwise by a logical field preserves the values where the criterion is true and zeros it elsewhere, which is exactly what is required here.

With this interpretation of the velocity field, the velocity given inside the circle by Equation (3.6) is irrelevant.

Listing 3.4   Octave code to generate Figure 3.7a, using direction (Listing 3.2).

```
direction (-2-2i, 2+2i, @ (z) (abs(z)>1) .* (1 - z .^ -2))
```

Listing 3.5   Octave code to generate Figure 3.7b, using isopotentials (Listing 3.3).

```
isopotentials (-2-2i, 2+2i, @ (z) (abs(z)>1) .* (z + 1 ./ z))
```

3.4 Transforming the Whole Velocity Field

3.4.1 Translating the Whole Velocity Field

If we have an analytic function $f(z)$ and form another by replacing z by $z - z_0$, the effect is to translate the whole velocity field.

3.4.2 Example: Doublet as the Sum of a Source and Sink

Consider for small real ϵ

$$\frac{1}{z^2} = \lim_{\epsilon \to 0} \frac{1}{z^2 - \epsilon^2} = \lim_{\epsilon \to 0} \frac{1}{(z + \epsilon)(z - \epsilon)}.$$

Now an expansion in partial fractions,

$$\begin{aligned}
\frac{1}{z^2 - \epsilon^2} = \frac{1}{(z + \epsilon)(z - \epsilon)} &= \frac{a}{z + \epsilon} + \frac{b}{z - \epsilon} \\
&= \frac{a(z - \epsilon) + b(z + \epsilon)}{z^2 - \epsilon^2} \\
&= \frac{(a + b)z + (b - a)\epsilon}{z^2 - \epsilon^2},
\end{aligned}$$

is valid if $a = -b$ (since the variable z doesn't appear in the numerator on the left-hand side) and then $-a = b = 1/2\epsilon$ so

$$\frac{1}{z^2} = \lim_{\epsilon \to 0} \left(\frac{1/2\epsilon}{z - \epsilon} - \frac{1/2\epsilon}{z + \epsilon} \right),$$

which is the superposition of a source at $z = +\epsilon$ and a sink of the same strength at $z = -\epsilon$; i.e. at the position of its mirror-image in the y-axis.

3.4.3 Rotating the Whole Velocity Field

If we have an analytic function $f(z)$ and form another by replacing z by $e^{-i\alpha}z$; i.e. $g(z) = f(e^{-i\alpha}z)$; and then form a third by $h(z) = e^{-i\alpha}g(z) = e^{-i\alpha}f(e^{-i\alpha}z)$, the velocity field $h(z)$ is that of $f(z)$ rotated through an anticlockwise angle α.

Verification of this using some of the examples above is left as an exercise. The vector field obtained from the circular obstacle example of Figure 3.7 using $\alpha = \pi/6$ is plotted in Figure 3.8 using Listing 3.6.

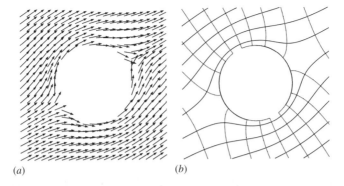

Figure 3.8 The complex velocity field $w = e^{-i\pi/6} \left\{ 1 - \left(e^{-i\pi/6} z \right)^{-2} \right\}$ for $|z| > 1$

Listing 3.6 Octave code to generate Figure 3.8a.

```
r = exp (-1i * pi / 6);
w = @ (z) (abs (z) > 1) .* (1 - z .^ -2);
direction (-2-2i, 2+2i, @ (z) r * w (r * z))
```

3.5 Circulation and Outflow

3.5.1 Curve-integrals in Plane Ideal Flow

We have seen in Section 2.7 how both Cartesian components $u = \Re w$ and $v = -\Im w$ can be recovered from either the real ($\phi = \Re W$)

$$u = \frac{\partial \phi}{\partial x}, \quad v = \frac{\partial \phi}{\partial y}$$

or imaginary ($\psi = \Im W$)

$$u = +\frac{\partial \psi}{\partial y}, \quad v = -\frac{\partial \psi}{\partial x}$$

parts of the complex potential W defined in Equation (2.26) by $dW/dz = w$. We now investigate recovering the scalar potential and the stream function from the velocity. This leads to important and useful physical interpretations of ϕ and ψ.

Integrating Equation (2.26) between two points in the fluid gives

$$W(z_2) - W(z_1) = \int_{z_1}^{z_2} w \, dz. \tag{3.7}$$

Expanding Equation (3.7) into real and imaginary parts gives

$$\phi(z_2) - \phi(z_1) + i\{\psi(z_2) - \psi(z_1)\} = \int_{z_1}^{z_2} (u - iv)\,(dx + i\,dy)$$

$$= \int_{z_1}^{z_2} (u\,dx + v\,dy) + i \int_{z_1}^{z_2} (u\,dy - v\,dx).$$

Now, as in the discussion of circulation in Section 2.5.3,

$$u\,dx + v\,dy = \boldsymbol{q} \cdot \hat{\boldsymbol{\tau}}\,ds;$$

i.e. the tangential component of velocity along an infinitesimal curve segment. And, as in the discussion of conservation of mass in Section 2.3,

$$u\,dy - v\,dx = \boldsymbol{q} \cdot \hat{\boldsymbol{n}}\,ds;$$

i.e. the component of velocity at right-angles to the same segment (reckoned positive when the velocity crosses the segment from left to right). Thus

$$\phi(z_2) - \phi(z_1) = \Re \int_{z_1}^{z_2} w\,dz = \int_{z_1}^{z_2} \boldsymbol{q} \cdot \hat{\boldsymbol{\tau}}\,ds \qquad (3.8a)$$

$$\psi(z_2) - \psi(z_1) = \Im \int_{z_1}^{z_2} w\,dz = \int_{z_1}^{z_2} \boldsymbol{q} \cdot \hat{\boldsymbol{n}}\,ds. \qquad (3.8b)$$

In words, change in scalar potential along a curve through the flow field gives minus the circulation along that curve, and the difference in the stream function between two ends of the curve gives the volumetric flow-rate (per unit span) across the curve from left to right.

Also, Equations (3.8) can be used to determine or measure either ϕ or ψ, up to an arbitrary additive constant: the value at some reference point z_1.

3.5.2   Example: Numerical Line-integrals for Circulation and Outflow

With these ideas, the example of Section 2.6.3 can now also be done numerically, using one of Octave's automatic quadrature routines for each of the legs of the loop. Again consider the quarter in the first quadrant of a circle centred on the origin, of radius a so that the legs are:

1. $z = x$, $dz = dx$ for $0 < x < a$, with $w \equiv \theta = 0$;
2. $z = ae^{i\theta}$, $dz = aie^{i\theta}\,d\theta$ for $0 < \theta < \pi/2$, with $w \equiv \theta = \theta$; and
3. $z = iy$, $dz = i\,dy$ for $a > y > 0$, with $w \equiv \theta = \pi/2$.

Since the integrand along the arc is complex (and in more general cases, all integrands will be complex), we cannot use the basic **quad** integration routine, which only accepts real integrands, but instead use **quadgk** which is not so restricted. To demonstrate that the closed-form answer is correct for different values of a, it is not specified in the input Listing 3.7 but generated randomly at run-time. From Equations (3.8), the real and imaginary parts of the integral $\int w\,dz$ are the circulation and outflow, respectively.

Listing 3.7 Octave code to integrate a complex velocity along a broken curve.

```
a = rand ()
W = [quadgk(@ (x) 0*x, 0, a);
     quadgk(@ (t) t .* exp (i*t), 0, pi/2) * 1i * a;
     quadgk(@ (y) ones (size (y)) * pi/2, a, 0) * 1i ]
sum (W)
```

Listing 3.8 Octave output on integrating complex velocity along a broken curve.

```
a =   0.37662
W =

   0.00000 + 0.00000i
  -0.37662 + 0.21497i
  -0.00000 - 0.59159i

ans = -0.37662 - 0.37662i
```

The output of this program is given in Listing 3.8, in which the real and imaginary parts of each of the three line-integrals indeed agree with the exact results obtained in Section 2.6.3.

3.5.3 Closed Circuits

Now say we take the final point, z_2, of the curve to coincide with the first point, z_1; i.e. close the curve to form a circuit C. Then the jump $[W]_C$ in W as it moves around the circuit C is

$$[W]_C = \oint_C w \, dz.$$

Of course, if W is a single-valued function of z, it will have the same value at the end of the loop as it did at the start, and so $[W]_C$ is necessarily zero for any circuit C; however, we will see shortly (in Section 3.7) that multivalued complex potentials are also of interest in aerodynamics.

The real and imaginary parts of the jump involve the circulation Γ_C around and outflow Q_C through the circuit C,

$$[W]_c \equiv [\phi]_C + i[\psi]_C = \oint_C (u \, dx + v \, dy) + i \oint_C (u \, dy - v \, dx)$$
$$= \oint_C \boldsymbol{q} \cdot \hat{\boldsymbol{\tau}} \, ds + i \oint_C \boldsymbol{q} \cdot \hat{\boldsymbol{n}} \, ds$$
$$\equiv -\Gamma_C + iQ_C. \tag{3.9}$$

If the circuit C is filled with fluid, the circuit integrals can be replaced with integrals of vorticity and divergence over the contained region R (see Section 2.5.3 for the circulation and Section 2.3 for the outflow):

$$[\phi]_C = -\Gamma_C = \iint_R \zeta \, dx \, dy$$

$$[\psi]_C = Q_C = \iint_R \Theta \, dx \, dy.$$

If the velocity field is irrotational in the region R, the integrand for the circulation Γ_C vanishes, and if it's divergence-free, the integrand for the outflow Q_C vanishes.

If the circuit is not filled with fluid in plane ideal flow, say, for example, because it encloses an obstacle such as an aerofoil, then the integrals over R are meaningless, the circulation and outflow need not vanish, and the complex potential might be multivalued. The same conclusions apply if the circuit contains a *singular point* at which the velocity is undefined.

3.5.4   Example: Powers of z and Circles around the Origin

Say we take $w = A_k z^k = A_k r^k e^{ik\theta}$, where A_k is a complex constant, and let the circuit C be a circle of radius a, centred on the origin; i.e. on C, $z = ae^{i\theta}$. Then around the circuit,

$$dz = d(ae^{i\theta}) = aie^{i\theta} \, d\theta$$

and $w = A_k a^k e^{ik\theta}$ with θ going from 0 to 2π, and the integral around the circuit is

$$\oint_C w \, dz = A_k \int_0^{2\pi} a^k e^{ik\theta} iae^{i\theta} \, d\theta$$

$$= ia^{k+1} A_k \int_0^{2\pi} e^{i(k+1)\theta} d\theta.$$

Now if $k \neq -1$, this is

$$A_k \oint_C z^k \, dz = ia^{k+1} A_k \left[\frac{e^{i(k+1)\theta}}{i(k+1)} \right]_0^{2\pi}$$

$$= 0,$$

since $e^{i\theta} = \cos\theta + i\sin\theta$ is periodic in θ with period 2π.

For the special case $k = -1$ though, we have

$$A_{-1} \oint_C \frac{dz}{z} = iA_{-1} \int_0^{2\pi} d\theta$$

$$= 2\pi iA_{-1}.$$

Notice that this result is independent of the radius of the circle used for integration. In fact, though not proven here, the circulation Γ and outflow Q are both independent of the shape of the circuit too, so long as the circuit encloses the same obstacles and singularities.

In summary, for $w = A_k z^k$, $\Gamma = Q = 0$ for $k \neq -1$. For $w = A_{-1}/z$, $\Gamma = 2\pi \mathrm{i} A_{-1}$ and $Q = 2\pi \Re A_{-1}$. For this reason we usually represent a point source at the origin as

$$w_{\mathrm{source}} = \frac{Q}{2\pi z}$$

and a point vortex as

$$w_{\mathrm{vortex}} = \mathrm{i}\, \frac{\Gamma}{2\pi z}, \tag{3.10}$$

and call Q and Γ the *strengths* of the source and vortex. For $Q < 0$, we call the source a *sink*; or, if the sign of Q is yet to be determined, a source of strength $-Q$ can be called a sink of strength Q.

The results obtained here correspond to the fact that the complex potentials for $w = A_k z^k$ are, as in Section 3.1.2,

$$W = \int A_k z^k \,\mathrm{d}z = \begin{cases} A_k(k+1)^{-1} z^{k+1}, & (k \neq -1); \\ A_{-1} \ln z, & (k = -1), \end{cases}$$

i.e. all single-valued except for when $k = -1$.

3.6 More on the Scalar Potential and Stream Function

3.6.1 The Scalar Potential and Irrotational Flow

We have seen that in an ideal flow, the difference in the scalar potential between two points is equal to minus the circulation along a path though the fluid joining them. In general, any plane velocity field derived from a scalar function ϕ by the operations

$$u = \frac{\partial \phi}{\partial x}, \qquad v = \frac{\partial \phi}{\partial y}$$

will be irrotational, since

$$\zeta \equiv \frac{\partial v}{\partial x} - \frac{\partial u}{\partial y} = \frac{\partial^2 \phi}{\partial x \partial y} - \frac{\partial^2 \phi}{\partial y \partial x} = 0.$$

On the other hand, if the scalar potential isn't the real part of an analytic function, the derived velocity field needn't be divergence-free; the continuity Equation (2.3), $\nabla \cdot \boldsymbol{q} = 0$, becomes

$$\frac{\partial^2 \phi}{\partial x^2} + \frac{\partial^2 \phi}{\partial y^2} = 0.$$

This second-order linear partial differential equation is called Laplace's equation, and is often written concisely as

$$\nabla^2 \phi = 0. \tag{3.11}$$

3.6.2   The Stream Function and Divergence-free Flow

If a velocity field is derived from a scalar function ψ by

$$u = +\frac{\partial \psi}{\partial y}, \quad v = -\frac{\partial \psi}{\partial x},$$

it will automatically be divergence-free since

$$\Theta \equiv \frac{\partial u}{\partial x} + \frac{\partial v}{\partial y} = \frac{\partial^2 \psi}{\partial x \partial y} - \frac{\partial^2 \psi}{\partial y \partial x} = 0.$$

On the other hand, it needn't be irrotational: the condition is

$$\zeta \equiv \frac{\partial v}{\partial x} - \frac{\partial u}{\partial y} = -\frac{\partial^2 \psi}{\partial x^2} - \frac{\partial^2 \psi}{\partial y^2} \equiv -\nabla^2 \psi = 0.$$

Thus, velocity fields derived from an analytic complex velocity are automatically divergence-free and irrotational, those derived from a scalar potential are irrotational, and those from a stream function divergence-free. Scalar potentials or stream functions can be used for ideal flow, but only if they satisfy Laplace's Equation (3.11).

For more general, nonideal, kinds of flow, it is still possible to define the complex velocity by $w = u - iv$, but it won't be analytic and so will be much less useful; however, these new representations in terms of scalar potential or stream function can still be useful in less restricted cases. Thus, the scalar potential can be used for irrotational flow of compressible fluids for which generally the divergence is nonzero, and the stream function can be used for viscous flow of incompressible fluids for which vorticity is generated at the flow boundaries. An example of the latter may be found in the treatment of Blasius's boundary layer in Section 16.4. An example of the former is the disturbance potential in Section 19.4.

3.7   Lift

In this section, we derive an expression for the aerodynamic force per unit span on an impermeable object in a plane ideal flow in terms of the complex velocity. We then show that the drag always vanishes while the lift is proportional to the circulation around the object.

3.7.1   Blasius's Theorem

The Cartesian components of the aerodynamic force per unit span on an obstacle with contour $\mathcal{C}$ are given by Equations (2.10). From them, form the complex quantity

$$a_x - ia_y = -\oint_{\mathcal{C}} p\,(\mathrm{d}y + i\,\mathrm{d}x) = -i\oint_{\mathcal{C}} p\,\mathrm{d}z^*$$

and use Bernoulli's Equation (2.18) for ideal flow, dropping the constant terms which integrate to zero,

$$
\begin{aligned}
a_x - \mathrm{i}a_y &= -\mathrm{i} \oint_C \left(p_\infty + \frac{\rho q_\infty^2}{2} - \frac{\rho q^2}{2} \right) \, \mathrm{d}z^* \\
&= \frac{\mathrm{i}\rho}{2} \oint_C q^2 \, \mathrm{d}z^* \\
&= \frac{\mathrm{i}\rho}{2} \oint_C ww^* \, \mathrm{d}z^*.
\end{aligned}
$$

Now

$$
w^* \, \mathrm{d}z^* = (w \, \mathrm{d}z)^* = \mathrm{d}W^* = \mathrm{d}\phi - \mathrm{i} \, \mathrm{d}\psi
$$

but $\mathrm{d}\psi = 0$ along the outline of an impermeable obstacle, so we can replace $\mathrm{d}W^*$ with its complex conjugate $\mathrm{d}\phi + \mathrm{i} \, \mathrm{d}\psi = \mathrm{d}W \equiv w \, \mathrm{d}z$ and

$$
a_x - \mathrm{i}a_y = \frac{\mathrm{i}\rho}{2} \oint_C w^2 \, \mathrm{d}z \tag{3.12}
$$

which is *Blasius's Theorem*.

3.7.2 The Kutta–Joukowsky Theorem

Now say, for example, the flow is composed of a uniform stream with speed q_∞ and incidence α and a vortex of strength Γ; i.e.

$$
w = q_\infty e^{-\mathrm{i}\alpha} + \mathrm{i} \frac{\Gamma}{2\pi z}. \tag{3.13}
$$

Then

$$
w^2 = q_\infty^2 e^{-\mathrm{i}2\alpha} + \mathrm{i} \frac{q_\infty \Gamma}{\pi z} e^{-\mathrm{i}\alpha} - \frac{\Gamma^2}{4\pi^2 z^2},
$$

and using the results on the circuit integrals of powers of z from Section 3.5.4

$$
\begin{aligned}
\oint_C w^2 \, \mathrm{d}z &= 2\pi \mathrm{i} \times \left(\mathrm{i} \frac{q_\infty \Gamma}{\pi} e^{-\mathrm{i}\alpha} \right) \\
&= -2q_\infty \Gamma e^{-\mathrm{i}\alpha}.
\end{aligned}
$$

Blasius's theorem, Equation (3.12), then gives

$$
a_x - \mathrm{i}a_y = -\mathrm{i}\rho q_\infty \Gamma e^{-\mathrm{i}\alpha}. \tag{3.14}
$$

Since the drag is measured in the direction of the airstream (α) and the lift is measured a right-angle anticlockwise from the drag ($\alpha + \pi/2$), we have (as in Section 2.6.2 for velocity components)

$$
\ell - \mathrm{i}(-d) = \ell + \mathrm{i}d = e^{\mathrm{i}(\alpha + \pi/2)}(a_x - \mathrm{i}a_y) = \rho q_\infty \Gamma;
$$

i.e.

$$\ell = \rho q_\infty \Gamma \tag{3.15a}$$

$$d = 0. \tag{3.15b}$$

If the vortex is clockwise $\Gamma > 0$ and the lift is positive (ρ and q_∞ are always positive).

These two results constitute the *Kutta–Joukowsky Theorem*, which actually holds for any flow over any aerofoil and is perhaps the most important result in two-dimensional aerodynamics. It means that instead of having to calculate the aerodynamic force by integrating contributions from stresses around the surface, we can just calculate the circulation and use Equations (3.15).

The Kutta–Joukowsky Theorem is easily verified for plane ideal flow over a circular cylinder.

More general flows than that described by Equation (3.13) could be considered by adding more terms to the series. Terms of the form z^k with positive k should be excluded, however, as they imply speeds increasing without bound in the far field ($q \sim |z|^k$). Terms of the form z^k with $k = -2, -3, \ldots$ on the other hand all only produce terms like z^m with $m \leqslant -2$ in w^2, and none of these contribute to the contour integral: all z^m terms vanish except those for $m = -1$.

The interpretation of the theorem is straightforward. We know that the pressure is related to the kinetic energy of the fluid (i.e. the square of the velocity) by Bernoulli's equation, and so it is pairwise products of velocities that will be of interest. Of all possible pairs of velocities of the type z^k, the only combination giving an average speed difference between the upper and lower surfaces is one varying like $\cos\theta$ or $\sin\theta$ or $e^{\pm i\theta}$, not $e^{in\theta}$ for any $n \neq \pm 1$. If the velocities are restricted to have $k \leqslant 0$ (by the requirement of bounded velocity in the far field) this leaves only the possibility $0 - 1 = -1$ or $-1 + 0 = -1$, the uniform stream–vortex pairing. In this pairing, the vortex on one side reinforces the stream and on the other retards it, and it is this that generates the net difference in speeds, kinetic energies, and pressures, and therefore generates the net lift. On the other hand, there exists no such mechanism for the generation of drag by plane ideal flow.

3.8 Exercises

1. Using the rule for dot product of a complex velocity and a complex vector in the exercise at the end of Chapter (2), rederive Equation (3.2a) from the complex velocity z^k and the unit vector directed away from the origin $e^{i\theta}$.

2. Write the Octave code to reproduce the silk-thread and streamline plots of the uniform stream with arbitrary direction in Figure 3.5.

3. Show that the doublet z^{-2} can also be expressed as the sum of counterrotating vortices on the y-axis, with equal strengths, equally displaced about the x-axis. [*Hint:* $z^2 + \epsilon^2 = z^2 - (i\epsilon)^2$.]

4. Show that, for $z = x + iy$ and positive real constants q_∞ and a, the complex velocity

$$w = q_\infty \left(1 - \frac{a^2}{z^2}\right)$$

gives a plane ideal ideal flow outside the solid circle of radius a centred on the origin $x = y = 0$.

(a) Show that a plane ideal line vortex at the origin of any strength can be added to the flow without violating the impermeability of the circle $|z| = a$.

(b) Compute the pressure around the surface of the circle, assuming the vortex strength is S.

(c) Compute the aerodynamic force on the circle.

(d) What is the circulation around the boundary? Does the direct evaluation of the aerodynamic force by integrating the pressure agree with the result implied by the Kutta–Joukowsky Theorem?

(e) Find the *stagnation points* (where $w = 0$). What is the pressure there?

5. Find the complex potential for the uniform stream with arbitrary direction plotted in Figure 3.5.

6. Investigate the plane ideal flow generated by placing a source of strength Q in a uniform flow of speed q_∞.

(a) Find the stagnation point and describe all the streamlines passing through it.

(b) Take the curved streamline through the stagnation point as the boundary of a (semi-infinite) solid. Find the pressure on this boundary.

(c) Discuss the aerodynamic force on the solid. (Warning: the integral $a_x = -\oint p \, dy$ is ugly.)

7. Employ the 'Waypoints' option of Octave's **quadgk** to repeat the numerical path integration of Listing 3.7 over the right-angle isoceles triangle with the same vertices; i.e.

```
quadgk (@angle, 0, 0, 'Waypoints', [1, 1i] * a)
```

Verify that this agrees with the exact answer obtained in the exercise at the end of Chapter 2.

8. Show that the complex conjugate of a product is the product of the complex conjugates of its factors.

9. Show that, following Exercise 2.8.15, both the real and imaginary parts of a differentiable function of $x - iy$ satisfy Laplace's Equation (3.11).

10. Show that if ϕ and ψ are the potential and stream function of the same plane flow then $W(x + iy) = \phi(x, y) + i\psi(x, y)$ satisfies the Cauchy–Riemann Equations (2.22).

3.9   Further Reading

For the governing equations in polar coordinates, see Thwaites (1987) and Bertin (2002).

Blasius's theorem is derived by Glauert (1926), Milne-Thomson (1973), Paraschivoiu (1998), and Pope (2009) using similar complex variable methods to those employed here; Karamcheti (1966) shows that it is a special case of a three-dimensional theorem derived using vector calculus methods more akin to those of Chapter 9.

References

Bertin, J.J. (2002) *Aerodynamics for Engineers* 4th edn. New Jersey: Prentice Hall.

Glauert, H. (1926) *The Elements of Aerofoil and Airscrew Theory*. Cambridge: Cambridge University Press.

Karamcheti, K. (1966) *Principles of Ideal-Fluid Aerodynamics*. Chichester: John Wiley & Sons, Ltd.

Milne-Thomson, L.M. (1973) *Theoretical Aerodynamics*, 4th edn. New York: Dover.

Paraschivoiu, I. (1998) *Aérodynamique Subsonique*. Éditions de l'École de Montréal.

Pope, A. (2009) *Basic Wing and Airfoil Theory*. New York: Dover.

Thwaites, B. (ed.) (1987) *Incompressible Aerodynamics*. New York: Dover.

4

Conformal Mapping

In Chapter 2 it was shown that plane ideal flows could be readily obtained from analytic complex functions and in Chapter 3 this was exploited to come up with several examples of two-dimensional flow-fields satisfying the Euler equations; however, none of them much resembled flow over an aerofoil.

In Section 4.1, a powerful means of combining and extending complex velocity fields by the composition of analytic functions is introduced which does yield flow-fields around shapes as complicated as aerofoils. In particular, it is used in Section 4.3 to find the exact solution for the flow over a thin flat plate at an angle to the free-stream, and this will be shown in the following chapters to contain many of the essential features of flow over a wing.

4.1 Composition of Analytic Functions

Another way to obtain more analytic functions is through composition of functions; e.g. $f(z(\zeta))$: the composition of two analytic functions, f and z, is analytic since by the chain rule

$$\frac{\mathrm{d}f}{\mathrm{d}\zeta} = \frac{\mathrm{d}f}{\mathrm{d}z}\frac{\mathrm{d}z}{\mathrm{d}\zeta}. \tag{4.1}$$

On the other hand, given $f(z(\zeta))$ as an analytic function of ζ, $f(z)$ is analytic too since

$$\frac{\mathrm{d}f}{\mathrm{d}z} = \frac{\mathrm{d}f}{\mathrm{d}\zeta}\frac{\mathrm{d}\zeta}{\mathrm{d}z} = \frac{\mathrm{d}f}{\mathrm{d}\zeta}\left(\frac{\mathrm{d}z}{\mathrm{d}\zeta}\right)^{-1} = \frac{\mathrm{d}f/\mathrm{d}\zeta}{\mathrm{d}z/\mathrm{d}\zeta}. \tag{4.2}$$

This could have been obtained more easily by dividing Equation (4.1) by $\mathrm{d}z/\mathrm{d}\zeta$.

If the point $z_1 = z(\zeta_1)$ is mapped from the point ζ_1, then the value of the function f at z_1 in the z-plane is the same as the value of f at ζ_1 in the ζ-plane, since $z_1 = z(\zeta_1)$ implies $f(z_1) = f(z(\zeta_1))$.

This is particularly useful if for f we consider the complex potential W. If $\mathcal{C}$ is a streamline of the flow in the ζ-plane, then $\psi = \Im W$ is constant along it, and this also applies to the imaginary part of W along the curve $z(\mathcal{C})$ in the z-plane to which $\mathcal{C}$ is mapped by $z = z(\zeta)$. It follows that $z(\mathcal{C})$ is a streamline in the ζ-plane.

Theory of Lift: Introductory Computational Aerodynamics in MATLAB®/Octave, First Edition. G. D. McBain.
© 2012 John Wiley & Sons, Ltd. Published 2012 by John Wiley & Sons, Ltd.

The complex velocities in the ζ- and z-planes are then given by applications of Equations (4.1) and (4.2):

$$w(\zeta) = \frac{dW}{d\zeta} = \frac{dW}{dz}\frac{dz}{d\zeta}$$

$$w(z) = \frac{dW}{dz} = \frac{dW/d\zeta}{dz/d\zeta}.$$

4.2   Mapping with Powers of ζ

A simple family of useful transformations is $z = \zeta^k$, with k a real constant.

4.2.1   Example: Square Mapping

The mapping $z = \zeta^2$ takes the first quadrant of the ζ-plane into the first two quadrants of the z-plane. Let us denote the coordinate-angle in the ζ-plane as

$$\chi \equiv \arg \zeta$$

so that ζ has the polar form

$$\zeta = |\zeta|e^{i\chi}.$$

Then, since $\zeta^2 = (|\zeta|e^{i\chi})^2 = |\zeta|^2 e^{2i\chi}$, an effect of the mapping is to double the angle each point makes with the positive real axis: $\arg z = 2 \arg \zeta$. Also, it is no surprise that the square of a positive real number is also real and positive, so that the positive real ζ-axis maps to the positive real z-axis, while the square of an imaginary number is negative, so the positive imaginary ζ-axis maps to the negative real z-axis.

Consider then the complex potential $W = z$ from Figure 3.2b, which represents a simple uniform horizontal velocity in the z-plane, with stream function $\psi(z) = \Im W(z) = \Im z = y$, and horizontal streamlines $\psi = y = $ const. The same complex potential represents the corner flow of Figure 3.1b in the ζ-plane (apart from a scale factor of 2): $\psi = \Im W = \Im \zeta^2$. In terms of the real and imaginary parts of ζ:

$$\zeta \equiv \xi + i\eta, \tag{4.3}$$

we have

$$\psi = \Im W = \Im \zeta^2 = \Im(\xi + i\eta)^2 = \Im(\xi^2 - \eta^2 + 2i\xi\eta) = 2\xi\eta.$$

Four streamlines are plotted in Figure 4.1: the same four in each case. The velocity in the ζ-plane is given by

$$w(\zeta) = \frac{dW}{d\zeta} = 2\zeta$$

which (apart from the scale factor 2) is the same as the complex velocity in Figure 3.1a.

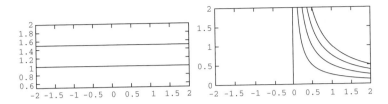

Figure 4.1   Mapping the upper-half z-plane (left) to the first quadrant of the ζ-plane (right) with $z = \zeta^2$, to get the corner flow in the ζ-plane from a uniform horizontal flow in the z-plane with the complex potential $W = z$

4.2.2   *Conforming Mapping by Contouring the Stream Function*

There are several different computational approaches to conformal mapping. One is to plot the level-sets of the imaginary part of the complex potential in the z- and ζ-planes, using Octave's **contour** function, as in Listing 4.1; this is a natural extension of Listing 3.3.

As an example of its use, the code used to produce Figure 4.1 is given in Listing 4.2.

4.2.3   *Example: Two-thirds Power Mapping*

Just as $z = \zeta^2$ takes the first ζ-quadrant to the first two z-quadrants, $z = \zeta^{2/3}$ takes the first three to the first two. If we consider the same flow in the z-plane as in the last example, i.e.

Listing 4.1   `conformal_streamlines`: plot the level sets v of the imaginary parts of two complex potential functions `W1` and `W2` in side-by-side copies of the same complex rectangle with defined lower-left and upper-right corners `ll` and `ur`.

```
function conformal_streamlines (ll, ur, W1, W2, v)
  Z = cmeshgrid (ll, ur, 1e3);

  subplot (1, 2, 1)
  contour (real (Z), imag (Z), imag (W1 (Z)), v)
  axis ('image')

  subplot (1, 2, 2)
  contour (real (Z), imag (Z), imag (W2 (Z)), v)
  axis ('image')
```

Listing 4.2   Octave code to generate Figure 4.1, using Listing 4.1.

```
conformal_streamlines (-2, 2+2i, @ (Z) Z, @ (Z) Z.^2, 0:0.5:2)
```

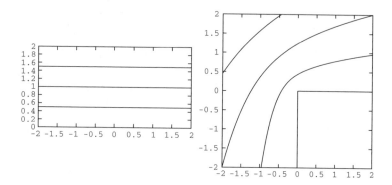

Figure 4.2 Mapping the upper-half z-plane (left) to the first three quadrants of the ζ-plane (right) with $z = \zeta^{2/3}$, to get the external corner flow in the ζ-plane from a uniform horizontal flow in the z-plane with the complex potential $W = z$

$W = z$, and apply $z = \zeta^{2/3}$, we get the flow around an external corner, as in Figure 4.2. The velocity in the ζ-plane is

$$w(\zeta) = \frac{\mathrm{d}W}{\mathrm{d}z}\frac{\mathrm{d}z}{\mathrm{d}\zeta} = 1 \times \frac{2}{3}\zeta^{-1/3} = \frac{2}{3}|\zeta|^{-1/3}\mathrm{e}^{-\mathrm{i}\chi/3},$$

so that the direction of flow at any point makes an angle $\beta = \frac{\chi}{3}$, one-third that of the line from the origin to the point; e.g. for all points on the positive η-axis, the flow is 'two o'clock', and for all points on the negative ξ-axis, it's one o'clock.

4.2.4 Branch Cuts

One difficulty with the above mapping functions is that they are not bijective. In the case of $z = \zeta^2$, the third quadrant of the ζ-plane also maps to the upper-half z-plane; it is therefore not injective. For $z = \zeta^{2/3}$, the mapping itself is multivalued; for example, should $\zeta = -1 = \mathrm{e}^{\mathrm{i}\pi}$ go to $\mathrm{e}^{2\mathrm{i}\pi/3}$ or should $\zeta = -1 = \mathrm{e}^{-\mathrm{i}\pi}$ go to $\mathrm{e}^{-2\mathrm{i}\pi/3}$? In Section 4.2.3, we chose the branch that maps $\zeta = |\zeta|\mathrm{e}^{\mathrm{i}\chi}$ with $0 \leqslant \chi \leqslant \frac{3\pi}{2}$ to $z = r\mathrm{e}^{\mathrm{i}\theta} = |\zeta|^{2/3}\mathrm{e}^{2\mathrm{i}\chi/3}$, since it's continuous in the first three quadrants, which constitute the domain of the flow while the fourth quadrant represents a solid obstacle.

Octave and many other computer programs, such as NumPy's numpy.**angle**, take the branch-cut in the power functions following from the rule that the argument of z is to be taken between $-\pi$ and $+\pi$. In Listing 4.3, this is got around by using **mod** (angle (z), 2***pi**) to get the argument between 0 and 2π; i.e. with the branch-cut along the positive real axis, which was suitable as it was a boundary in the example, whereas the negative real-axis is inside the flow-field in the ζ-plane.

Another small numerical trick used in that program is using a very small value (1e-6) of the stream function instead of zero, since the zero-streamline would lie 'exactly' on the branch-cut, and 'exactly' is not a well-defined concept in floating-point arithmetic. The ambiguity is avoided by slightly but definitely displacing the streamline away from the branch-cut.

Listing 4.3   Octave script used to generate Figure 4.2.

```
powmag = @ (z, k) abs (z) .^ k;
powarg = @ (z, k) k * mod (angle (z), 2 * pi);
pow = @ (z, k) powmag (z, k) .* exp (1i * powarg (z, k));
conformal_streamlines (-2-2i, 2+2i, ...
        @ (Z) Z, @ (Z) pow (Z, 2/3), ...
        [1e-6, 0.5:0.5:2])
```

4.2.5   Other Powers

Other shapes can be obtained using other powers. Also, one can use a different complex potential.

4.3   Joukowsky's Transformation

The mapping function $z = \zeta + a^2/\zeta$ for a real constant a is called *Joukowsky's transformation*.

4.3.1   Unit Circle from a Straight Line Segment

A shifted special case is

$$z = \zeta + \frac{c}{2} + \frac{c^2}{16\zeta}. \tag{4.4}$$

Notice that both ζ and $\frac{c^2}{16\zeta}$ map to the same point: the mapping Equation (4.4) is not injective. If $4|\zeta| = c$ then both these points are on the circle in the ζ-plane, otherwise one is inside and one outside, and both the interior and exterior of the circle in the ζ-plane map to the entire z-plane. This means care must be taken in constructing the inverse transformation

$$\zeta = \frac{c}{4}\left\{\frac{2z}{c} - 1 \pm 2\sqrt{\frac{z}{c}\left(\frac{z}{c} - 1\right)}\right\} = \frac{z - \frac{c}{2} \pm \sqrt{z(z - c)}}{2}. \tag{4.5}$$

Usually for aerodynamical applications it is the exterior, $4|\zeta| > c$, that is of interest so this condition should be checked to determine the appropriate sign in Equation (4.5).

The mapping of Equation (4.4) takes the circle $4|\zeta| = c$ in the ζ-plane to the straight-line segment joining zero and c in the z-plane. To see this, put $4\zeta = ce^{i\chi}$ and let χ run from zero to 2π:

$$z = \frac{ce^{i\chi}}{4} + \frac{c}{2} + \frac{c}{4e^{i\chi}} = \frac{c}{2}(1 + \cos\chi) \equiv c\cos^2\frac{\chi}{2}, \tag{4.6}$$

so that z is real and confined to $0 \leqslant z \leqslant c$.

As χ runs from zero to π, ζ moves anticlockwise around the upper unit semicircle and z moves left from c to zero, and then as χ increases to 2π, ζ moves anticlockwise around the lower unit semicircle and z moves right from zero back to c.

In this double-tracing of the segment, it can help to think of the leftward tracing for $0 < \chi < \pi$ as corresponding to the upper side of the segment and the right tracing for $\pi < \chi < 2\pi$ as corresponding to the lower surface. If the segment $0 < z < c$ is a streamline, there is no flow across it and it can be replaced by an impermeable thin flat plate, and then differences in the velocity or pressure between the upper and lower surfaces can be tolerated, even though the plate thickness is neglected. Thus in talking about the pressure at $x = \frac{c}{2}$, say, we should specify whether we mean the upper or lower surface; i.e. $z = \frac{c}{2} \pm i\epsilon$ as $\epsilon \downarrow 0$.

4.3.2 Uniform Flow and Flow over a Circle

Consider the complex potential $W = q_\infty z$ in the z-plane (with q_∞ a positive constant). Its stream function is $\psi(z) = \Im W(z) = q_\infty \Im z = q_\infty y$; which is constant along horizontal lines. This means the real z-axis is a streamline, and so is the segment from $z = 0$ to $z = c$. This further means that the corresponding complex potential in the ζ-plane will have the circle $4|\zeta| = c$ as a streamline. That potential is

$$W = q_\infty z = q_\infty \left(\zeta + \frac{c}{2} + \frac{c^2}{16\zeta} \right).$$

Instead, let's shift some constants and consider

$$W = q_\infty \left(z - \frac{c}{2} \right) = q_\infty \left(\zeta + \frac{c^2}{16\zeta} \right). \tag{4.7}$$

The corresponding stream functions

$$\psi = q_\infty y = q_\infty \eta \left(1 - \frac{c^2}{16|\zeta|^2} \right), \tag{4.8}$$

where $\eta \equiv \Im \zeta$, from Equation (4.3), which are evidently constant along both $y = 0$ and $4|\zeta| = c$, respectively, are plotted in Figure 4.3 using the Octave code of Listings 4.4 and 4.5.

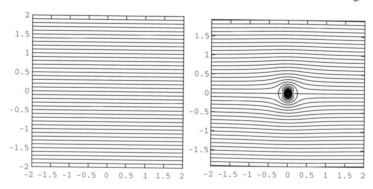

Figure 4.3 Flow over a thin plate at zero incidence (left), and flow over a circle (right), being z- and ζ-plane representations of the complex potential $W = q_\infty z$ with the Joukowsky transformation of Equation (4.4)

Listing 4.4   Octave code used to generate Figure 4.3, using Listing 4.5.

```
c = 1;
W = @ (zeta) zeta + (c/4)^2 ./ zeta;
conformal_streamlines (-2-2i, 2+2i, ...
         @ (z) W (joukowsky_inverse (z, c)), ...
         W, -2:0.1:2 )
```

Listing 4.5   `joukowsky_inverse`: compute the point in the ζ plane (outside the circle $4|\zeta| = c$) corresponding to z in the Joukowsky transformation, implementing Equation (4.5).

```
function zeta = joukowsky_inverse (z, c)
  zeta = (z - c/2 + sqrt (z .* (z - c))) / 2;
  flip = abs (zeta) < c/4;
  zeta(flip) = c^2 ./ (16 * zeta(flip));
```

4.3.3   Thin Flat Plate at Nonzero Incidence

Nontrivial flows over the thin flat plate in the z-plane can now be obtained by modifying the complex potential in the ζ-plane, provided the circle $4|\zeta| = c$ remains a streamline. First, rotate the whole streamline pattern in the ζ-plane by replacing ζ with $e^{-i\alpha}\zeta$:

$$W = q_\infty \left(\frac{\zeta}{e^{i\alpha}} + \frac{c^2 e^{i\alpha}}{16\zeta} \right), \tag{4.9}$$

and then recover the modified flow over the flat plate from the Joukowsky transformation Equations (4.4)–(4.5).

Notice that as $|\zeta| \to \infty$, $W \sim q_\infty \zeta e^{-i\alpha}$, which is the complex potential of a uniform stream with speed q_∞ inclined at an angle α to the positive real axis. Because the Joukowsky transformation Equation (4.4) far from the origin asymptotically approaches $z \sim \zeta + \frac{c}{2}$, the velocity far from the thin flat plate

$$w(z) = \frac{\mathrm{d}W/\mathrm{d}\zeta}{\mathrm{d}z/\mathrm{d}\zeta} \sim q_\infty e^{-i\alpha} \qquad (|z| \to \infty)$$

is a uniform stream with the same speed q_∞ and the same direction α; this is plausible in Figure 4.4.

Since the complex velocity for the thin flat plate is recovered from Equation (4.2), the conformal transformation can introduce a singularity in $w(z)$ if the transformation has *stationary points* for which $\mathrm{d}z/\mathrm{d}\zeta$ vanishes. This occurs for the Joukowsky mapping in Equation (4.4) at

$$\frac{\mathrm{d}}{\mathrm{d}\zeta} \left\{ \zeta + \frac{c}{2} + \frac{(c/4)^2}{\zeta} \right\} = 1 - \left(\frac{c}{4\zeta} \right)^2 = 0; \tag{4.10}$$

i.e. $4\zeta = \pm c$. These points map to $z = c$ and $z = 0$, the ends of the thin flat plate.

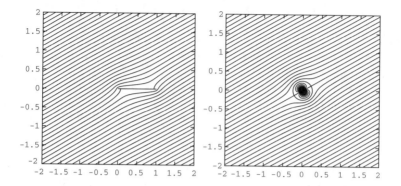

Figure 4.4 Flow over a thin plate at nonzero incidence (left), and flow over a circle (right), being z- and ζ-plane representations of the complex potential of Equation (4.9) with the Joukowsky transformation Equations (4.4)–(4.5)

Implemented in Octave as shown in Listing 4.6, this produces Figure 4.4. Changing the incidence to $\alpha = \pi/2$ produces Figure 4.5.

4.3.4 Flow over the Thin Flat Plate with Circulation

The flow can be further modified by adding a vortex from Equation (3.10) at the origin in the ζ-plane, since this still preserves $|\zeta| = c/4$ as a streamline. This changes the complex potential of Equation (4.9) to

$$W = \frac{q_\infty c}{4}\left(\frac{4\zeta}{ce^{i\alpha}} + \frac{ce^{i\alpha}}{4\zeta}\right) + \frac{i\Gamma}{2\pi}\ln\frac{4\zeta}{ce^{i\alpha}}. \tag{4.11}$$

Note that $\ln\zeta$ and $\ln(4\zeta/ce^{i\alpha}) = \ln\zeta - (i\alpha + \ln c - \ln 4)$ differ only by a physically irrelevant complex constant.

The results are plotted in Figure 4.6 for the choice $\alpha = \frac{\pi}{6}$ and $\Gamma = q_\infty c\pi \sin\alpha$. The latter choice happens to be exactly that which causes the flow in the z-plane to leave the trailing

Listing 4.6 Octave code to generate Figure 4.4.

```
alpha = pi/6; c = 1;
W = @ (zeta) zeta + (c/4)^2 ./ zeta;
V = @ (zeta) W (zeta / exp (1i * alpha));
conformal_streamlines (-2-2i, 2+2i, ...
        @ (z) V (joukowsky_inverse (z, c)), ...
        V, -3:0.1:3 )
subplot (1, 2, 1)
hold ('on')
plot ([0, 1], [0, 0], '-')
```

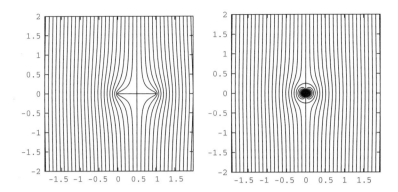

Figure 4.5   Flow over a thin plate at incidence $\alpha = \frac{\pi}{2}$ with zero circulation (z), and flow over a circle (ζ), being z- and ζ-plane representations of the complex potential of Equation (4.11) with the Joukowsky transformation of Equation (4.4)

Listing 4.7   Octave code to generate Figure 4.6.

```
alpha = pi/6; c = 1; Gamma = c * pi * sin (alpha);
W = @ (zeta) zeta + (c/4)^2 ./ zeta;
V = @ (zeta) W (zeta) - Gamma * log (4 * zeta / c) / pi / 2i;
U = @ (zeta) V (zeta / exp (1i * alpha));
conformal_streamlines (-2-2i, 2+2i, ...
        @ (z) U (joukowsky_inverse (z, c)), ...
        U, -3:0.1:4 )
subplot (1, 2, 1)
hold ('on')
plot ([0, 1], [0, 0], '-')
```

edge ($z = c$) of the plate smoothly, as can be seen in Figure 4.6, generated by Listing 4.7; the significance of this choice will be explored in Chapter 5.

4.3.5   Joukowsky Aerofoils

The Joukowsky transformation Equation (4.4) maps the circle $4|\zeta| = c$ to the thin flat plate $0 < z < c$, but it also maps circles not centred on $\zeta = 0$ to some very interesting shapes, including some that look like aerofoils.

4.4   Exercises

1. Derive Equation (4.8) for the stream functions over an aligned flat plate and around a circular obstacle in plane ideal flow without circulation by taking the imaginary parts of Equation (4.7) in the z- and ζ-planes.

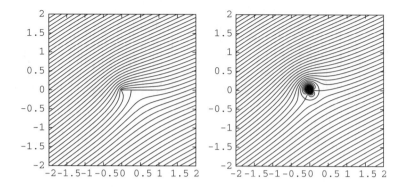

Figure 4.6   Flow with circulation over a thin plate at nonzero incidence with circulation (left), and over a circle (right), being z- and ζ-plane, respectively, representations of the complex potential of Equation (4.11) with the Joukowsky transformation of Equation (4.4)

2. (a) Show that the conformal transformation

$$W = z = i(\zeta - 1)^k \tag{4.12}$$

   where

$$k = \frac{\pi}{2\pi - \kappa}$$

   and $0 < \kappa < \pi$ is a small positive angle, illustrated in Figure 4.7 for $\kappa = \frac{\pi}{6}$, has a streamline with a corner at the point $\zeta = 1$, and which lies on the two rays from it making angles $\frac{\kappa}{2}$ with the negative real axis in the ζ-plane.
   (b) Produce the silk-thread velocity plot (of Section 3.1.3) for the ζ-plane.
   (c) Show that velocity at the vertex $\zeta = 1$ is infinite if $\kappa < \pi$. Show that the pressure is negatively infinite.
   (d) Describe the limiting case $\kappa \to \pi$.
   (e) What value of κ is required here to produce a flow in the ζ-plane congruent to (a possibly rotated and translated) Figure 4.2b. What subsequent conformal transformation would accomplish the required rotation and translation?
   (f) Beginning from Figure 4.2, what subsequent conformal transformation would modiy it so that internal bisector of the solid wedge lay along the negative real axis of the ζ-plane?

3. (a) For the same value of k as in the previous exercise, consider the flow in the ζ-plane described by

$$W = z = (\zeta - 1)^{2k}. \tag{4.13}$$

   (b) Show that the two rays from $\zeta = 1$ making angles $\pm\frac{\kappa}{2}$ with the negative real axis in the ζ-plane are again streamlines. See Figure 4.8.

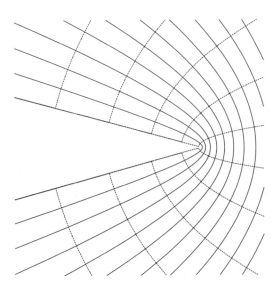

Figure 4.7   Antisymmetrical plane ideal flow around a wedge of angle $\kappa = \frac{\pi}{6}$ lying along the negative real axis, as given by Equation (4.12)

(c) Show that a third ray emanates from $\zeta = 1$ and that the stream function has the same value along it. What angle does it make with the positive real axis in the ζ-plane?

(d) Show that $\zeta = 1$ is a stagnation point.

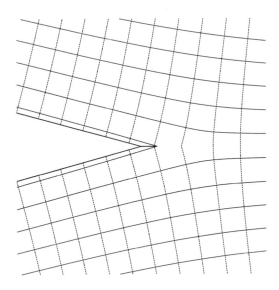

Figure 4.8   Symmetrical plane ideal flow around a wedge of angle $\kappa = \frac{\pi}{6}$ lying along the negative real axis, as given by Equation (4.13)

4. (a) Show that any linear combination of the flows in the ζ-plane in the preceding two
   exercises also treats as impermeable the boundaries of the wedge.
   (b) Under what conditions is there a stagnation point on the surface of the wedge? How
   does its location depend on the relative contributions of the two basic flows in the
   linear combination?
   (c) Under what conditions are the velocity and pressure bounded everywhere along the
   surface of the wedge in the vicinity of the vertex?

4.5 Further Reading

The theory of conformal mapping and the Joukowsky transformation in particular are much
discussed in classical works on aerodynamics such as those by Glauert (1926), Prandtl and
Tietjens (1957), Abbott and von Doenhoff (1959), Milne-Thomson (1973), Ashley and Landahl
(1985), Paraschivoiu (1998), Katz and Plotkin (2001), and Moran (2003); most of these also
go on to describe the Joukowsky aerofoils.

The matrix computation library NumPy is described by Oliphant (2006).

A flow-pattern similar to Figure 4.5 was derived slightly differently by Lamb (1932).

References

Abbott, I.H. and von Doenhoff, A.E. (1959) *Theory of Wing Sections*. New York: Dover.
Ashley, H. and Landahl, M. (1985) *Aerodynamics of Wings and Bodies*. New York: Dover.
Glauert, H. (1926) *The Elements of Aerofoil and Airscrew Theory*. Cambridge: Cambridge University Press.
Katz, J. and Plotkin, A. (2001) *Low-Speed Aerodynamics*, 2nd edn. Cambridge: Cambridge University Press.
Lamb, H. (1932) *Hydrodynamics*, 6th edn. Cambridge: Cambridge University Press.
Milne-Thomson, L.M. (1973) *Theoretical Aerodynamics*, 4th edn. New York: Dover.
Moran, J. (2003) *An Introduction to Theoretical and Computational Aerodynamics*. New York: Dover.
Oliphant, T.E. (2006) *Guide to NumPy*. Trelgol.
Paraschivoiu, I. (1998) *Aérodynamique Subsonique*. Éditions de l'École de Montréal.
Prandtl, L. and Tietjens, O.G. (1957) *Applied Hydro- and Aeromechanics*. New York: Dover.

5

Flat Plate Aerodynamics

5.1 Plane Ideal Flow over a Thin Flat Plate

From the conformal mapping of Section 4.3.4, the complex velocity (in the z-plane) for a stream of speed q_∞ at an angle α to a thin flat plate of length c is given by

$$w(z) = \frac{dW}{dz} = \frac{dW/d\zeta}{dz/d\zeta} \tag{5.1}$$

where the complex potential W is given by Equation (4.11)

$$W = \frac{q_\infty c}{4} \left(\frac{4\zeta}{ce^{i\alpha}} + \frac{ce^{i\alpha}}{4\zeta} \right) + \frac{i\Gamma}{2\pi} \ln \frac{4\zeta}{ce^{i\alpha}},$$

and the mapping by Equation (4.4)

$$z = \zeta + \frac{c}{2} + \frac{c^2}{16\zeta}.$$

The numerator of Equation (5.1) is

$$\frac{dW}{d\zeta} = \frac{q_\infty c}{4} \left(\frac{4}{ce^{i\alpha}} - \frac{ce^{i\alpha}}{4\zeta^2} \right) + \frac{i\Gamma}{2\pi\zeta}$$

and the denominator is

$$\frac{dz}{d\zeta} = 1 - \frac{c^2}{16\zeta^2}.$$

Hence

$$w(z) = \frac{\frac{q_\infty c}{4} \left(\frac{4}{ce^{i\alpha}} - \frac{ce^{i\alpha}}{4\zeta^2} \right) + \frac{i\Gamma}{2\pi\zeta}}{1 - \frac{c^2}{16\zeta^2}} = \frac{q_\infty \left(\frac{4\zeta}{ce^{i\alpha}} - \frac{ce^{i\alpha}}{4\zeta} \right) + \frac{2i\Gamma}{\pi c}}{\frac{4\zeta}{c} - \frac{c}{4\zeta}}. \tag{5.2}$$

Notice that the denominator vanishes at $4\zeta = \pm c$; i.e. $z = 0$ and c; so that the velocity will be unbounded at both edges of the plate unless the numerator vanishes there too.

Theory of Lift: Introductory Computational Aerodynamics in MATLAB®/Octave, First Edition. G. D. McBain.
© 2012 John Wiley & Sons, Ltd. Published 2012 by John Wiley & Sons, Ltd.

On the segment $0 < z < c$ (or in terms of ζ the circle $|\zeta| = \frac{c}{4}$) we have from Equation (4.6) $z = \frac{c}{2}(1 + \cos \chi)$, which is real and so equal to x. In a sense, χ can be thought of as an alternative coordinate x for points along the plate; the forward and inverse coordinate transformations are

$$x(\chi) = \frac{c}{2}(1 + \cos \chi) \tag{5.3a}$$

$$\chi(x) = \arccos\left(\frac{2x}{c} - 1\right). \tag{5.3b}$$

In this context, χ is called the *eccentric angle*.

On the plate then the complex velocity is

$$w\,(x(\chi)) = \frac{q_\infty\left\{e^{i(\chi-\alpha)} - e^{i(\alpha-\chi)}\right\} + \frac{2i\Gamma}{\pi c}}{e^{i\chi} - e^{-i\chi}} = \frac{q_\infty \sin(\chi - \alpha) + \frac{\Gamma}{\pi c}}{\sin \chi}. \tag{5.4}$$

This is real which means $w = u$ and $v = 0$ and corresponds to no flow through the plate, v being the component of velocity normal to the horizontal plate which lies in the x-axis.

5.1.1 Stagnation Points

The velocity given by Equation (5.4) on the plate vanishes when

$$\pi c q_\infty \sin(\chi - \alpha) + \Gamma = 0. \tag{5.5}$$

For example, if $\alpha = \frac{\pi}{2}$ and $\Gamma = 0$ (which corresponds to flow, without circulation, broadside on to the plate as in Figure 4.5), the stagnation points occur at $\chi = \frac{\pm\pi}{2}$; i.e. $x = \frac{c}{2}$ on the upper and lower sides of the plate; which makes sense on symmetry grounds.

As another example, in Figure 4.4 we had $\alpha = \frac{\pi}{6}$ and $\Gamma = 0$ so that the stagnation points were $\chi = \frac{\pi}{6}$ and $\chi = \frac{-5\pi}{6}$; i.e.

$$\frac{x}{c} = \frac{1 + \cos \chi}{2} = \begin{cases} \frac{2+\sqrt{3}}{4} \doteq 0.933, & \text{on the upper side} \\ \frac{2-\sqrt{3}}{4} \doteq 0.067, & \text{on the lower side.} \end{cases}$$

This is plausible from the streamline pattern in the (left) z-plane in Figure 4.4.

5.1.2 The Kutta–Joukowsky Condition

Also, the velocity on the surface of the plate is undefined at the points for which $\sin \chi = 0$; i.e. $\chi = 0$ or π; i.e. $x = c$ or 0, the trailing and leading edges. See the tilted and broadside configurations of Figures 4.4 and 4.5. If the numerator is nonzero at these points, the velocity is infinite; however, if the numerator vanishes at one of these points, as in Equation (5.5), a finite velocity might be obtained—the limit would have to be investigated more carefully, using l'Hôpital's rule for example. This could be used as a condition to determine Γ:

$$\Gamma = \pi c q_\infty \sin \alpha, \quad \text{(trailing edge)} \tag{5.6}$$

or $\Gamma = -\pi c q_\infty \sin\alpha$ if the condition were enforced at the leading edge instead. The first of these—the condition that the singularity at the trailing edge be cancelled—is known as the *Kutta–Joukowsky condition*.

The streamlines for the thin flat plate with the Kutta–Joukowsky condition were shown in Figure 4.6, where it is seen that the flow leaves the trailing edge smoothly.

Notice that the Kutta–Joukowsky condition leads to a positive circulation, and therefore, by the Kutta–Joukowsky Equation (3.15a), a positive lift. Since this is a desirable thing, it suggests that wing sections should have sharp trailing edges; to eliminate the leading edge velocity singularity, wing section leading edges are rounded. Thus, in this treatment of the thin plate, we are only interested in the Kutta–Joukowsky condition, and ignore the difficulty at the leading edge, knowing that it doesn't arise for real wing sections.

Reversed Aerofoils

Actually while the equation for circulation following from the leading-edge counterpart of the Kutta–Joukowsky condition ($\Gamma = -\pi c q_\infty \sin\alpha$) would lead to negative circulation at positive incidence, it would give positive circulation and therefore positive lift at negative incidence.

Thus one might be led to wonder whether this phenomenon could be observed physically, if one had a wing section with leading edge sharp and trailing edge rounded, to promote the reverse Kutta–Joukowsky condition. Of course, this configuration does apply regularly in practice: whenever a regular rounded-leading sharp-trailing aerofoil is subject to reverse flow, as can happen for rotary-wing craft. The answer is negative, experiments still show a positive lift/incidence slope; despite the sharpness of the leading edge, the overall flow pattern appears much as in the usual case: at small positive incidence, the air hits the underside not too far from the leading edge, moves forwards against the main flow and rounds the leading edge before moving back along the upper side with the main flow towards the trailing edge which it leaves smoothly. Merely sharpening the leading edge appears to be insufficient to attract the forward stagnation point (from its usual position at small positive incidence) forwards along the underside to the leading edge; similarly, a slight rounding the trailing edge does not stop the flow from leaving the trailing edge smoothly.

Thus although it is consistent with the Kutta–Joukowsky condition, the conventional rounded-leading sharp-trailing geometry is not necessary for it to apply. The reason for the success of the geometry is that it maintains this overall flow pattern to higher incidence before *stall* sets in. A detailed discussion of that lies outside beyond ideal flow theory and will be deferred to Chapter 17. This is an early hint that flow contains an upstream–downstream asymmetry which is not reflected in the ideal flow model. Nevertheless, since wings are generally operated at such incidences as do not lead to stall, it remains useful to pursue a relatively simple theory which is effective in the absence of stall.

5.1.3 Lift on a Thin Flat Plate

Notice that Γ is the circulation around any loop enclosing the thin flat plate. To see this, use the rule in Equation (3.9)

$$\Gamma_C = -\Re[W]_C$$

and the expression in Equation (4.11) for W; as the outline of the plate is traced anticlockwise, $\zeta = ae^{i\chi}$ goes from ae^{i0} to $ae^{i2\pi}$ which decreases W by Γ, because of the logarithmic term.

Now that the circulation is related to the incidence by the Kutta–Joukowsky condition, Equation (5.6), the lift per unit span can be calculated from the Kutta–Joukowsky Equation (3.15a) as

$$\ell = \rho q_\infty \Gamma = \pi \rho q_\infty^2 c \sin \alpha$$

and the two-dimensional lift coefficient from Equation (1.10) as

$$C_\ell \equiv \frac{\ell}{\frac{1}{2}\rho q_\infty^2 c} = 2\pi \sin \alpha. \tag{5.7}$$

Thus we have a simple and important result for the lift coefficient of a thin flat plate as a function of incidence. The result agrees quite well with the experimentally measured lift for flat plates for angles of incidence up to about 3°. At higher incidence, the flow separates from the upper surface of the plate and the vortex sheet model is no longer accurate. This arises because of the infinite suction at the leading edge, and can be avoided till higher incidence by rounding the leading edge; thus the characteristic shape of wing sections: rounded at the leading edge and sharp at the trailing edge.

Since Equation (5.7) only holds for small values of α, it's often convenient to invoke the *small-angle approximation* for the sine (where α is expressed in radians)

$$\sin \alpha \sim \alpha \qquad (\alpha \to 0). \tag{5.8}$$

5.1.4   Surface Speed Distribution

Accepting the Kutta–Joukowsky condition Equation (5.6), the velocity over the plate is

$$u = \frac{q_\infty \{\sin(\chi - \alpha) + \sin \alpha\}}{\sin \chi} \tag{5.9a}$$

$$= \frac{q_\infty \{\cos \alpha \sin \chi - \sin \alpha \cos \chi + \sin \alpha\}}{\sin \chi}$$

$$= q_\infty \cos \alpha \left\{ 1 + \tan \alpha \frac{1 - \cos \chi}{\sin \chi} \right\}. \tag{5.9b}$$

The first term, $q_\infty \cos \alpha$, is just the x-component of velocity of the free-stream; the other is the correction for the flat plate. The correction factor $(1 - \cos \chi)/\sin \chi$ does become infinite at $\chi = \pi$ (the leading edge), but on the trailing edge the numerator vanishes too so, using l'Hôpital's rule,

$$\lim_{\chi \to 0} \frac{1 - \cos \chi}{\sin \chi} \left(= \frac{0}{0} \right) = \lim_{\chi \to 0} \frac{\sin \chi}{\cos \chi} = 0.$$

Thus, with the Kutta–Joukowsky condition, the velocity at the trailing edge should have a horizontal component equal to that of the free-stream and a zero vertical component. The surface speed distribution along the plate is plotted in Figure 5.1 for $\alpha = \frac{\pi}{6}$ (the same case as Figure 4.6). In the figure can be seen:

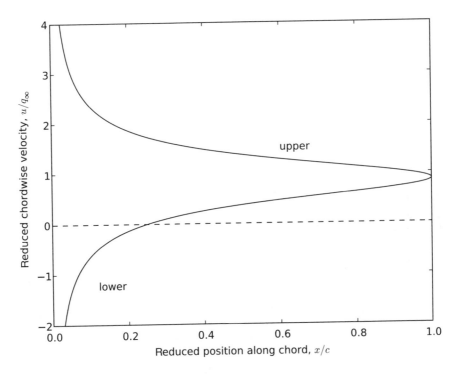

Figure 5.1 The speed distribution Equation (5.9b) along a thin flat plate, with the Kutta–Joukowsky condition, for incidence $\alpha = \frac{\pi}{6}$

- just one stagnation point, on the underside at $\chi = 2\alpha - \pi$, $x = \frac{c}{2}(1 - \cos 2\alpha) = \frac{c}{4}$, which is consistent with Equation (5.9b)
- infinite speed at the leading edge, left underneath and right above, as the fluid flows backwards along the underside from the stagnation point to the leading edge and then around to the upper surface; and
- a speed $u = q_\infty \cos \alpha = \sqrt{3}q_\infty/2 \doteq 0.866q_\infty$ at the trailing edge, equal to the horizontal component of the free-stream.

5.1.5 Pressure Distribution

Knowing the speed on the surface, the pressure distribution can be calculated using Bernoulli's Equation (2.18) for irrotational flow. With this, the pressure coefficient is

$$C_p \equiv \frac{p - p_\infty}{\frac{1}{2}\rho q_\infty^2} = 1 - \frac{q^2}{q_\infty^2}$$

which on the surface (where $q = |u|$) is

$$C_p = 1 - \frac{u^2}{q_\infty^2} = 1 - \cos^2 \alpha \left\{ 1 + \tan \alpha \frac{1 - \cos \chi}{\sin \chi} \right\}^2 \tag{5.10}$$

and is plotted in Figure 5.2.

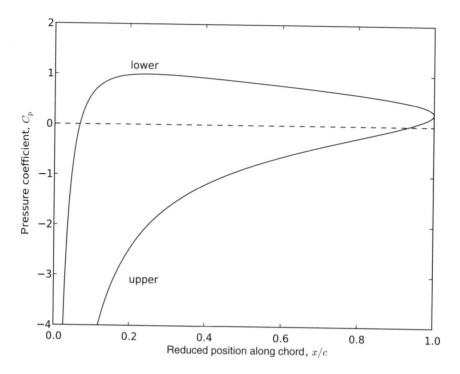

Figure 5.2 The pressure-coefficient distribution of Equation (5.10) along a thin flat plate, with the Kutta–Joukowsky condition, for incidence $\alpha = \frac{\pi}{6}$

5.1.6 Distribution of Circulation

Because of the infinite suction at the leading edge ($C_p \to -\infty$ as $\chi \to \pi$), it's difficult to calculate the aerodynamic force from the pressure distribution in this case, but we can use the Kutta–Joukowsky theorem and get the lift from the circulation.

Consider an infinitesimal rectangular circuit around the point $x = x'$ on the plate, as in Figure 5.3. There is no contribution to the circulation from the sides of the circuit piercing the plate, since the plate is impermeable. If the sides parallel to the plate have length δx, the

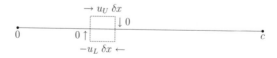

Figure 5.3 Infinitesimal circuit around a point on the thin flat plate, showing that local vortex strength is proportional to the difference in the tangential velocity across the plate; $\delta \Gamma = (u_U - u_L)\delta x$

contribution from them is $\delta\Gamma = (u_U - u_L)\delta x$, where u_U and u_L are the horizontal components of velocity on the upper and lower surfaces. Then

$$\begin{aligned}
\Gamma &= \int_0^c (u_U - u_L)\,dx \\
&= \int_\pi^0 \{u(\chi) - u(-\chi)\}\frac{-c\sin\chi}{2}\,d\chi \\
&= \frac{c}{2}\int_0^\pi \{u(\chi) - u(-\chi)\}\sin\chi\,d\chi.
\end{aligned}$$

From Equation (5.9a), the integrand is

$$\begin{aligned}
\{u(\chi) - u(-\chi)\}\sin\chi &= q_\infty\{\sin(\chi - \alpha) - \sin(\chi + \alpha) + 2\sin\alpha\} \\
&= q_\infty\{-2\sin\alpha\cos\chi + 2\sin\alpha\} \\
&= 2q_\infty\sin\alpha\{1 - \cos\chi\}
\end{aligned}$$

so

$$\Gamma = q_\infty c\sin\alpha\int_0^\pi(1 - \cos\chi)\,d\chi = \pi q_\infty c\sin\alpha \tag{5.11}$$

$$\ell = \rho q_\infty\Gamma = \pi\rho q_\infty^2 c\sin\alpha$$

$$C_\ell \equiv \frac{\ell}{\frac{1}{2}\rho q_\infty^2 c} = 2\pi\sin\alpha,$$

as obtained in Equation (5.7) from the jump in the complex potential around the plate.

5.1.7 Thin Flat Plate as Vortex Sheet

One way of interpreting the results of Section 5.1.6 is that the circulation arises from a distribution of circulation along the plate. Since circulation is associated with vortices (Section 3.5.4), and we have amounts of circulation

$$\begin{aligned}
\delta\Gamma &= -cq_\infty\sin\alpha(1 - \cos\chi)\,\delta\chi \\
&= 2q_\infty\sin\alpha\frac{1 - \cos\chi}{\sin\chi}\,\delta x \tag{5.12} \\
&= 2q_\infty\sin\alpha\frac{1 - \cos\chi}{\sqrt{1 - \cos^2\chi}}\,\delta x \\
&= 2q_\infty\sin\alpha\sqrt{\frac{1 - \cos\chi}{1 + \cos\chi}}\,\delta x \\
&= 2q_\infty\sin\alpha\sqrt{\frac{c}{x} - 1}\,\delta x \tag{5.13}
\end{aligned}$$

along the plate, we might think of this as being due to a vortex between $z = x$ and $z = x + \delta x$ of strength $\delta\Gamma = \gamma(x)\delta x$. This vortex strength distribution is plotted in Figure 5.4.

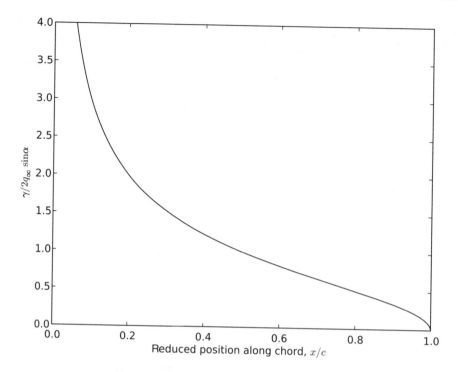

Figure 5.4 The distribution of circulation along a thin flat plate with the Kutta–Joukowsky condition at arbitrary nonzero incidence, as given by Equation (5.13)

The vortex at x' would have the complex velocity (at some other point z)

$$\delta w(z) = \frac{2iq_\infty \sin \alpha}{2\pi(z-x')} \sqrt{\frac{c}{x'} - 1} \, \delta x'$$

or, since $x = z$ on the plate, where all the vortices must be

$$\delta w(z) = \frac{iq_\infty \sin \alpha}{\pi(z-z')} \sqrt{\frac{c}{z'} - 1} \, \delta z'.$$

Adding in the free-stream $q_\infty e^{-i\alpha}$, the total flow field due to the free-stream and the vortex sheet along the plate then has complex velocity

$$w(z) = q_\infty e^{-i\alpha} + \frac{iq_\infty \sin \alpha}{\pi} \int_0^c \frac{1}{z-z'} \sqrt{\frac{c}{z'} - 1} \, dz'.$$

With a change of variables to facilitate the integration, this would eventually get us back to a representation of the velocity field the same as Equation (5.2); however, the details of that do not concern us here. What is of interest is the concept that a thin plate can be represented by a *vortex sheet*.

5.2 Application of Thin Aerofoil Theory to the Flat Plate

5.2.1 Thin Aerofoil Theory

For aerofoils typically in use, the camber and thickness are an order of magnitude smaller than the chord. This separation of length scales is exploited in *thin aerofoil theory*. It is also assumed in thin aerofoil theory that the incidence is small; i.e. the flight is more or less parallel to the chord.

The basic idea of thin aerofoil theory is to model the flow over the aerofoil by adding a uniform stream and a continuous distribution of vortices along the chord or camber line (these two becoming equivalent in the thin aerofoil approximation).

The strengths of the vortices are determined by:

- the obvious requirement that there is no flow through the aerofoil; and
- the more subtle *Kutta–Joukowsky condition*, which stipulates that the flow leave the trailing edge smoothly.

The lift on the aerofoil is then obtained from the Kutta–Joukowsky Equation (3.15a). The circulation around the aerofoil is the integral of the strengths of the vortices distributed along it.

In Section 5.2, thin aerofoil theory is introduced by its application to the flow over a thin flat plate.

5.2.2 Vortex Sheet along the Chord

If vortices are distributed with (yet to be determined) strength $\gamma(z')$ at z' along the chord of the aerofoil, $0 \leqslant z' \leqslant c$, the complex velocity at z is

$$w(z) = q_\infty e^{-i\alpha} + \frac{i}{2\pi} \int_C \frac{\gamma \, dz'}{z - z'} \tag{5.14}$$

and along the chord is (noting that $z = x + iy = x$ on the chord)

$$u(x) - iv(x) = q_\infty e^{-i\alpha} + \frac{i}{2\pi} \int_C \frac{\gamma \, dx'}{x - x'}.$$

Thus the normal component of velocity to the surface of the aerofoil ($y = 0$) is minus the imaginary part;

$$v(x) = q_\infty \sin \alpha - \frac{1}{2\pi} \int_0^c \frac{\gamma \, dx'}{x - x'},$$

and the condition that there be no flow through the aerofoil is $v(x) = 0$:

$$\int_0^c \frac{\gamma \, dx'}{x - x'} = 2\pi q_\infty \sin \alpha.$$

To be consistent with the other approximations of thin aerofoil theory, the small-angle approximation Equation (5.8) should be applied; thus

$$\int_0^c \frac{\gamma \, dx'}{x - x'} = 2\pi q_\infty \alpha. \tag{5.15}$$

5.2.3   Changing the Variable of Integration

The integral in Equation (5.15) is fairly nasty and is more easily treated after the transformation via Equation (5.3a) to the eccentric angle with a similar transformation for x' to χ'. The differential in the integral becomes

$$dx' = \frac{-c \sin \chi' \, d\chi'}{2},$$

and Equation (5.15) becomes

$$\int_0^\pi \frac{\gamma \sin \chi' \, d\chi'}{\cos \chi - \cos \chi'} = 2\pi q_\infty \alpha. \tag{5.16}$$

This is the basic integral equation of thin aerofoil theory, as applied to a flat plate.

5.2.4   Glauert's Integral

The integral in Equation (5.16) is still nasty, but can be solved by reference to Glauert's integral

$$\int_0^\pi \frac{\cos n\chi' \, d\chi'}{\cos \chi - \cos \chi'} = \frac{-\pi \sin n\chi}{\sin \chi}, \tag{5.17}$$

which holds for $n = 0, 1, 2, \ldots$.
Setting $n = 0$ and $n = 1$ in Equation (5.17) gives

$$\int_0^\pi \frac{d\chi'}{\cos \chi - \cos \chi'} = 0,$$

$$\int_0^\pi \frac{\cos \chi' \, d\chi'}{\cos \chi - \cos \chi'} = -\pi.$$

Thus, for any constant C,

$$\int_0^\pi \frac{C + \cos \chi'}{\cos \chi - \cos \chi'} \, d\chi' = -\pi.$$

Comparing this with Equation (5.16), we see that the solution for the distribution of vortex strength is

$$\gamma(\chi) = -2q_\infty \alpha \, \frac{C + \cos \chi}{\sin \chi}. \tag{5.18}$$

5.2.5 The Kutta–Joukowsky Condition

At the trailing edge $x = c$, $\chi = 0$, and the vortex strength in Equation (5.18) is infinite unless $C = -1$. Therefore this special value of the constant is chosen. This is how the Kutta–Joukowsky condition enters the problem in this formulation.

With the condition, the vortex strength distribution is

$$\gamma(\chi) = 2q_\infty\alpha\, \frac{1 - \cos\chi}{\sin\chi}, \qquad (5.19)$$

which is just as in Equation (5.12), with the small-angle approximation Equation (5.8).

5.2.6 Circulation and Lift

It follows that the circulation, lift (per unit span), and two-dimensional lift coefficient are

$$\Gamma = \pi c q_\infty \alpha \qquad (5.20)$$
$$\ell = \pi c \rho q_\infty^2 \alpha$$
$$C_\ell = 2\pi\alpha.$$

Again, just as in Equation (5.7), with the small-angle approximation.

5.3 Aerodynamic Moment

Up till now we have mostly been concerned with the aerodynamic force on the aerofoil, but the resulting line of action or moment associated with the distribution of this force is important too.

Since the lift (per unit span) is given by

$$\ell = \rho q_\infty \Gamma = \rho q_\infty \int_0^c \gamma\, dx,$$

we can think of

$$\delta\ell = \rho q_\infty \gamma\, \delta x$$

as the distribution of the lift along the aerofoil. Each element of lift contributes to the moment about the leading edge in proportion to its distance from that edge:

$$\delta m_{\text{l.e.}} = -x\delta\ell = -\rho q_\infty x \gamma \delta x,$$

where the moment is reckoned positive when it tends to increase the *pitch* of the aerofoil; i.e. to raise the leading edge and depress the trailing edge. Thus the pitching moment (per unit length) is given by

$$m_{\text{l.e.}} = -\rho q_\infty \int_0^c x\gamma\, dx.$$

For the flat plate, using Equation (5.19),

$$
\begin{aligned}
m_{\text{l.e.}} &= -2\rho q_\infty^2 \alpha \int_0^c x \frac{1 - \cos \chi}{\sin \chi}\, dx \\
&= \frac{-\rho q_\infty^2 \alpha c^2}{2} \int_0^\pi (1 - \cos \chi)(1 + \cos \chi)\, d\chi \\
&= \frac{-\rho q_\infty^2 c^2 \alpha}{4}.
\end{aligned}
$$

Defining the leading-edge pitching moment coefficient as

$$
C_{m_{\text{l.e.}}} \equiv \frac{m_{\text{l.e.}}}{\frac{1}{2}\rho q_\infty^2 c^2}, \tag{5.21}
$$

we have the result

$$
C_{m_{\text{l.e.}}} = -\frac{\pi \alpha}{2} = \frac{-C_\ell}{4}.
$$

5.3.1 Centre of Pressure and Aerodynamic Centre

The *centre of pressure* is the point through which the action of the resultant aerodynamic force is equivalent to that of its total distribution. If the lift is placed at a point x, it produces a pitching moment about the leading edge

$$
-x\ell,
$$

and this balances the actual leading edge pitching moment if $-x\ell = m_{\text{l.e.}}$, or

$$
\frac{x}{c} = \frac{-C_{m_{\text{l.e.}}}}{C_\ell}.
$$

For the flat plate this gives $x = \frac{c}{4}$; i.e. the centre of pressure is at the quarter-chord point. Note that this result is independent of the incidence.

It is left as an exercise to show that the only point about which the pitching moment is independent of incidence—i.e. the *aerodynamic centre*—also occurs at the quarter-chord point for the flat plate.

5.4 Exercises

1. (a) *The aerodynamic centre of a flat plate.* Derive the formula for pitching moment $m(x)$ about an arbitary point x on a flat plate is in terms of lift ℓ, position x, and chord c.
   (b) Show that since the lift ℓ varies with the incidence α, the only point x about which pitching moment is independent of incidence is the quarter-chord point; i.e. $x_{\text{a.c.}} = c/4$.

5.5 Further Reading

For the Kutta–Joukowksy condition (Section 5.1.2), see: Goldstein (1938), Abbott and von Doenhoff (1959), Bertin (2002), and Houghton and Carpenter (2003).

An analysis of the leading edge suction, mentioned in Section 5.1.6, may be found in Milne-Thomson (1973).

Thin aerofoil theory is a fundamental component of aerodynamics, and is treated extensively in Chapter 6.

The definition of the eccentric angle given here in Equation (5.3b) is taken from Milne-Thomson (1973). Glauert's integral is treated (in varying levels of detail) by Glauert (1926), Abbott and von Doenhoff (1959), Milne-Thomson (1973), Kuethe and Chow (1998), Bertin (2002), Moran (2003, appendix A), and Anderson (2007); for a direct solution of the thin aerofoil integral equation, not using Glauert's integral, see Ashley and Landahl (1985).

For references to experimentally measured lift on thin flat plates, see Hoerner and Borst (1985) and Anderson (2007); for reversed aerofoils, see Hoerner and Borst (1985, figure 14).

References

Abbott, I.H. and von Doenhoff, A.E. (1959) *Theory of Wing Sections*. New York: Dover.

Anderson, J.D. (2007) *Fundamentals of Aerodynamics*, 4th edn. New York: McGraw-Hill.

Ashley, H. and Landahl, M. (1985) *Aerodynamics of Wings and Bodies*. New York: Dover.

Bertin, J.J. (2002) *Aerodynamics for Engineers*, 4th edn. New Jersey: Prentice Hall.

Glauert, H. (1926) *The Elements of Aerofoil and Airscrew Theory*. Cambridge: Cambridge University Press.

Goldstein, S. (1938) *Modern Developments in Fluid Dynamics*. Oxford: Clarendon Press.

Hoerner, S.F. and Borst, H.V. (1985) *Fluid-Dynamic Lift*, 2nd edn. Bakersfield, CA: Hoerner Fluid Dynamics.

Houghton, E.L. and Carpenter, P.W. (2003) *Aerodynamics for Engineering Students*, 5th edn. Oxford: Butterworth Heinemann.

Kuethe, A.M. and Chow, C.Y. (1998) *Foundations of Aerodynamics*, 5th edn. Chichester: John Wiley & Sons, Ltd.

Milne-Thomson, L.M. (1973) *Theoretical Aerodynamics*, 4th edn. New York: Dover.

Moran, J. (2003) *An Introduction to Theoretical and Computational Aerodynamics*. New York: Dover.

6

Thin Wing Sections

In Chapter 5 we found the lift and pitching moment on a thin tilted flat plate by modelling the flow as the sum of a uniform stream and a vortex sheet along the plate. By applying similar ideas to a thin curved plate, we obtain a theory of thin wing sections. The basic notions are:

- thickness is ignored – the aerofoil is reduced to its camber line;
- the camber is assumed to be much less than the chord length;
- the incidence is small.

6.1 Thin Aerofoil Analysis

6.1.1 Vortex Sheet along the Camber Line

Spread a vortex sheet along the camber line; the complex velocity at a general point $z = x + iy$ is

$$w(z) = q_\infty e^{-i\alpha} + \int_0^c \frac{i\gamma(z')\, dz'}{2\pi(z - z')}.$$

The integration here is along the camber line and so, unlike in the case of a flat plate, departs from the real axis.

6.1.2 The Boundary Condition

The boundary condition on the flow is that the aerofoil, represented by its camber line, should be impermeable; i.e. that the component of velocity normal to the camber line should vanish. Recalling Equation (2.24) that the component of velocity normal to a line making an angle λ with the positive x-axis is minus the imaginary part of $e^{i\lambda} w$, the impermeability condition is

$$-\Im\left\{ e^{i\lambda} w(z) \right\} = 0 \tag{6.1}$$

Theory of Lift: Introductory Computational Aerodynamics in MATLAB®/Octave, First Edition. G. D. McBain.
© 2012 John Wiley & Sons, Ltd. Published 2012 by John Wiley & Sons, Ltd.

for z on the camber line and where the camber line slope is

$$\tan \lambda = \frac{dy}{dx}.$$

The free-stream contribution is

$$-\Im \left(e^{i\lambda} q_\infty e^{-i\alpha} \right) = q_\infty \sin(\alpha - \lambda) \tag{6.2}$$

and the contribution from the infinitesimal vortex at z' is

$$-\Im \left\{ e^{i\lambda} \frac{i\gamma\delta z'}{2\pi(z - z')} \right\}. \tag{6.3}$$

6.1.3 Linearization

Since it is assumed that the incidence is small, Equation (5.8) may be used; since it is further assumed that the aerofoil is thin and the incidence and slope are small, the following approximations may be added:

$$z \sim x$$
$$\delta z \sim \delta x$$
$$\tan \alpha \sim \sin \alpha \sim \alpha$$
$$\sin \lambda \sim \tan \lambda \sim \lambda \sim \frac{dy}{dx}$$
$$\cos \alpha, \cos \lambda \sim 1$$
$$e^{i\lambda} \sim 1 + i\lambda.$$

This means the free-stream contribution to the normal velocity component in Equation (6.2) is approximately

$$q_\infty \sin(\alpha - \lambda) \equiv q_\infty \sin\left(\alpha - \arctan \frac{dy}{dx} \right) \sim q_\infty \left(\alpha - \frac{dy}{dx} \right)$$

and the infinitesimal vortex contribution of Equation (6.3) is approximately

$$\frac{-\gamma\delta x'}{2\pi(x - x')}$$

Thus the condition of Equation (6.1) becomes

$$\frac{1}{2\pi q_\infty} \int_0^c \frac{\gamma \, dx'}{x - x'} = \alpha - \frac{dy}{dx}. \tag{6.4}$$

This is the *thin aerofoil integral equation*.

6.1.4   Glauert's Transformation

As in the case of the flat plate, we introduce the transformation to the eccentric angle $x = \frac{c}{2}(1 + \cos \chi)$, so that the thin aerofoil integral Equation (6.4) becomes

$$\frac{1}{2\pi q_\infty} \int_0^\pi \frac{\gamma \sin \chi' \, d\chi'}{\cos \chi - \cos \chi'} = \alpha - \frac{dy}{dx}. \tag{6.5}$$

6.1.5   Glauert's Expansion

Comparing Equation (6.5) with the special case of the flat plate in Equation (5.16)

$$\frac{1}{2\pi q_\infty} \int_0^\pi \frac{\gamma_{\text{f.p.}} \sin \chi' \, d\chi'}{\cos \chi - \cos \chi'} = \alpha$$

which had the solution Equation (5.19)

$$\gamma_{\text{f.p.}} = 2q_\infty \alpha \, \frac{1 - \cos \chi}{\sin \chi}$$

suggests trying an expansion of the form

$$\frac{\gamma}{2q_\infty} = A_0 \frac{1 - \cos \chi}{\sin \chi} + \sum_{n=1}^\infty A_n \sin n\chi \tag{6.6}$$

for the general cambered wing section. Note that, since $\sin(n \times 0) = 0$, the circulation at the trailing edge $\chi = 0$ remains zero, which is the Kutta–Joukowsky condition.

With the trigonometric identity

$$2 \sin n\chi \sin \chi \equiv \cos\{(n-1)\chi\} - \cos\{(n+1)\chi\},$$

the numerator of the integrand of Equation (6.5) can be written as a cosine series:

$$\frac{\gamma \sin \chi'}{2q_\infty} = A_0 (1 - \cos \chi') + \sum_{n=1}^{\infty} A_n \sin n\chi' \sin \chi'$$

$$= A_0 (1 - \cos \chi') + \sum_{n=1}^{\infty} \frac{A_n}{2} \left[\cos\{(n-1)\chi'\} - \cos\{(n+1)\chi'\}\right]$$

$$= A_0 (1 - \cos \chi') + \sum_{n=1}^{\infty} \frac{A_n}{2} \cos\{(n-1)\chi'\} - \sum_{n=1}^{\infty} \frac{A_n}{2} \cos\{(n+1)\chi'\}$$

$$= A_0 (1 - \cos \chi') + \sum_{n=0}^{\infty} \frac{A_{n+1}}{2} \cos n\chi' - \sum_{n=2}^{\infty} \frac{A_{n-1}}{2} \cos n\chi'$$

$$= A_0 (1 - \cos \chi') + \frac{A_1}{2} + \frac{A_2}{2} \cos \chi' + \sum_{n=2}^{\infty} \frac{A_{n+1} - A_{n-1}}{2} \cos n\chi'$$

$$= A_0 + \frac{A_1}{2} + \left(\frac{A_2}{2} - A_0\right) \cos \chi' + \sum_{n=2}^{\infty} \frac{A_{n+1} - A_{n-1}}{2} \cos n\chi'. \qquad (6.7)$$

This means that Glauert's integral (Equation 5.17)

$$\int_0^\pi \frac{\cos n\chi' \, d\chi'}{\cos \chi - \cos \chi'} = \frac{-\pi \sin n\chi}{\sin \chi}$$

can be used to integrate Equation (6.5) term by term:

$$\frac{1}{2\pi q_\infty} \int_0^\pi \frac{\gamma \sin \chi' \, d\chi'}{\cos \chi - \cos \chi'} = A_0 - \frac{A_2}{2} - \sum_{n=2}^{\infty} (A_{n+1} - A_{n-1}) \frac{\sin n\chi}{2 \sin \chi}.$$

This solves Equation (6.5) provided

$$\alpha - \frac{dy}{dx} = A_0 - \frac{A_2}{2} - \sum_{n=2}^{\infty} (A_{n+1} - A_{n-1}) \frac{\sin n\chi}{2 \sin \chi}$$

$$= A_0 - \frac{A_2}{2} - \sum_{n=2}^{\infty} A_{n+1} \frac{\sin n\chi}{2 \sin \chi} + \sum_{n=2}^{\infty} A_{n-1} \frac{\sin n\chi}{2 \sin \chi}$$

$$= A_0 - \frac{A_2}{2} - \sum_{n=3}^{\infty} A_n \frac{\sin(n-1)\chi}{2 \sin \chi} + \sum_{n=1}^{\infty} A_n \frac{\sin(n+1)\chi}{2 \sin \chi}$$

$$= A_0 - \frac{A_2}{2} + \frac{A_1 \sin 2\chi + A_2 \sin 3\chi}{2 \sin \chi}$$

$$+ \sum_{n=3}^{\infty} A_n \frac{\sin(n+1)\chi - \sin(n-1)\chi}{2 \sin \chi}.$$

Which, using the trigonometric identities

$$\frac{\sin 2\chi}{2\sin\chi} \equiv \cos\chi$$

$$\frac{\sin 3\chi}{2\sin\chi} \equiv \cos 2\chi + \frac{1}{2}$$

$$\frac{\sin(n+1)\chi - \sin(n-1)\chi}{2\sin\chi} \equiv \cos n\chi,$$

can be converted back to a cosine series

$$\alpha - \frac{dy}{dx} = A_0 - \frac{A_2}{2} + A_1 \cos\chi + A_2 \left(\cos 2\chi + \frac{1}{2}\right) + \sum_{n=3}^{\infty} A_n \cos n\chi$$

$$= A_0 + A_1 \cos\chi + A_2 \cos 2\chi + \sum_{n=3}^{\infty} A_n \cos n\chi$$

$$= \sum_{n=0}^{\infty} A_n \cos n\chi. \tag{6.8}$$

6.1.6 Fourier Cosine Decomposition of the Camber Line Slope

The decomposition of the camber line slope implied by Equation (6.8) can be carried out by multiplying each side by $\cos m\chi$ and integrating over the chord:

$$\int_0^\pi \left(\alpha - \frac{dy}{dx}\right) \cos m\chi \, d\chi = \sum_{n=0}^{\infty} A_n \int_0^\pi \cos n\chi \cos m\chi \, d\chi,$$

since the cosine integrals are, for integer m and n,

$$\int_0^\pi \cos n\chi \cos m\chi \, d\chi = \begin{cases} \pi, & (m=n=0) \\ \frac{\pi}{2}, & (m=n\neq 0) \\ 0, & (m\neq n); \end{cases}$$

thus

$$\int_0^\pi \left(\alpha - \frac{dy}{dx}\right) d\chi = \pi A_0$$

$$\int_0^\pi \left(\alpha - \frac{dy}{dx}\right) \cos m\chi \, d\chi = \frac{\pi A_m}{2}, \qquad (m = 1, 2, \ldots);$$

i.e. noting that the incidence α is independent of χ,

$$A_0 = \alpha - \frac{1}{\pi} \int_0^\pi \frac{dy}{dx} \, d\chi \tag{6.9a}$$

$$A_m = \frac{-2}{\pi} \int_0^\pi \frac{dy}{dx} \cos m\chi \, d\chi, \qquad (m = 1, 2, \ldots). \tag{6.9b}$$

We will return to the evaluation of these integrals later (Sections 6.3 and 6.4), after considering in Section 6.2 the general aerodynamic properties of a thin aerofoil with a vortex strength distribution γ of the assumed type (Equation 6.6).

6.2 Thin Aerofoil Aerodynamics

6.2.1 Circulation and Lift

The circulation around the wing section is just the integral of the circulation density over the vortex sheet:

$$\Gamma = \int_0^c \gamma \, dx = \frac{c}{2} \int_0^\pi \gamma \sin \chi \, d\chi$$

(here and in the following, where we are concerned with global properties such as circulation and lift, we drop the primes on x and χ) so on using Equation (6.7),

$$\Gamma = q_\infty c \int_0^\pi \left\{ A_0 + \frac{A_1}{2} + \left(\frac{A_2}{2} - A_0 \right) \cos \chi + \sum_{n=2}^\infty \frac{A_{n+1} - A_{n-1}}{2} \cos n\chi \right\} d\chi$$

$$= \pi q_\infty c \left(A_0 + \frac{A_1}{2} \right).$$

Thus, using the Kutta–Joukowsky Equation (3.15a), the lift per unit span is

$$\ell = \rho q_\infty \Gamma = \pi \rho q_\infty^2 c \left(A_0 + \frac{A_1}{2} \right)$$

and the wing section lift coefficient is

$$C_\ell \equiv \frac{\ell}{\frac{1}{2} \rho q_\infty^2 c} = 2\pi \left(A_0 + \frac{A_1}{2} \right).$$

In terms of the incidence and the camber line slope distribution, using Equations (6.9a) and (6.9b),

$$C_\ell = 2\pi \left\{ \alpha - \frac{1}{\pi} \int_0^\pi \frac{dy}{dx} (1 + \cos \chi) \, d\chi \right\}.$$

Introducing the *zero-lift incidence*

$$\alpha_0 = \frac{1}{\pi} \int_0^\pi \frac{dy}{dx} (1 + \cos \chi) \, d\chi, \tag{6.10}$$

this can be written as

$$C_\ell = 2\pi(\alpha - \alpha_0), \tag{6.11}$$

which is a simple linearized and offset generalization of the Equation (5.7) for a flat plate.

Thus, in thin aerofoil theory, all thin aerofoils have the same lift–incidence slope, and only differ in the offset. The latter is given in Equation (6.10) by a weighted integral of the camber line slope.

6.2.2 Pitching Moment about the Leading Edge

Since each element of lift $\delta\ell$ contributes $-x\cos\alpha\,\delta\ell \approx -x\delta\ell$ to the pitching moment about the leading edge, the total (per unit span) is

$$m_{\text{l.e.}} = -\int_0^c x\frac{\mathrm{d}\ell}{\mathrm{d}x}\,\mathrm{d}x$$

$$= -\rho q_\infty \int_0^c x\gamma\,\mathrm{d}x$$

$$= \frac{-\rho q_\infty c^2}{4}\int_0^\pi (1+\cos\chi)\gamma\sin\chi\,\mathrm{d}\chi.$$

In terms of the thin aerofoil theory vortex strength distribution of Equation (6.7)

$$m_{\text{l.e.}} = \frac{-\rho q_\infty^2 c^2}{2}\int_0^\pi (1+\cos\chi)\left\{A_0 + \frac{A_1}{2} + \left(\frac{A_2}{2}-A_0\right)\cos\chi \right.$$
$$\left. + \sum_{n=2}^\infty \frac{A_{n+1}-A_{n-1}}{2}\cos n\chi \right\}\mathrm{d}\chi.$$

This nondimensionalizes to give the leading-edge pitching moment coefficient defined by Equation (5.21) as

$$C_{m_{\text{l.e.}}} = -\int_0^\pi (1+\cos\chi)\left\{A_0 + \frac{A_1}{2} + \left(\frac{A_2}{2}-A_0\right)\cos\chi \right.$$
$$\left. + \sum_{n=2}^\infty \frac{A_{n+1}-A_{n-1}}{2}\cos n\chi \right\}\mathrm{d}\chi.$$

Of the trigonometrical integrals here, all vanish except the one involving the square of $\cos\chi$ which gives

$$\int_0^\pi \cos^2\chi\,\mathrm{d}\chi = \frac{\pi}{2}$$

and the one not involving cosines at all; thus

$$C_{m_{\text{l.e.}}} = -\pi\left(A_0 + \frac{A_1}{2}\right) - \frac{\pi}{2}\left(\frac{A_2}{2}-A_0\right) = \frac{-\pi}{2}\left(A_0 + A_1 + \frac{A_2}{2}\right).$$

Notice a very convenient feature of Glauert's expansion Equation (6.6): only the first three expansion coefficients are required to compute the lift and pitching moment.

6.2.3 Aerodynamic Centre

To obtain instead the pitching moment about another point x_1, we would have used

$$
m_1 = \int_0^c (x_1 - x)\frac{d\ell}{dx}\,dx
$$
$$
= x_1 \int_0^c \frac{d\ell}{dx}\,dx - \int_0^c x\frac{d\ell}{dx}\,dx
$$
$$
= x_1\ell + m_{\text{l.e.}},
$$

which is the usual formula for the transferral of moments from point to point. In dimensionless terms,

$$
C_{m_1} = \frac{x_1}{c}C_\ell + C_{m_{\text{l.e.}}},
$$

and so for the thin aerofoil,

$$
C_{m_1} = \frac{2\pi x_1}{c}\left(A_0 + \frac{A_1}{2}\right) - \frac{\pi}{2}\left(A_0 + A_1 + \frac{A_2}{2}\right)
$$
$$
= \pi\left\{\left(\frac{2x_1}{c} - \frac{1}{2}\right)A_0 + \left(\frac{x_1}{c} - \frac{1}{2}\right)A_1 - \frac{A_2}{4}\right\}.
$$

Recalling that only A_0 and not A_1 nor A_2 depends on the incidence α, we can find a point x_1 about which the pitching moment is independent of incidence by setting the coefficient of A_0 to zero. Thus the thin aerofoil has an aerodynamic centre at

$$
x_{\text{a.c.}} = \frac{c}{4}.
$$

The corresponding pitching moment coefficient is

$$
C_{m_{c/4}} = \frac{-\pi}{4}(A_1 + A_2).
$$

In terms of the incidence and the camber line slope distribution, using Equations (6.9a)–(6.9b),

$$
C_{m_{c/4}} = \frac{-\pi}{4}\left(\frac{-2}{\pi}\int_0^\pi \frac{dy}{dx}\cos\chi\,d\chi - \frac{2}{\pi}\int_0^\pi \frac{dy}{dx}\cos 2\chi\,d\chi\right)
$$
$$
= \frac{1}{2}\int_0^\pi \frac{dy}{dx}(\cos\chi + \cos 2\chi)\,d\chi, \tag{6.12}
$$

which is independent of α, as claimed.

6.2.4   Summary

The results of thin aerofoil theory may be summarized as follows:

- All thin aerofoils have lift coefficient $C_\ell = 2\pi(\alpha - \alpha_0)$, where the zero-lift incidence α_0 is given by a weighted integral of the camber line slope distribution Equation (6.10)

$$\alpha_0 = \frac{1}{\pi} \int_0^\pi \frac{dy}{dx}(1 + \cos \chi)\, d\chi.$$

- All thin aerofoils have an aerodynamic centre at the quarter-chord point.
- The pitching moment coefficient about the aerodynamic centre is given by another weighted integral of the camber line slope distribution Equation (6.12)

$$C_{m_{c/4}} = \frac{1}{2} \int_0^\pi \frac{dy}{dx}(\cos \chi + \cos 2\chi)\, d\chi.$$

- For an uncambered aerofoil (for which $dy/dx \equiv 0$), both the zero-lift incidence and aerodynamic centre pitching moment coefficient vanish (as expected by symmetry).

6.3   Analytical Evaluation of Thin Aerofoil Integrals

Some aerofoils have camber lines defined by an explicit formula; e.g. the NACA four- and five-digit wing sections have a polynomial with coefficients that jump at a specified value $x = pc$ but are constant over $0 < x < pc$ and $pc < x < c$. For such camber lines, the slope distribution can be expressed by a pair of cosine series of the form

$$\frac{dy}{dx} = \begin{cases} f_0 + f_1 \cos \chi + f_2 \cos 2\chi + \cdots, & \frac{x}{c} < p \\ a_0 + a_1 \cos \chi + a_2 \cos 2\chi + \cdots, & p < \frac{x}{c} \end{cases}$$

and the zero-lift incidence integral in Equation (6.10) as

$$\alpha_0 = \frac{1}{\pi} \sum_{k=0}^\infty \left\{ f_k \int_{\chi_p}^\pi \cos k\chi\,(1 + \cos \chi)\, d\chi + a_k \int_0^{\chi_p} \cos k\chi\,(1 + \cos \chi)\, d\chi \right\}$$

where

$$\chi_p = \arccos(2p - 1).$$

Using the trigonometric identity

$$2 \cos k\chi \cos \chi \equiv \cos(k + 1)\chi + \cos(k - 1)\chi,$$

the indefinite integral that appears twice above is

$$\int \cos k\chi(1 + \cos \chi)\, d\chi = \begin{cases} \chi + \sin \chi + \text{const.}, & k = 0 \\ \sin \chi + \frac{\sin 2\chi}{4} + \frac{\chi}{2} + \text{const.}, & k = 1 \\ \frac{\sin k\chi}{k} + \frac{\sin(k+1)\chi}{2(k+1)} + \frac{\sin(k-1)\chi}{2(k-1)} + \text{const.}, & k > 1 \end{cases}$$

and hence the definite integrals are

$$
\int_{\chi_p}^{\pi} \cos k\chi (1 + \cos \chi)\, d\chi =
\begin{cases}
\pi - \chi_p - \sin \chi_p, & k = 0 \\[2mm]
-\sin \chi_p - \frac{\sin 2\chi_p}{4} + \frac{\pi - \chi_p}{2}, & k = 1 \\[2mm]
-\frac{\sin k\chi_p}{k} - \frac{\sin(k+1)\chi_p}{2(k+1)} - \frac{\sin(k-1)\chi_p}{2(k-1)}, & k > 1
\end{cases}
$$

$$
\int_{0}^{\chi_p} \cos k\chi (1 + \cos \chi)\, d\chi =
\begin{cases}
\chi_p + \sin \chi_p, & k = 0 \\[2mm]
\sin \chi_p + \frac{\sin 2\chi_p}{4} + \frac{\chi_p}{2}, & k = 1 \\[2mm]
\frac{\sin k\chi_p}{k} + \frac{\sin(k+1)\chi_p}{2(k+1)} + \frac{\sin(k-1)\chi_p}{2(k-1)}, & k > 1
\end{cases}
$$

so

$$
\begin{aligned}
\alpha_0 \pi &= f_0 \pi + (a_0 - f_0)(\chi_p + \sin \chi_p) + \frac{f_1 \pi}{2} \\
&\quad + (a_1 - f_1)\left(\sin \chi_p + \frac{\sin 2\chi_p}{4} + \frac{\chi_p}{2} \right) \\
&\quad + (a_2 - f_2)\left(\frac{\sin 2\chi_p}{2} + \frac{\sin 3\chi_p}{6} + \frac{\sin \chi_p}{2} \right) + \cdots \\
&\quad + (a_k - f_k)\left\{ \frac{\sin k\chi_p}{k} + \frac{\sin(k+1)\chi_p}{2(k+1)} + \frac{\sin(k-1)\chi_p}{2(k-1)} \right\} + \cdots \\
&= \left(f_0 + \frac{f_1}{2} \right)\pi + \left\{ a_0 - f_0 + \frac{a_1 - f_1}{2} \right\}\chi_p \\
&\quad + \left\{ a_0 - f_0 + a_1 - f_1 + \frac{a_2 - f_2}{2} \right\}\sin \chi_p \\
&\quad + \left\{ \frac{a_1 - f_1}{4} + \frac{a_2 - f_2}{2} + \frac{a_3 - f_3}{4} \right\}\sin 2\chi_p + \cdots \\
&\quad + \left\{ \frac{a_{k-1} - f_{k-1}}{2k} + \frac{a_k - f_k}{k} + \frac{a_{k+1} - f_{k+1}}{2k} \right\}\sin k\chi_p + \cdots
\end{aligned}
$$

The same approach could be applied to compute the pitching moment:

$$
C_{m_{c/4}} = \frac{1}{2} \sum_{k=0}^{\infty} \left\{ f_k \int_{\chi_p}^{\pi} \cos k\chi (\cos \chi + \cos 2\chi)\, d\chi \right. \\
\left. + a_k \int_{0}^{\chi_p} \cos k\chi (\cos \chi + \cos 2\chi)\, d\chi \right\}.
$$

The indefinite integrals are:

$$\int (\cos \chi + \cos 2\chi)\, d\chi = \sin \chi + \frac{\sin 2\chi}{2}$$

$$\int \cos \chi(\cos \chi + \cos 2\chi)\, d\chi = \frac{\chi}{2} + \frac{\sin \chi}{2} + \frac{\sin 2\chi}{4}$$

$$\int \cos 2\chi(\cos \chi + \cos 2\chi)\, d\chi = \frac{\chi}{2} + \frac{\sin \chi}{2} + \frac{\sin 3\chi}{6} + \frac{\sin 4\chi}{8}$$

and for $k = 3, 4, \ldots$

$$\int \cos k\chi(\cos \chi + \cos 2\chi)\, d\chi$$

$$= \frac{\sin(k-2)\chi}{2(k-2)} + \frac{\sin(k-1)\chi}{2(k-1)} + \frac{\sin(k+1)\chi}{2(k+1)} + \frac{\sin(k+2)\chi}{2(k+2)}.$$

The definite integrals are

$$\int_{\chi_p}^{\pi} (\cos \chi + \cos 2\chi)\, d\chi = -\sin \chi_p - \frac{\sin 2\chi_p}{2}$$

$$\int_{\chi_p}^{\pi} \cos \chi(\cos \chi + \cos 2\chi)\, d\chi = \frac{\pi - \chi_p}{2} - \frac{\sin \chi_p}{2} - \frac{\sin 2\chi_p}{4} - \frac{\sin 3\chi_p}{6}$$

$$\int_{\chi_p}^{\pi} \cos 2\chi(\cos \chi + \cos 2\chi)\, d\chi = \frac{\pi - \chi_p}{2} - \frac{\sin \chi_p}{2} - \frac{\sin 3\chi_p}{6} - \frac{\sin 4\chi_p}{8}$$

and for $k = 3, 4, \ldots$

$$\int_{\chi_p}^{\pi} \cos k\chi(\cos \chi + \cos 2\chi)\, d\chi$$

$$= -\frac{\sin(k-2)\chi_p}{2(k-2)} - \frac{\sin(k-1)\chi_p}{2(k-1)} - \frac{\sin(k+1)\chi_p}{2(k+1)} - \frac{\sin(k+2)\chi_p}{2(k+2)}$$

and aft

$$\int_{0}^{\chi_p} (\cos \chi + \cos 2\chi)\, d\chi = \sin \chi_p + \frac{\sin 2\chi_p}{2}$$

$$\int_{0}^{\chi_p} \cos \chi(\cos \chi + \cos 2\chi)\, d\chi = \frac{\chi_p}{2} + \frac{\sin \chi_p}{2} + \frac{\sin 2\chi_p}{4} + \frac{\sin 3\chi_p}{6}$$

$$\int_{0}^{\chi_p} \cos 2\chi(\cos \chi + \cos 2\chi)\, d\chi = \frac{\chi_p}{2} + \frac{\sin \chi_p}{2} + \frac{\sin 3\chi_p}{6} + \frac{\sin 4\chi_p}{8}$$

and for $k = 3, 4, \ldots$

$$\int_{0}^{\chi_p} \cos k\chi(\cos \chi + \cos 2\chi)\, d\chi$$

$$= \frac{\sin(k-2)\chi_p}{2(k-2)} + \frac{\sin(k-1)\chi_p}{2(k-1)} + \frac{\sin(k+1)\chi_p}{2(k+1)} + \frac{\sin(k+2)\chi_p}{2(k+2)}.$$

So

$$
\begin{aligned}
2C_{m_{c/4}} = {} & \frac{(f_1 + f_2)\pi}{2} + (a_0 - f_0)\left(\sin\chi_p + \frac{\sin 2\chi_p}{2}\right) \\
& + (a_1 - f_1)\left(\frac{\chi_p + \sin\chi_p}{2} + \frac{\sin 2\chi_p}{4} + \frac{\sin 3\chi_p}{6}\right) \\
& + (a_2 - f_2)\left(\frac{\chi_p + \sin\chi_p}{2} + \frac{\sin 3\chi_p}{6} + \frac{\sin 4\chi_p}{8}\right) \\
& + \sum_{k=3}^{\infty}(a_k - f_k)\left\{\frac{\sin(k-2)\chi_p}{2(k-2)} + \frac{\sin(k-1)\chi_p}{2(k-1)}\right. \\
& \left. \qquad\qquad\quad + \frac{\sin(k+1)\chi_p}{2(k+1)} + \frac{\sin(k+2)\chi_p}{2(k+2)}\right\}
\end{aligned}
$$

$$
\begin{aligned}
= {} & \frac{(f_1 + f_2)\pi}{2} + (a_1 - f_1 + a_2 - f_2)\frac{\chi_p}{2} \\
& + \{2(a_0 - f_0) + a_1 - f_1 + a_2 - f_2 + a_3 - f_3\}\frac{\sin\chi_p}{2} \\
& + \{2(a_0 - f_0) + a_1 - f_1 + a_3 - f_3 + a_4 - f_4\}\frac{\sin 2\chi_p}{4} \\
& + \sum_{k=3}^{\infty}(a_{k-2} - f_{k-2} + a_{k-1} - f_{k-1} + a_{k+1} - f_{k+1} + a_{k+2} - f_{k+2})\frac{\sin k\chi_p}{2k}
\end{aligned}
$$

6.3.1 Example: the NACA Four-digit Wing Sections

The camber lines of the NACA four-digit wing sections have the formula (Jacobs *et al.* 1933; Abbott and von Doenhoff 1959):

$$
\frac{y}{c} = \begin{cases} \frac{m}{p^2}\left(\frac{2px}{c} - \frac{x^2}{c^2}\right), & \frac{x}{c} < p \\ \frac{m}{(1-p)^2}\left(1 - 2p + \frac{2px}{c} - \frac{x^2}{c^2}\right), & p < \frac{x}{c} \end{cases} \tag{6.13}
$$

so that the slope is

$$
\frac{dy}{dx} = \begin{cases} \frac{2m}{p^2}\left(p - \frac{x}{c}\right), & \frac{x}{c} < p \\ \frac{2m}{(1-p)^2}\left(p - \frac{x}{c}\right), & p < \frac{x}{c} \end{cases} \tag{6.14}
$$

or in terms of the eccentric angle

$$
\frac{dy}{dx} = \begin{cases} \frac{2m}{p^2}\left(p - \frac{1}{2}\{1 + \cos\chi\}\right), & \frac{x}{c} < p \\ \frac{2m}{(1-p)^2}\left(p - \frac{1}{2}\{1 + \cos\chi\}\right), & p < \frac{x}{c} \end{cases}
$$

$$
= \begin{cases} \frac{m}{p^2}(2p - 1 - \cos\chi), & \frac{x}{c} < p \\ \frac{m}{(1-p)^2}(2p - 1 - \cos\chi), & p < \frac{x}{c} \end{cases}
$$

so that the coefficients are

$$
f_0 = \frac{m(2p-1)}{p^2}, \qquad\qquad a_0 = \frac{m(2p-1)}{(1-p)^2}
$$

$$
f_1 = \frac{-m}{p^2}, \qquad\qquad a_1 = \frac{-m}{(1-p)^2}
$$

and their jumps are

$$
a_0 - f_0 = \frac{m(2p-1)^2}{p^2(1-p)^2}
$$

$$
a_1 - f_1 = \frac{-m(2p-1)}{p^2(1-p)^2}
$$

and the zero-lift incidence is given by

$$
\frac{\alpha_0}{m} = \frac{4p-3}{2p^2} + \frac{(2p-1)\left\{\left(2p - \frac{3}{2}\right)\chi_p - 2(1-p)\sin\chi_p - \frac{\sin 2\chi_p}{4}\right\}}{p^2(1-p)^2\pi}; \tag{6.15}
$$

that it is proportional to the maximum camber m is a necessary consequence of the linearization. This result is plotted in Figure 6.1. Note that the NACA 2412 example of Bertin (2002) gives $\alpha_0 \doteq -2.095°$, which is slightly out; the correct theoretical value for the 24 camber line to three decimal places is $\alpha_0 \doteq -2.077°$.

Similarly, the quarter-chord pitching moment coefficient is

$$
\frac{C_{m_{c/4}}}{m} = \frac{-\pi}{4p^2} - \frac{(2p-1)\left\{\chi_p - (4p-3)\left(\sin\chi_p + \frac{\sin 2\chi_p}{2}\right) + \frac{\sin 3\chi_p}{3}\right\}}{4p^2(1-p)^2}. \tag{6.16}
$$

This is compared with the wind-tunnel measurements of Jacobs *et al.* (1933, table V) in Figure 6.2. In this case, the value given by Bertin (2002) $C_{m_{c/4}} \doteq -0.05309$ is closer to the correct value $C_{m_{c/4}} = -0.05312$.

6.4 Numerical Thin Aerofoil Theory

The analytical approach demonstrated in Section 6.3 is somewhat involved, and can't be used unless an explicit differentiable formula for the camber line is available. A more generally applicable approach is to approximate the weighted integrals of the camber line slope distribution using a numerical quadrature rule, such as the trapezoidal rule.

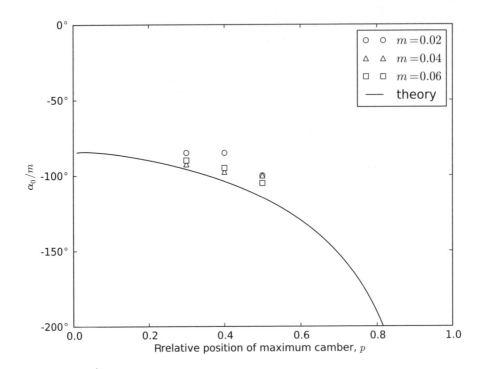

Figure 6.1   Movement of the zero-lift incidence α_0 with position p of maximum camber m for the NACA four-digit camber lines: comparison of theoretical Equation (6.15) with experiments at 12% thickness from Jacobs *et al.* (1933, table III)

Consider the thin aerofoil theoretical zero-lift incidence integral in Equation (6.10)

$$\alpha_0 = \frac{1}{\pi} \int_0^\pi \frac{dy}{dx} (1 + \cos \chi) \, d\chi.$$

Say we partitioned the domain with $0 = x_1 < x_2 < \ldots x_k < \ldots < x_n = c$ and corresponding eccentric angles $\pi = \chi_1 > \chi_2 > \ldots > \chi_k > \ldots > \chi_n = 0$

$$\alpha_0 = \frac{-1}{\pi} \sum_{k=1}^{n-1} \int_{\chi_k}^{\chi_{k+1}} \frac{dy}{dx} (1 + \cos \chi) \, d\chi$$

and treated the camber line slope as constant on each segment

$$\alpha_0 \doteq \frac{-1}{\pi} \sum_{k=1}^{n-1} \int_{\chi_k}^{\chi_{k+1}} \frac{y_{k+1} - y_k}{x_{k+1} - x_k} (1 + \cos \chi) \, d\chi$$

$$= \frac{-1}{\pi} \sum_{k=1}^{n-1} \frac{y_{k+1} - y_k}{x_{k+1} - x_k} [\chi + \sin \chi]_{\chi_k}^{\chi_{k+1}} .$$

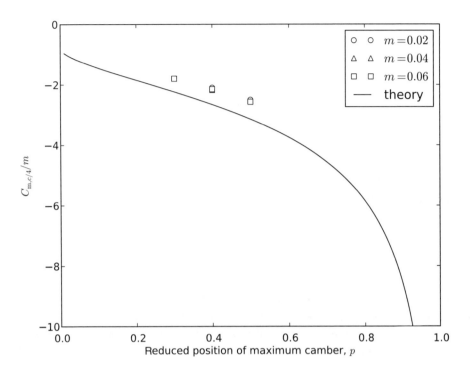

Figure 6.2 Variation of the pitching moment with position of maximum camber for the NACA four-digit camber lines: comparison of theoretical Equation (6.16), with experiments at 12% thickness from Jacobs *et al.* (1933, table III) and camber as marked

Introducing the *forward difference* notation

$$\Delta y_k \equiv y_{k+1} - y_k,$$

this is

$$\alpha_0 \doteq \frac{-1}{\pi} \sum_{k=1}^{n-1} \frac{\Delta y_k}{\Delta x_k} \Delta(\chi + \sin \chi)_k.$$

For the pitching moment, the corresponding numerical approximation is

$$C_{m_{c/4}} \doteq \frac{-1}{2} \sum_{k=1}^{n-1} \frac{\Delta y_k}{\Delta x_k} \Delta \left(\sin \chi + \frac{\sin 2\chi}{2} \right)_k.$$

These numerical approximation formulae can be conveniently calculated manually using a table or semi-manually with spreadsheet, or better yet a matrix-oriented programming language such as Octave (in which the forward difference operator is called **diff**). By exploiting the matrix-oriented features, the program thin (Listing 6.1) turns out to be both very compact and very similar in form to the mathematical formulae.

Listing 6.1 thin.m.

```
function [alpha0, Cmac] = thin (x, y)
  chi = acos (2 * x - 1);
  slope = diff (y) ./ diff (x);
  alpha0 = -slope' * diff (chi + sin (chi)) / pi;
  Cmac = -slope' * diff (sin (chi) + sin (2 * chi) / 2) / 2;
```

Listing 6.2 naca4pars.m.

```
function [m, p, t] = naca4pars (n)
  m = 0.01 * round (n / 1e3);
  p = 0.1 * round (rem (n, 1e3) / 1e2);
  t = 0.01 * rem (n, 1e2);
```

One such feature has been used to write the sum almost implicitly using matrix multiplication, as described in Appendix A.2.8. Here the inputs are assumed to be column-vectors and the left factor is converted into row-vector by transposition with the apostrophe operator.

To demonstrate the program, we need a sample geometry; to this end we provide naca4pars and naca4meanline in Listings 6.2 and 6.3.

A transcript demonstrating the use of these procedures is given in Listing 6.4.

The procedure, as implemented above and using the indicated x-values, predicts $\alpha_0 \doteq -2.052°$ and $C_{mc/4} \doteq -0.0524$, which compare well with the exact thin aerofoil theoretic values $\alpha_0 \doteq -2.08°$ and $C_{mc/4} \doteq -0.0531$ (Section 6.3); indeed, the agreement between the numerical and exact thin aerofoil answers is better than the agreement of the exact values with the experimental results $\alpha_0 \doteq -1.7°$, $C_{mc/4} \doteq -0.042$ (Jacobs *et al.* 1933).

Listing 6.3 naca4meanline.m.

```
function y = naca4meanline (x, m, p)
  aft = x(:) >= p;
  y(~aft,1) = m / p^2 * (2*p*x(~aft) - x(~aft) .^2);
  y(aft,1) = m / (1-p)^2 * (1-2*p + 2*p*x(aft) - x(aft).^2);
```

Listing 6.4 Octave instructions to demonstrate thin.m from Listing 6.1.

```
x = [0, 1.25, 2.5:2.5:10, 15:5:25, 30:10:90, 95, 100]'/1e2;
[m, p] = naca4pars (2412);
y = naca4meanline (x, 0.02, 0.4);
[alpha0, Cmc4] = thin (x, y);
alpha0*180/pi, Cmc4
```

6.5 Exercises

1. For the NACA 4-digit wing sections
   (a) 0012,
   (b) 1412,
   (c) 2412, and
   (d) 4412

   apply the theory of thin wing sections to compute the distribution of circulation density along the camber line as a function of incidence, and then the circulation, lift, and pitching moment.

   For the NACA 2412, compute the required integrals both analytically and numerically; for the others, use either approach.

6.6 Further Reading

Thin aerofoil theory is discussed in most textbooks on aerodynamics (Glauert 1926; Abbott and von Doenhoff 1959; Karamcheti 1966; Batchelor 1967; Milne-Thomson 1973; Ashley and Landahl 1985; McCormick 1995; Kuethe and Chow 1998; Paraschivoiu 1998; Bertin 2002; Houghton and Carpenter 2003; Moran 2003; Anderson 2007).

The formulae for the NACA four- and five-digit camber lines are given by Jacobs *et al.* (1933) and Jacobs and Pinkerton (1936); see also Abbott and von Doenhoff (1959).

References

Abbott, I.H. and von Doenhoff, A.E. (1959) *Theory of Wing Sections*. New York: Dover.

Anderson, J.D. (2007) *Fundamentals of Aerodynamics*, 4th edn. New York: McGraw-Hill.

Ashley, H. and Landahl, M. (1985) *Aerodynamics of Wings and Bodies*. New York: Dover.

Batchelor, G.K. (1967) *An Introduction to Fluid Dynamics*. Cambridge: Cambridge University Press.

Bertin, J.J. (2002) *Aerodynamics for Engineers*, 4th edn. New Jersey: Prentice Hall.

Glauert, H. (1926) *The Elements of Aerofoil and Airscrew Theory*. Cambridge: Cambridge University Press.

Houghton, E.L. and Carpenter, P.W. (2003) *Aerodynamics for Engineering Students*, 5th edn. Oxford: Butterworth Heinemann.

Jacobs, E.N. and Pinkerton, R.M. (1936) Tests in the Variable-Density Wind Tunnel of Related Airfoils having the Maximum Camber Unusually Far Forward. Report 537, NACA.

Jacobs, E.N., Ward, K.E. and Pinkerton, R.M. (1933) The Characteristics of 78 Related Airfoil Sections from Tests in the Variable-Density Wind Tunnel. Report 460, NACA.

Karamcheti, K. (1966) *Principles of Ideal-Fluid Aerodynamics*. Chichester: John Wiley & Sons, Ltd.

Kuethe, A.M. and Chow, C.Y. (1998) *Foundations of Aerodynamics*, 5th edn. Chichester: John Wiley & Sons, Ltd.

McCormick, B.W. (1995) *Aerodynamics, Aeronautics, and Flight Mechanics*, 2nd edn. Chichester: John Wiley & Sons, Ltd.

Milne-Thomson, L.M. (1973) *Theoretical Aerodynamics*, 4th edn. New York: Dover.

Moran, J. (2003) *An Introduction to Theoretical and Computational Aerodynamics*. New York: Dover.

Paraschivoiu, I. (1998) *Aérodynamique Subsonique*. Éditions de l'École de Montréal.

7

Lumped Vortex Elements

7.1 The Thin Flat Plate at Arbitrary Incidence, Again

Recall that for a flat plate we obtained the solution to the flow field as Equation (5.14)

$$w(z) = q_\infty e^{-i\alpha} + \frac{i}{2\pi} \int_0^c \frac{\gamma(x') \, dx'}{z - x'}$$

$$\text{where } \gamma = 2q_\infty \sin\alpha \, \frac{1 - \cos\chi}{\sin\chi}$$

$$\text{and } \cos\chi = \frac{2x}{c} - 1.$$

The total circulation is, by Equation (5.6),

$$\Gamma = \pi q_\infty c \sin\alpha.$$

7.1.1 Single Vortex

Using our knowledge of this solution, we attempt to model the flat plate using a single vortex. Recalling (Section 5.3.1) that the aerodynamic force acts through the quarter-chord point $x = c/4$, we place the vortex there. The complex velocity field due to the free-stream and the single vortex is

$$w(z) = q_\infty e^{-i\alpha} + \frac{i\Gamma}{2\pi(z - c/4)} = q_\infty \left(e^{-i\alpha} + \frac{i \sin\alpha}{\frac{2z}{c} - \frac{1}{2}} \right).$$

7.1.2 The Collocation Point

The question then arises: how well does this single vortex model the flow? The criterion to apply is that there should be no flow through the plate. On the plate $y = 0$ and the normal

---

Theory of Lift: Introductory Computational Aerodynamics in MATLAB®/Octave, First Edition. G. D. McBain.
© 2012 John Wiley & Sons, Ltd. Published 2012 by John Wiley & Sons, Ltd.

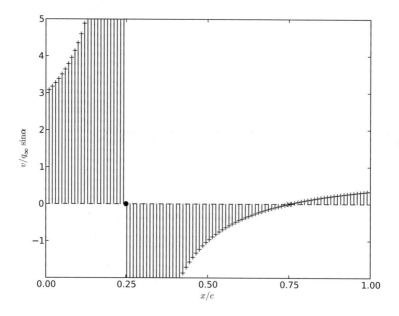

Figure 7.1 Normal component of velocity on a flat plate due to a nonzero incidence and a corresponding vortex lumped at the quarter-point, as computed from Equation (7.1)

component is

$$v(x) = -\Im w = q_\infty \left\{ \sin\alpha - \frac{\sin\alpha}{\frac{2x}{c} - \frac{1}{2}} \right\} = q_\infty \left(1 - \frac{1}{\frac{2x}{c} - \frac{1}{2}} \right) \sin\alpha, \qquad (7.1)$$

which, as shown in Figure 7.1, vanishes only at

$$x = \frac{3c}{4}.$$

We call this special point at which the boundary condition is satisfied the *collocation point*; note that it is independent of the incidence α.

7.1.3 Lumped Vortex Model of the Thin Flat Plate

As an artificial and trivial application of the lumped vortex method, say we knew the correct location for the vortex ($x = c/4$) and the collocation point ($x = 3c/4$), but didn't know the vortex strength Γ. The configuration of this single lumped vortex model of the plate is shown in Figure 7.2.

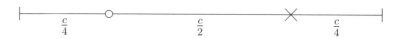

Figure 7.2 Model of a flat plate consisting of a single lumped vortex at the quarter-chord and a collocation point at the three-quarter-chord

The complex velocity would be

$$w(z) = q_\infty e^{-i\alpha} + \frac{i\Gamma}{2\pi(z - c/4)},$$

and the normal velocity at the collocation point would be

$$v(3c/4) = q_\infty \sin\alpha - \frac{\frac{c}{2}\Gamma}{2\pi\left(\frac{c}{2}\right)^2} = q_\infty \sin\alpha - \frac{\Gamma}{\pi c}.$$

Therefore, setting this to zero gives the circulation as

$$\Gamma = \pi c q_\infty \sin\alpha,$$

in agreement with the results of conformal mapping in Equation (5.6) and the continuous vortex sheet model in Equation (5.11).

The flow produced by the single lumped vortex is compared with the full solution of Section 4.3.4 obtained from conformal mapping in Figure 7.3. The figures were produced by plotting the level-sets of the stream function

$$\psi = \Im W = \Im \int w\, dz = \Im \left\{ q_\infty e^{-i\alpha} z + \frac{i\Gamma \ln\left(z - \frac{c}{4}\right)}{2\pi} \right\}$$

with the Octave functions in Listings 7.1 and 7.2.

Notice that far from the plate the agreement is very good, but it's less good in the immediate vicinity of points like the leading edge, $z = 0$, or the lumped vortex point, $z = \frac{c}{4}$. Observe too the horizontal streamlines near the collocation point, $z = \frac{3c}{4}$.

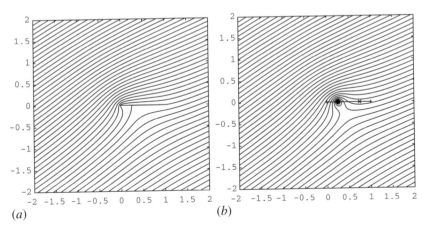

(a) (b)

Figure 7.3 Streamlines over a thin plate at nonzero incidence with circulation, exactly from conformal mapping (a) and approximately with a single lumped vortex (b)

Listing 7.1 `streamlines_plate.m`.

```
alpha = pi/6; c = 1; Gamma = c * pi * sin (alpha);
W = @ (zeta) zeta + (c/4)^2 ./ zeta;
W1 = @ (zeta) W (zeta) - Gamma * log (4 * zeta / c) / pi / 2i;
W2 = @ (zeta) W1 (zeta / exp (1i * alpha));
W3 = @ (z) W2 (joukowsky_inverse (z, c));
streamlines_W (-2-2i, 2+2i, W3, -3:0.1:4 )
hold ('on')
plot ([0, 1], [0, 0], 'LineWidth', 3, 'LineStyle', '-')
```

Listing 7.2 `streamlines_lvm1.m`.

```
alpha = pi/6; c = 1; Gamma = c * pi * sin (alpha);
W0 = @ (z) exp (-1i * alpha) * z;
W = @ (z) W0 (z) - Gamma * log (z - c/4) / pi / 2i;
streamlines_W (-2-2i, 2+2i, W, -3:0.1:4)
hold ('on')
plot ([0, 1], [0, 0], '+-', 'LineWidth', 3, ...
      1/4, 0, 'o', 3/4, 0, 'x')
axis ('image')
```

7.2   Using Two Lumped Vortices along the Chord

As a slightly more realistic example, say the plate is split at $x = c/2$ and each half represented with a lumped vortex panel, as shown in Figure 7.4. This problem is still simple enough to be worked exactly by hand, but begins to exhibit some of the structure of a general numerical panel method.

The complex velocity field is

$$w(z) = q_\infty e^{-i\alpha} + \frac{i\Gamma_1}{2\pi\left(z - z_1^{(v)}\right)} + \frac{i\Gamma_2}{2\pi\left(z - z_2^{(v)}\right)}. \tag{7.2}$$

Here the vortex points are the quarter-points of each panel:

$$z_1^{(v)} = \frac{c}{8}$$

$$z_2^{(v)} = \frac{5c}{8}.$$

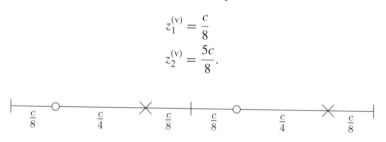

Figure 7.4   Model of a flat plate consisting of two lumped vortex panels end-to-end

On the plate $z = x$ and the normal component of velocity reduces to

$$v(x) = q_\infty \sin \alpha - \frac{1}{2\pi} \left(\frac{\Gamma_1}{x - x_1^{(v)}} + \frac{\Gamma_2}{x - x_2^{(v)}} \right),$$

where $x_i^{(v)}$ is the real part of $z_i^{(v)}$.

The two unknown vortex strengths can be determined by enforcing the impermeability condition (here $v = 0$) at the collocation points $x_i^{(c)}$: the three-quarter points of the panels. Thus:

$$q_\infty \sin \alpha - \frac{1}{2\pi} \left(\frac{\Gamma_1}{x_1^{(c)} - x_1^{(v)}} + \frac{\Gamma_2}{x_1^{(c)} - x_2^{(v)}} \right) = 0$$

$$q_\infty \sin \alpha - \frac{1}{2\pi} \left(\frac{\Gamma_1}{x_2^{(c)} - x_1^{(v)}} + \frac{\Gamma_2}{x_2^{(c)} - x_2^{(v)}} \right) = 0,$$

where the collocation points are

$$x_1^{(c)} = \frac{3c}{8}$$

$$x_2^{(c)} = \frac{7c}{8}.$$

This system can be written in matrix–vector form as

$$\frac{1}{2\pi} \begin{bmatrix} \frac{1}{x_1^{(c)} - x_1^{(v)}} & \frac{1}{x_1^{(c)} - x_2^{(v)}} \\ \frac{1}{x_2^{(c)} - x_1^{(v)}} & \frac{1}{x_2^{(c)} - x_2^{(v)}} \end{bmatrix} \begin{bmatrix} \Gamma_1 \\ \Gamma_2 \end{bmatrix} = q_\infty \sin \alpha \begin{bmatrix} 1 \\ 1 \end{bmatrix}. \tag{7.3}$$

Inserting the numerical values for the coordinates of the vortex and collocation points gives

$$\frac{1}{2\pi c} \begin{bmatrix} 4 & -4 \\ \frac{4}{3} & 4 \end{bmatrix} \begin{bmatrix} \Gamma_1 \\ \Gamma_2 \end{bmatrix} = q_\infty \sin \alpha \begin{bmatrix} 1 \\ 1 \end{bmatrix}.$$

Solving the linear system gives

$$\begin{bmatrix} \Gamma_1 \\ \Gamma_2 \end{bmatrix} = 2\pi q_\infty c \sin \alpha \begin{bmatrix} \frac{3}{8} \\ \frac{1}{8} \end{bmatrix}.$$

The streamlines are plotted in Figure 7.5.

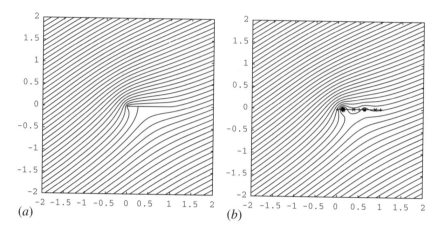

Figure 7.5 Streamlines over a thin plate at nonzero incidence with circulation, exactly from conformal mapping (*a*) and approximately with two lumped vortices (*b*)

7.2.1 Postprocessing

Having calculated the strengths of the lumped vortices, the lift can be deduced, either:

- from the total circulation and the Kutta–Joukowsky theorem or
- by aggregrating the forces on individual vortices.

The total circulation is

$$\Gamma = \Gamma_1 + \Gamma_2 = \pi q_\infty c \sin \alpha,$$

the lift (per unit span) is

$$\ell = \rho q_\infty \Gamma = \rho \pi q_\infty^2 c \sin \alpha,$$

and the lift coefficient is

$$C_\ell = \frac{\ell}{\frac{1}{2}\rho q_\infty^2 c} = 2\pi \sin \alpha.$$

These three results agree with the exact results.

If we compute the pitching moment about the quarter-chord from the two-panel model, we again get the correct answer, zero (Section 5.3.1):

$$m_{c/4} = \left(\frac{c}{4} - \frac{c}{8}\right)\rho q_\infty \Gamma_1 + \left(\frac{c}{4} - \frac{5c}{8}\right)\rho q_\infty \Gamma_2 = \frac{\rho q_\infty c}{8}(\Gamma_1 - 3\Gamma_2) = 0.$$

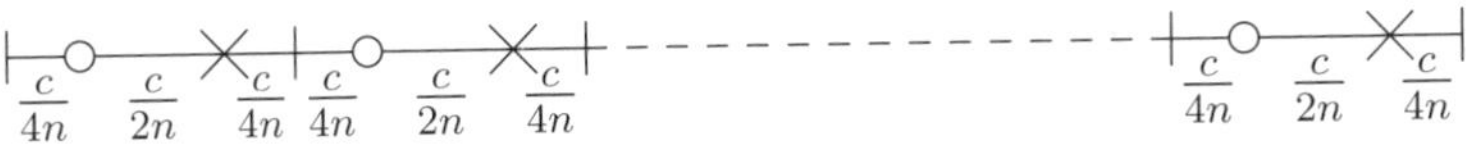

Figure 7.6 Model of a flat plate consisting of n lumped vortex panels end-to-end

7.3 Generalization to Multiple Lumped Vortex Panels

If instead of two, we used n vortex panels, as shown in Figure 7.6 the matrix–vector Equation (7.3) would be:

$$
\frac{1}{2\pi}
\begin{bmatrix}
\frac{1}{x_1^{(c)}-x_1^{(v)}} & \frac{1}{x_1^{(c)}-x_2^{(v)}} & \cdots & \frac{1}{x_1^{(c)}-x_n^{(v)}} \\
\frac{1}{x_2^{(c)}-x_1^{(v)}} & \frac{1}{x_2^{(c)}-x_2^{(v)}} & \cdots & \frac{1}{x_2^{(c)}-x_n^{(v)}} \\
\vdots & & & \\
\frac{1}{x_n^{(c)}-x_1^{(v)}} & \frac{1}{x_n^{(c)}-x_2^{(v)}} & \cdots & \frac{1}{x_n^{(c)}-x_n^{(v)}}
\end{bmatrix}
\begin{bmatrix}
\Gamma_1 \\ \Gamma_2 \\ \vdots \\ \Gamma_n
\end{bmatrix}
= q_\infty \sin\alpha
\begin{bmatrix}
1 \\ 1 \\ \vdots \\ 1
\end{bmatrix},
\tag{7.4}
$$

where $x_i^{(c)}$ is the collocation (three-quarter) point of the ith panel and $x_j^{(v)}$ is the lumped vortex (one-quarter) point of the jth panel.

7.3.1 Postprocessing

Having obtained the lumped vortex strengths, the complex velocity at an arbitrary point z (not coinciding with any of the vortices) is

$$
w(z) = q_\infty e^{-i\alpha} + \sum_{j=1}^{n} \frac{i\Gamma_j}{2\pi(z - z_j^{(v)})}
$$

and the lift and quarter-chord pitching moment coefficients are

$$
C_\ell = 2 \sum_{j=1}^{n} \frac{\Gamma_j}{q_\infty c}
\tag{7.5}
$$

$$
C_{mc/4} = 2\cos\alpha \sum_{j=1}^{n} \left(\frac{1}{4} - \frac{x_j^{(v)}}{c} \right) \frac{\Gamma_j}{q_\infty c}.
\tag{7.6}
$$

7.4 General Considerations on Discrete Singularity Methods

- The use of two panels gave the same result for the lift coefficient as the use of one panel. Both agreed with the exact result for this problem, since the lumped vortex panels are based on the exact solution.
- The use of one panel gave only a result for the overall circulation, but the use of two panels began to give information about the distribution of vortex strength along the plate.

- The two-panel example exhibits the basic steps common to all numerical panel methods

  1. select singularity element;
  2. discretize geometry and generate grid;
  3. compute matrix of influence coefficients;
  4. establish right-hand side vector;
  5. solve linear set of equations;
  6. compute derived quantities (e.g. lift, pitching moment).

Here, these steps were:

1. Select singularity element: we chose to generate the flow perturbation with lumped vortex panels.
2. Discretize geometry and generate grid: we split the plate into two, fore and aft, and put vortices and collocation points at the quarter and three-quarter points of each panel.
3. Compute matrix of influence coefficients: enforcing the impermeability condition at the collocation points required the calculation of the contribution to the normal velocity at the ith collocation point from the jth lumped vortex. This appears as the coefficient in the ith row and jth column of the matrix, and represents the influence of the jth panel on the ith.
4. Establish right-hand side vector: here there were constant terms in the impermeability condition equations due to the free-stream velocity; these formed the right-hand side of the matrix–vector equation.
5. Solve linear set of equations: here the example was small (2×2) so that we could solve it exactly by hand. For larger examples, it would be difficult to use exact fractions and easier to use approximate decimals. For much larger examples (say bigger than 8×8), the matrix–vector equation should be solved by a specialized computer program such as Octave.
6. Compute derived quantities (e.g. lift): here we just added the vortex strengths to obtain the circulation; however, we could also have gone back to the expression of Equation (7.2) and computed the velocity field, then used Bernoulli's equation to compute the pressure field, etc.

- The method for two panels can easily be generalized to any number of panels and is well suited to computer implementation, e.g. in Octave.
- With some modifications, the method can be generalized to cambered aerofoils. The advantage of this over the use of thin aerofoil theory is that there is no need to assume either the camber or incidence small. This is developed in Section 7.5.
- Let's consider the main differences between the thin aerofoil type solution and the lumped vortex panel solution to the flat plate problem. The thin aerofoil solution required a number of difficult and slightly mysterious mathematical steps; e.g. Glauert's transformation and Glauert's integral. The lumped vortex method is conceptually much simpler, but involves the solution of a matrix–vector equation: this requires a computer if a realistic number of panels are used.
- The real advantage of the lumped vortex panel method is that it can be systematically generalized to handle much more complicated problems. Some of the generalizations are:

- Instead of using lumped vortex panels, we can use panels with point sources or point doublets, or other kinds of singularity.
- Instead of concentrating the singularity at a point on the panel, we can spread it out over the panel. This must be done fairly simply so that the method remains suitable for computer implementation. Common approaches include constant strength panels, panels with strength varying linearly over the panel, panels with strength varying quadratically, etc. The difficulty of using these distributed-strength panels as opposed to the lumped-strength panels of the problem worked above is that a number of complicated integrals (of the singularity strength over the panel) have to be solved to compute the influence coefficients; however, these can usually be found in reference books and can be solved once and for all when the computer program is written.
- Instead of using straight line segments as panels, we could use curved panels. Again, the curvature must be kept simple if the method is to be implemented on a computer. One approach is to allow each panel to be parabolic; another is to represent the camber line with a spline.
- The method can be generalized to three-dimensional problems. The two-dimensional surface of a three-dimensional body can be covered in quadrilateral panels and a variety of flow singularities (source, doublets, vortices) attached to each.
- The computation of subsidiary quantities is often accompanied by the production of graphical output, such as pressure plots or streamlines. This can be computationally intensive, but is an excellent means of visualizing the aerodynamical problem and building intuitive understanding.

7.5   Lumped Vortex Elements for Thin Aerofoils

7.5.1   Panel Chains for Camber Lines

Here we use a chain of lumped vortex elements to model a thin aerofoil. This provides an alternative to thin aerofoil theory. The results obtained are similar.

Say we have the complex coordinates for $n + 1$ points along the camber line of an aerofoil: $z = z_1, z_2, \ldots, z_{n+1}$, with z_1 at the leading edge and z_{n+1} at the trailing edge. The camber line can be modelled by putting a lumped vortex element on each line segment joining these points; i.e. the jth element runs from z_j to z_{j+1}. This is illustrated in Figure 7.7.

With the forward difference notation

$$\Delta z_j \equiv z_{j+1} - z_j,$$

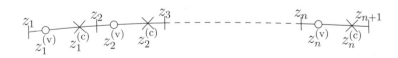

Figure 7.7   Model of a slightly cambered mean line consisting of n lumped vortex panels. As in Figures 7.2, 7.4, and 7.6, circles and crosses represent vortices and collocation points, respectively; the notches are the points defining the geometry

the vortex points (quarter-points) are

$$z_j^{(v)} = z_j + \frac{1}{4}\Delta z_j$$

and the collocation points (three-quarter-points) are

$$z_i^{(c)} = z_i + \frac{3}{4}\Delta z_i.$$

We place vortices of strength Γ_j at the n vortex points. With a free-stream at incidence α, the complex velocity field is then:

$$w(z) = q_\infty e^{-i\alpha} + \frac{i}{2\pi}\sum_{j=1}^{n}\frac{\Gamma_j}{z - z_j^{(v)}}.$$

Evaluated at the ith collocation point, the complex velocity is

$$w\left(z_i^{(c)}\right) = q_\infty e^{-i\alpha} + \frac{i}{2\pi}\sum_{j=1}^{n}\frac{\Gamma_j}{z_i^{(c)} - z_j^{(v)}}.$$

Write this as

$$w\left(z_i^{(c)}\right) = q_\infty e^{-i\alpha} + \sum_{j=1}^{n} i\omega_{ij}\Gamma_j,$$

where the

$$\omega_{ij} = \frac{1}{2\pi\left\{z_i^{(c)} - z_j^{(v)}\right\}}$$

are the *influence coefficients*, being the contribution of a unit point source at $z_j^{(v)}$ to the complex velocity at $z_i^{(c)}$.

The n unknown vortex strengths Γ_j are determined by requiring that the n panels are impermeable at their collocation points; that is, that the velocity normal to the panel vanish at the three-quarter point. The slope of the ith panel is

$$\tan \lambda_i = \tan \arg \Delta z_i \equiv \frac{\Delta y_i}{\Delta x_i},$$

and the normal component of velocity at the collocation point of the ith panel, which has direction λ_i, is

$$-\Im\left(e^{i\lambda_i}w(z_i^{(c)})\right) = q_\infty \sin(\alpha - \lambda_i) + \sum_{j=1}^{n}\left\{-\Im\left(e^{i\lambda_i} i\omega_{ij}\right)\right\}\Gamma_j.$$

Setting this to zero results in a system of n linear equations in n unknowns, and can therefore be written as a matrix–vector equation

$$\sum_{j=1}^{n} a_{ij} x_j = b_j \tag{7.7}$$

where

$$a_{ij} = -\Im\left(e^{i\lambda_i} i \omega_{ij} c\right)$$
$$x_j = \frac{\Gamma_j}{c q_\infty}$$
$$b_i = \sin(\lambda_i - \alpha).$$

Once this has been solved for the unknown vortex strenths Γ_j, the lift and pitching moment coefficients can be obtained from Equations (7.5) and (7.6).

7.5.2 Implementation in Octave

The lumped vortex method for thin aerofoils can be directly implemented in a computer programming language that handles complex matrices such as Octave.

It's convenient to first define a function lvmi.m (Listing 7.3) to compute the influence factors from one point to another (or array of points to another), since as well as being required in the assembly of the problem, this is used in postprocessing to evaluate the solution at points of interest (e.g. on a grid for a silk-thread arrow plot).

Once this one-line influence coefficient subroutine is defined, the lumped-vortex method program itself lvm.m (Listing 7.4) is only half a dozen lines.

Listing 7.3 lvmi: compute the lumped vortex influence coefficients for lvm of Listing 7.4.

```
function w = lvmi (z, zeta)
  w = 1.0 ./ (2 * pi * bsxfun (@minus, z(:), zeta));
```

Listing 7.4 lvm: solve the lumped vortex model of a thin wing section with complex vertices z at incidence alpha radians; requires lvmi from Listing 7.3.

```
function [zc, omega, Gamma] = lvm (z, alpha)
  dz = diff (z);
  lambda = angle (dz)';
  zc = (z(1:(end-1)) + dz*3/4).';
  omega = lvmi (zc, zc.' - dz / 2);
  Gamma = -imag (diag (exp (1i*lambda) * 1i) ...
                * omega) \ sin (lambda-alpha);
```

Listing 7.5   lvm_demo: demonstrating naca4meanline and lvm of Listings 6.3 and 7.4.

```
x = linspace (0, 1, 21);
y = naca4meanline (x, 0.02, 0.4);
alpha = 0;
[zc, omega, Gamma] = lvm (complex (x, y'), alpha);
Cl = 2 * sum (Gamma)
Cmc4 = 2 * sum (Gamma .* real (1/4 - zc)) * cos (alpha)
```

The matrix–vector Equation (7.7) is solved there using Octave's backslash operator, as in Listing 1.4 in Section 1.5.2. With naca4meanline from Listing 6.3, the method can be applied to the NACA 2400 camber line, as shown in Listing 7.5.

The zero-incidence lift, 0.21758, is within 10% of the exact thin aerofoil result $C_\ell = -2\pi\alpha_0 = 0.228$ (recalling $\alpha_0 = -2.077°$, and $C_{m_{c/4}} = -0.053$ from Section 6.3.1). The agreement of the result for the pitching moment, $C_{m_{c/4}} = -0.059078$, with the exact theoretical value is less good, but still reasonable.

Better results can be obtained from the method by using more panels. Another possibility for improvement is opened by not necessarily taking all panels of equal length—there is nothing in either the lumped vortex element method or its implementation in Listing 7.4 that depends on the abscissae being equally spaced. An opportunity to explore these avenues is given in the exercises at the end of the chapter.

7.5.3   Comparison with Thin Aerofoil Theory

In the NACA 2412 example, we found that the lumped vortex panel method gave very similar results to thin aerofoil theory. Let's explore this further.

The matrix equation of the lumped vortex panel method is Equation (7.7). If we make the thin aerofoil approximations:

$$\sin \lambda \sim \tan \lambda \sim \lambda \sim \frac{dy}{dx}$$

$$\cos \lambda \sim 1$$

$$e^{i\lambda} = \cos \lambda + i \sin \lambda \sim 1 + i\lambda \sim 1$$

$$z \sim x,$$

the matrix Equation (7.7) becomes

$$\sum_{j=1}^{n} \frac{-c}{2\pi \left\{ x_i^{(c)} - x_j^{(v)} \right\}} \frac{\Gamma_j}{cq_\infty} = \lambda_i - \alpha$$

$$\sum_{j=1}^{n} \frac{\Gamma_j}{2\pi q_\infty \left\{ x_i^{(c)} - x_j^{(v)} \right\}} = \alpha - \lambda_i$$

which may be compared with the thin aerofoil integral equation,

$$\int_0^c \frac{\gamma(x')\,dx'}{2\pi q_\infty(x - x')} = \alpha - \frac{dy}{dx}.$$

Thus, when applied to the thin aerofoil problem, the lumped vortex panel method is equivalent to a numerical approximation to the thin aerofoil integral equation, with Γ_j being the vortex strength equivalent to $\gamma(x')\delta x'$.

7.6 Disconnected Aerofoils

If the lumped vortex panel method served only as an alternative discretization of the thin aerofoil theory it would still be of some interest, since it avoids the difficulty of having to solve the integral equation via Glauert's integral; however, it is much more important than that: it permits generalization to problems that would be very much more difficult or impossible with the analytic approach.

One important direction that the method can be generalized in is to allow multiple disconnected aerofoils, such as leading-edge slats and slotted flaps, and biplanes. We exemplify the method by modelling a biplane with two lumped vortex panels. In particular, we choose the simplest example of a biplane: equal aerofoils with no *stagger*; i.e. both aerofoils have chord length c, with one running from $(0, 0)$ to $(c, 0)$ and the other from $(0, h)$ to (c, h).

As shown in Figure 7.8, the vortex points are:

$$z_1^{(v)} = \frac{c}{4} + i0$$

$$z_2^{(v)} = \frac{c}{4} + ih$$

and the collocation points are

$$z_1^{(c)} = \frac{3c}{4} + i0$$

$$z_2^{(c)} = \frac{3c}{4} + ih.$$

Figure 7.8 Model of a biplane with two lumped vortex elements

With vortices of strength Γ_1 and Γ_2 at $z = z_1^{(v)}$ and $z_2^{(v)}$, respectively, the complex velocity field is

$$w(z) = q_\infty e^{-i\alpha} + \frac{i}{2\pi} \left\{ \frac{(x - \frac{c}{4}) - iy}{(x - \frac{c}{4})^2 + y^2} \Gamma_1 + \frac{(x - \frac{c}{4}) - i(y - h)}{(x - \frac{c}{4})^2 + (y - h)^2} \Gamma_2 \right\}$$

and so the vertical component of velocity is

$$v(z) = q_\infty \sin\alpha - \frac{1}{2\pi} \left\{ \frac{x - \frac{c}{4}}{(x - \frac{c}{4})^2 + y^2} \Gamma_1 + \frac{x - \frac{c}{4}}{(x - \frac{c}{4})^2 + (y - h)^2} \Gamma_2 \right\}.$$

By requiring this to vanish at the two collocation points the unknown vortex strengths can be determined. This leads to the matrix equation

$$\begin{bmatrix} \frac{2}{c} & \frac{2}{c + \frac{4h^2}{c}} \\ \frac{2}{c + \frac{4h^2}{c}} & \frac{2}{c} \end{bmatrix} \begin{bmatrix} \Gamma_1 \\ \Gamma_2 \end{bmatrix} = 2\pi q_\infty \sin\alpha \begin{bmatrix} 1 \\ 1 \end{bmatrix},$$

which has the solution

$$\Gamma_1 = \Gamma_2 = \frac{4\left(\frac{h}{c}\right)^2 + 1}{4\left(\frac{h}{c}\right)^2 + 2} \pi q_\infty c \sin\alpha.$$

This may be compared with the circulation for a single thin flat plate aerofoil given by Equation (5.6):

$$\Gamma = \pi q_\infty c \sin\alpha.$$

It can be seen that each aerofoil is inhibited by the other for finite $\frac{h}{c}$, with the results tending to the value for isolated aerofoils when the separation h becomes much larger than the chord length c. Further, as $h \to 0$, the circulation around each aerofoil tends to half the value for an isolated aerofoil, which is also a logical result.

A comparison with more detailed theoretical solutions for the biplane shows quite good agreement. In practice, however, three-dimensional effects (the interference between the wing-tip vortices) are important in biplane performance, so we don't pursue the details of the comparison here; see Chapter 13.

7.6.1   Other Applications

Other simple applications of the lumped vortex panel method include:

- modelling tandem aerofoils;
- modelling ground effect;
- calculating wind-tunnel corrections.

7.7   Exercises

1. The position of the vortex and the collocation point may be derived in another way: by forcing the lumped vortex panel to give the correct thin-aerofoil theory result for a parabolic camber line $y = mx(c - x)$. Show that this gives the same result.

2. Model two wing sections with single lumped vortex panels. Formulate the general problem of flow over them, assuming they don't move relative to each other. (Don't solve.)

3. Specialize the results of the previous question to the case where the two panels are parallel and of equal length.

4. Specialize the results of the previous question to the cases where the panels are aligned
   (a) horizontally
   (b) vertically.
   Solve the flow problem. How does the presence of the second wing section affect the aerodynamics of the first?

5. Using panels of equal sizes, how does the error decrease as the number of panels increases? Prepare a table and a graph, for both the lift and pitching moment coefficients. Does the error decrease like one over the number of panels? Like one over its square?

6. Repeat the previous question, but instead of defining the panels in terms of equally spaced values of the linear coordinate x, try using equally spaced values of the eccentric angle as given by the Equation (11.11). Alternatively, try making the panels in the aft- or forward-third half the length of those in the other two-thirds. Is it more efficient to cluster panels nearer the leading or trailing edge? What other schemes for spacing the panels can you come up with?

7. (a) Let the free-stream be parallel to the x-axis and place the lumped vortex panel entirely above it, at arbitrary inclination, in the half-space $y > 0$. Then place the 'mirror image' of the panel below the x-axis with equal and opposite vortex strength. Show that the x-axis is a streamline.
   (b) How does the presence of this mirror image affect the force on the upper panel?

7.8   Further Reading

The single lumped vortex method with the vortex at the quarter chord and the collocation point at the three-quarter chord was applied to the problem of ground effect by Pistolesi (1937) and Katz and Plotkin (2001), as in Exercise 7. The importance and usefulness of the three-quarter point is discussed by Weissinger (1947) and Ashley and Landahl (1985), among many others.

Milne-Thomson (1973) called the single lumped vortex approximation for an aerofoil the 'substitution vortex' and applied it to estimate the lift generated by biplanes, as did later Katz and Plotkin (2001), who also similarly analysed tandem aerofoils and wind-tunnel corrections. Modelling a slatted and flapped aerofoil with three disconnected lumped vortex panels was suggested as an exercise by Moran (2003); for more detailed discussion of slats and experimental measurements, see Abbott and von Doenhoff (1959) and Hoerner and Borst (1985, Chapters 5–6).

References

Abbott, I.H. and von Doenhoff, A.E. (1959) *Theory of Wing Sections*. New York: Dover.

Ashley, H. and Landahl, M. (1985) *Aerodynamics of Wings and Bodies*. New York: Dover.

Hoerner, S.F. and Borst, H.V. (1985) *Fluid-Dynamic Lift*, 2nd edn. Bakersfield, CA: Hoerner Fluid Dynamics.

Katz, J. and Plotkin, A. (2001) *Low-Speed Aerodynamics*, 2nd edn. Cambridge: Cambridge University Press.

Milne-Thomson, L.M. (1973) *Theoretical Aerodynamics*, 4th edn. New York: Dover.

Moran, J. (2003) *An Introduction to Theoretical and Computational Aerodynamics*. New York: Dover.

Pistolesi, E. (1937) Ground effect—theory and practice. Technical Memorandum 828, NACA.

Weissinger, J. (1947) The lift-distribution on swept-back wings. Technical Memorandum 1126, NACA.

8

Panel Methods for Plane Flow

The thin wing section theory of Chapter 6 and the lumped vortex panel method of Chapter 7 both neglect any effects of thickness. Arbitrarily shaped wing sections can be modelled using panel methods that distribute the singularities (sources and vortices) around the profile, instead of along the camber line.

Here we describe a method that breaks the profile up into a chain of line segments and along each segment distributes a different uniform source strength density. For lift, there must also be circulation and therefore vortices, so each segment is also given a uniform vortex strength distribution. For n panels, the n unknown source strength densities are determined by the n conditions that the panels should be impermeable, as approximated by collocation at their midpoints. The additional condition that the flow leave the trailing edge smoothly—the Kutta–Joukowsky condition of Section 5.1.2—is used to determine the vortex strength of the panels; as this is a single condition, all panels are given the same vortex strength. This method is essentially equivalent to that of Hess and Smith (1967). Here we derive the method using the ideas of complex velocity and conformal mapping, and also present and discuss its implementation in Octave. As the program is based on Complex Uniform Source Strength Segment Panels, we call it CUSSSP.

8.1 Development of the CUSSSP Program

In the development of the method, we follow the six steps for any panel method from Section 7.4.

8.1.1 The Singularity Elements

The panels of Hess and Smith (1967) are line segments, with each point on the segment being an infinitesimal source and vortex such that the outflow and circulation of the whole panel is finite. The source and vortex strength density is uniform along each panel. Since a vortex is a source with an imaginary strength, it's very convenient to treat sources and vortices together, as 'complex sources'.

Theory of Lift: Introductory Computational Aerodynamics in MATLAB®/Octave, First Edition. G. D. McBain.
© 2012 John Wiley & Sons, Ltd. Published 2012 by John Wiley & Sons, Ltd.

The Canonical Uniform Source Strength Segment

A source of strength Q (a real number) at a point z' induces a complex velocity field

$$w(z) = \frac{Q}{2\pi(z - z')}.$$

A vortex of strength Γ (another real number) at a point z' induces a complex velocity

$$w(z) = \frac{i\Gamma}{2\pi(z - z')}.$$

Thus a complex source of strength $Q + i\Gamma$ induces

$$w(z) = \frac{Q + i\Gamma}{2\pi(z - z')}.$$

If we allow Q to be complex rather than real, Γ isn't needed and we have $\Re Q$ as the (real) source strength or outflow and $\Im Q$ as the vortex strength or circulation.

Consider first uniformly distributing source strength along the real axis from $z = 0$ to $z = 1$. Each infinitesimal segment from z' to $z' + \delta z'$, with unit strength density (and therefore unit strength, as the segment has unit length) will induce a complex velocity field

$$\delta w(z) = \frac{\delta z'}{2\pi(z - z')}.$$

Integrating this along the segment gives

$$\begin{aligned} w(z) &= \int_0^1 \frac{dz'}{2\pi(z - z')} \\ &= \frac{-1}{2\pi} \left[\ln(z - z') \right]_0^1 \\ &= \frac{\ln z - \ln(z - 1)}{2\pi}. \end{aligned} \qquad (8.1)$$

There is a difficulty with this integral, as the logarithm of a complex number is not single valued. Recall from Equation (3.3) that

$$\ln z = \ln \left(r e^{i\theta} \right) = \ln r + i\theta.$$

It is rendered single-valued by restricting $\theta = \arg z$ to some range, e.g. $-\pi < \theta < \pi$; i.e. making a branch cut as in Section 4.2.4. This means that the complex velocity field $\ln z$ is discontinuous along the line $\theta = \pm\pi$ (the negative real axis); specifically, its imaginary part (corresponding to the component of velocity in the negative y-direction) jumps by 2π. Wherever the branch cut was laid, the logarithm $\ln z$ would be discontinuous across it. Also, similarly, the complex velocity field $\ln(z - 1)$ is discontinuous along the part of the real axis less than $z = 1$; however, in the combination of Equation (8.1), the jump discontinuities are such as to cancel over the negative real axis. Thus, the velocity field of Equation (8.1) is only discontinuous along the segment $0 < z < 1$ itself.

Conformal Mapping for Arbitrary Segments

The line segment from zero to one in the ζ-plane is mapped to the line segment from z_1 to z_2 in the z-plane with

$$\zeta = \frac{z - z_0}{z_1 - z_0}.$$

So if we have the complex velocity

$$w(\zeta) = \frac{\ln \zeta - \ln(\zeta - 1)}{2\pi}$$

in the ζ-plane, due to a unit uniform strength source segment on $0 < \zeta < 1$, this has the complex potential

$$W(\zeta) = \int w(\zeta)\, d\zeta = \frac{\zeta \ln \zeta - (\zeta - 1) \ln(\zeta - 1)}{2\pi} + \text{const.}$$

The corresponding velocity in the z-plane is

$$
\begin{aligned}
w(z) &= \frac{dW}{dz} = \frac{dW}{d\zeta} \frac{d\zeta}{dz} \\[6pt]
&= \frac{\ln \zeta - \ln(\zeta - 1)}{2\pi} \times \frac{1}{z_1 - z_0} \\[6pt]
&= \frac{\ln \frac{z - z_0}{z_1 - z_0} - \ln \frac{z - z_1}{z_1 - z_0}}{2\pi(z_1 - z_0)}.
\end{aligned}
\tag{8.2}
$$

This arrangement is such that provided the branch-cut of the logarithm function is taken along the negative real axis, the velocity field is only discontinuous across the line segment from z_0 to z_1, and only unbounded at the ends of the segment.

8.1.2 *Discretizing the Geometry*

Say the profile is described by a sequence of $n + 1$ points from trailing edge anticlockwise around to the trailing edge again.

Extensive collections of such descriptions can be found on the Internet (UIUC n.d., e.g.). They are also easily generated for the NACA four- and five-digit wing sections.

The NACA 4-digit Sections

The NACA 4-digit formulae for the mean line and its slope have already been given in Equations (6.13) and (6.14) and the former was implemented in Octave in Listings 6.2 and 6.3; the latter is given here in Listing 8.1.

In order to obtain the complete two-dimensional wing section shapes, the thickness must be added. This is specified by the formulae in the references listed in the Further Reading section

Listing 8.1 `naca4slope`: return the slope at the chordwise stations x for the NACA 4-digit wing section with maximum camber m at p.

```
function s = naca4slope (x, m, p)
  s = 2 * m / p ^ 2 * (p - x);
  s(x > p) = s(x > p) * (p / (1 - p)) ^ 2;
```

Listing 8.2 `naca4thickness`: return the (dimensionless) half-thickness from the camber line of a NACA 4-digit wing section at chordwise stations x.

```
function y = naca4thickness (x)
  p = [0.10150, -0.28430, 0.35160, 0.12600, 0];
  y = (0.29690 * sqrt (x) - polyval (p, x)) / 0.20;
```

at the end of the chapter and which in complex terms can be expressed for the upper and lower surfaces

$$z_U \equiv z_c + ir_t e^{i\lambda}$$

$$z_L \equiv z_c - ir_t e^{i\lambda},$$

where $z_c \equiv x_c + iy_c$ is the complex coordinate along the camber line, λ is related to its local slope by

$$\tan \lambda \equiv \frac{dy_c}{dx_c}$$

and the thickness function is given by

$$\frac{r_t(x)}{t} \equiv \frac{0.29690\sqrt{x} - 0.12600x - 0.35160x^2 + 0.28430x^3 - 0.10150x^4}{0.20},$$

which is implemented in `naca4thickness` in Listing 8.2.

Resolving the Trailing Edge

It turns out that a very fine resolution is needed near the trailing edge, where the Kutta–Joukowsky condition is applied, to obtain good results for the lift. This is achieved for NACA 4-digit sections, by **function** `naca4z` in Listing 8.3.

An example of the use of this code is given in Listing 8.4; it produces Figure 8.1.

Figure 8.1   The points describing the NACA 2412, as generated by the Octave code in Listing 8.4

Listing 8.3   naca4z: return a vector of pts plus a few to better resolve the trailing edge complex
coordinates describing the NACA 4-digit wing section mptt.

```
function z = naca4z (pts, mptt)

  [m, p, t] = naca4pars (mptt);

  chi = linspace (0, 2*pi, pts);
  x = (1 + cos (chi)) / 2;
  xte = fsolve (@naca4thickness, 1.0);
  nte = max (2, ceil ((xte - x(1)) / (x(1) - x(2))));
  xx = linspace (xte, x(1), nte);
  x = [xx, x(2:(end-1)), fliplr(xx)];
  chi = [zeros(1, nte-1), chi, 2*pi*ones(1, nte - 1)];

  z = (complex (x, naca4meanline (x, m, p)') + ...
       ((2 * (chi < pi) - 1) * 1i * t .* ...
        naca4thickness (x) .* ...
        exp (1i * atan (naca4slope (x, m, p))) ) );
```

Listing 8.4   naca4z_naca2412: demonstrate naca4z of Listing 8.3 by plotting a NACA 2412
wing section; see Figure 8.1 for output.

```
zp = naca4z (32, 2412);
plot (real (zp), imag (zp), 'o-');
axis ('off', 'image');
```

Computing the Collocation Points

Given the complex row-vector of points $\{z_1, z_2 \ldots, z_n\}$, the complex length of the kth panel
is $d_k \equiv z_{k+1} - z_k$ and so it makes an angle $\arg d_k$ with the positive x-axis.

The impermeability of each panel is enforced at its midpoint, which can be computed as
$z_k^{(c)} = z_k + \frac{1}{2}d_k$, or in Octave as (using zp for the ends of the panels and z for their midpoints):

```
dz = diff (zp);
n = length (dz);
lambda = angle (dz)';
z = zp(1:n) + 0.5*dz;
```

8.1.3   The Influence Matrix

The velocity induced at the ith collocation point $z_i^{(c)}$ by the jth panel, i.e. the panel running
from z_j to $z_j + d_j$, is $\omega_{ij}(Q_j + i\Gamma_j)$ where the influence coefficient ω_{ij} is obtained by putting

Listing 8.5 cussspi: return the influence coefficients at complex collocation points zc for panels from z with complex lengths d.

```
function w = cussspi (zc, z, d)
  w = bsxfun (@ (x,y) (log(x./y)-log((x-y)./y))./(2*pi*y), ...
       bsxfun (@minus, zc(:), z), d );
```

z_j for z_0 and $z_j + d_j$ for z_1 in Equation (8.2); i.e.

$$\omega_{ij} \equiv \frac{\ln \frac{z_i^{(c)}-z_j}{d_j} - \ln \frac{z_i^{(c)}-z_j-d_j}{d_j}}{2\pi d_j}.$$

However, this needs correcting when the point at which the velocity is being evaluated lies on the panel, since the velocity field induced by a panel is discontinuous across the segment of the panel. The magnitude is predicted correctly whichever side of the panel the computer thinks the midpoint is on, but if we force it to be considered as the right side (outside), the direction due to pure source strength will then be simply the direction of the plate minus ninety degrees; i.e. the diagonal influence coefficients satisify

$$\omega_{ii} = |\omega_{ii}|e^{-i(\lambda_i-\pi/2)}$$

Thus, the influence matrix may be computed by calling cussspi from Listing 8.5, and subsequently correcting the angle of the coefficients along the diagonal.

8.1.4 The Right-hand Side

The Impermeability Conditions

The total complex velocity at the ith collocation point, due to the free-stream and all n panels is

$$w_i = w\left(z_i^{(c)}\right) = q_\infty e^{-i\alpha} + \sum_{j=1}^{n} \omega_{ij}(Q_j + i\Gamma_j). \tag{8.3}$$

If the ith panel has slope λ_i, then the component of velocity normal to the panel is $-\Im e^{i\lambda_i}w$ and this should be zero for the plate to be impermeable; i.e. for $i = 1, 2, \ldots, n$:

$$\sum_{j=1}^{n} -\Im\left\{e^{i\lambda_i}\omega_{ij}(Q_j + i\Gamma_j)\right\} = q_\infty \sin(\lambda_i - \alpha)$$

and as all the panels are taken to have equal vortex strength, Γ,

$$\sum_{j=1}^{n} -\Im\left(e^{i\lambda_i}\omega_{ij}\right)Q_j + \left\{\sum_{j=1}^{n} -\Im\left(ie^{i\lambda_i}\omega_{ij}\right)\right\}\Gamma = q_\infty \sin(\lambda_i - \alpha).$$

The Kutta–Joukowsky Condition

The Kutta–Joukowsky condition that the flow leave the trailing edge smoothly is enforced by requiring that the tangential components of velocity on the first and last panels are equal, tracing each panel in the same direction. But as the panels are traced in opposite directions—the first forwards, the last aftwards, as the profile is described anticlockwise, beginning at the trailing edge—the condition is that the tangential components should be opposite. The tangential component at the ith collocation point is

$$\Re e^{i\lambda_i} w_i$$

so the condition is

$$\Re e^{i\lambda_1} w_1 + \Re e^{i\lambda_n} w_n = 0;$$

i.e.

$$\Re \sum_{i=1}^{n} e^{i\lambda_i} \left(\sum_{j=1}^{n} \omega_{ij} Q_j + \left[\sum_{j=1}^{n} \omega_{ij} i \right] \Gamma \right) = -\Re q_\infty \left\{ e^{i(\lambda_1 - \alpha)} + e^{i(\lambda_n - \alpha)} \right\}.$$

Thus the whole problem can be written in block-matrix form as

$$\begin{bmatrix} -\Im \left[e^{i\lambda_i} \omega_{ij} \right]_{(n \times n)} & -\Im \left[i e^{i\lambda_i} \sum_{j=1}^{n} \omega_{ij} \right]_{(n \times 1)} \\ \left[\sum_{i=1}^{n} \Re e^{i\lambda_i} \omega_{ij} \right]_{(1 \times n)} & \left[\sum_{i=1}^{n} \Re i e^{i\lambda_i} \sum_{j=1}^{n} \omega_{ij} \right]_{(1 \times 1)} \end{bmatrix} \begin{Bmatrix} \{Q_j\}_{(n \times 1)} \\ \Gamma \end{Bmatrix}$$

$$= q_\infty \begin{Bmatrix} \{\sin(\lambda_i - \alpha)\}_{(n \times 1)} \\ -\{\cos(\lambda_1 - \alpha) + \cos(\lambda_n - \alpha)\} \end{Bmatrix}$$

Having computed the influence matrix `omega`, the matrix equation is assembled by computing `A`, `k`, `B`, and `rhs`, in sequence:

```
A = diag (exp (1i*lambda)) * omega;
```

```
k = A(1,:) + A(n,:);
```

```
B = [-imag(A),  -sum(imag (1i * A), 2);
     +real(k),   sum(real (1i * k)) ];
```

```
rhs = [sin(lambda - alpha);
      -(cos(lambda(1) - alpha) + cos(lambda(n) - alpha))];
```

8.1.5 Solving the Linear System

Solving the linear system in Octave is easy; one just uses the backslash operator:

```
Q = B \ rhs;
```

This solution vector Q actually has the n values of the source strengths and the one value of the vortex strengths, so we convert it to the complex source strengths with

```
Q = Q(1:n) + 1i*Q(n+1);
```

Summary of the CUSSSP Program

The complete pair of functions is given in Listings 8.6 and 8.5.

The input geometry will typically be prepared by a program like naca4z of Listing 8.3. Examples interpreting the output will be discussed next.

Listing 8.6 cusssp: return the midpoints, influence matrix, and panel strengths for the CUSSSP model of a wing section zp at incidence alpha; see Listing 8.7 for a demo.

```
function [z, omega, Q] = cusssp (zp, alpha)

  dz = diff (zp);
  n = length (dz);
  lambda = angle (dz)';
  z = zp(1:n) + 0.5*dz;

  omega = cussspi (z, zp(1:n), dz);

  self = logical (eye (n));
  omega(self) = abs (omega(self)) .* exp (-1i*(lambda-pi/2));

  A = diag (exp (1i*lambda)) * omega;
  k = A(1,:) + A(n,:);

  B = [-imag(A), -sum(imag (1i * A), 2);
       +real(k),  sum(real (1i * k)) ];

  rhs = [sin(lambda - alpha);
         -(cos(lambda(1) - alpha) + cos(lambda(n) - alpha))];

  Q = B \ rhs;
  Q = Q(1:n) + 1i*Q(n+1);
```

8.1.6 Postprocessing

Velocity at the Panel Midpoints

The complex velocity at each of the midpoints is given by Equation (8.3) as the sum of the free-stream and the product of the influence matrix with the column-matrix of complex source strengths; in Octave, and taking the free-stream velocity q_∞ as the speed scale:

```
w = exp (-1i*alpha) + omega * Q;
```

Thus with the Octave code in Listing 8.7 we can produce Figure 8.2.

Notice that the velocity is tangential to the profile at each point, so that the panels are indeed treated as impermeable.

Velocity at Other Points

If we want to calculate the velocity at points other than the collocation points, we need to calculate a new influence matrix, or rather new rows of the influence matrix, since each row of the influence matrix corresponds to an evaluation point while each column corresponds to a panel. The Octave code in Listing 8.8 produces Figure 8.3.

Listing 8.7 cusssp_midpoint_w: simple graphical demonstration of cusssp of Listing 8.6; see Figure 8.2 for output.

```
zp = naca4z (32, 2412);
alpha = 8 * pi / 180;
[z, omega, Q] = cusssp (zp, alpha);
w = exp (-1i*alpha) + omega * Q;
quiver (real (z), imag (z), real (w'), imag (w'), 9)
hold ('on')
plot (real (z), imag (z), 'o', real (zp), imag (zp), '-')
axis ('off', 'image')
```

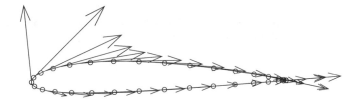

Figure 8.2 The velocity vectors at the midpoints of the panels on the NACA 2412 discretized in Figure 8.1, as computed by cusssp (Listing 8.6)

Listing 8.8 cusssp_w_field: another graphical demonstration of cusssp of Listing 8.6; see Figure 8.3 for output.

```
zp = naca4z (256, 2412);
alpha = 8 * pi / 180;
[z, omega, Q] = cusssp (zp, alpha);
ZP = cmeshgrid (-0.5-0.3i, 1.5+0.3i, 24);
ZP = ZP(:);
ZP = ZP(~inpolygon (real (ZP), imag (ZP), ...
                    real (zp), imag (zp) ));
w = exp (-1i * alpha);
w = w + cussspi (ZP, zp(1:(end-1)), diff (zp)) * Q;
quiver (real (ZP), imag (ZP), real (w), -imag (w), 12);
hold ('on')
plot (real (zp), imag (zp), '-')
axis ('off', 'image')
```

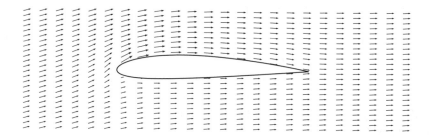

Figure 8.3   The velocity field around the NACA 2412 of Figures 8.1 and 8.2; plot generated by Listing 8.8

Lift

The lift is given by the Kutta–Joukowsky theorem: $\ell = \rho q_\infty \Gamma$, where the circulation Γ is the sum of the vortex strengths for each panel; i.e. the sum of the imaginary part of the complex source strength. The lift coefficient is

$$C_\ell = \frac{\ell}{\frac{1}{2}\rho q_\infty c} = \frac{2\Gamma}{q_\infty c},$$

so that in Octave, assuming all speeds are nondimensionalized by q_∞ and all lengths and coordinates by c, the lift coefficient is computed by

```
Cl = 2 * sum (imag (Q))
```

For example, the code in Listing 8.9 produces the data for the graph in Figure 8.4. The lift–incidence results in Figure 8.4 are compared with the prediction of thin wing section theory, $C_\ell = 2\pi(\alpha - \alpha_0)$, where $\alpha_0 \doteq -2.077°$, as in Chapter 6; notice:

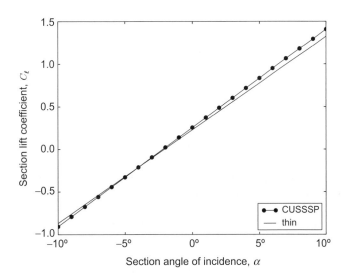

Figure 8.4 The lift–incidence curve for the NACA 2412, as computed by CUSSSP in Listing 8.9 and thin wing section theory (taking the zero-lift incidence as $\alpha_0 \doteq -2.077°$)

Listing 8.9 `cusssp_demo` demonstrate Listing 8.6 by computing a lift–incidence table; the output is graphed in Figure 8.4.

```
zp = naca4z (256, 2412);
alpha0d = -2.077;
alphad = -10
[z, omega, Q] = cusssp (zp, alphad*pi/180);
Cl = 2 * sum (imag (Q))
Cl_thin = 2 * pi * (alphad - alpha0d) * pi / 180
```

- the two predictions of zero-lift incidence agree closely; and
- the thickness increases the slope slightly.

The results may also be compared with experimental data (Abbott and von Doenhoff 1959). In fact the slope of the experimental data is closer to the 2π of thin wing section theory; this is due to a cancellation of errors: thickness does increase the slope but viscous effects, ignored in both thin wing section theory and this simple panel method, reduce it.

8.2 Exercises

1. If a single line segment panel with real uniform source strength Q is immersed in a uniform velocity field, what must Q be to stop the field passing through the centre of the panel. Show that (except when the external field is parallel to the panel) the velocity will not vanish at the midpoint on the other side of the panel; what is the velocity there (in terms of the length of the panel and the speed and relative direction of the free-stream)?

2. Modify the panel program CUSSSP to solve for the flow over an obstacle with specified circulation.

    (a) Check that for $\Gamma = 0$ it gives the same results as `function cusssp` for a (symmetrically panelled) symmetric aerofoil at zero incidence, for which the Kutta–Joukowsky condition should result in zero circulation.

    (b) Use the modified program to compute the flow over a circle. Compare the results with the known analytical solution (Section 3.3.1).

        i. What is the pressure coefficient halfway around the perimeter between the stagnation points? Is a more accurate answer for this obtained by using an odd or even number of panels on the upper and lower semicircles? (That is, so that the midpoint falls either on the joint or midpoint of a panel, respectively.)

3. (a) Extract the pressure coefficient at the collocation points associated with the velocities computed by CUSSSP. Plot the pressure coefficient around the NACA 2412 as a function of x.

    (b) Modify Listing 8.7 to draw instead the pressure force on each collocation point, as a vector acting on the midpoint and in its normal direction.

    (c) Extend Listing 8.8 to produce the pressure (or pressure coefficient) field around the aerofoil, e.g. using `contour` to plot the isobars.

4. Apply CUSSSP to three more aerofoils from Abbott *et al.* (1945) or Abbott and von Doenhoff (1959).

5. Apply CUSSSP the Clark Y aerofoil, with shape as given by Silverstein (1935) and quoted in Table 1.1. Compare with the original experimental results in the report.

6. Apply CUSSSP to three aerofoils from the UIUC database.

8.2.1   Projects

1. The above modified program of Exercise 2 is useful for symmetric aerofoils, for which the circulation is known to be zero. Modify it further by exploiting this symmetry; i.e. only model the flow in the half-plane $y > 0$.

    (a) How fine must the discretization of an obstacle be for this version to run noticeably faster?

    (b) For numbers of panels above this threshold, how do the run-times scale with the number of panels?

2. Modify the panel program CUSSSP to use different kinds of panels.

    (a) Let the source strength vary linearly along the panel, being continuous from panel to panel, but with a jump in tangential derivative.

    (b) Instead of distributing sources along the panels, distribute vortices.

    (c) Consider a line segment panel with a uniform strength distribution of doublets. Show that this has an identical influence to a panel with a combined point source–vortex, i.e. a point source of complex strength, at each end (Bousquet 1990).

8.3   Further Reading

The original journal paper on panel methods by Hess and Smith (1967) is still well worth reading. An excellent—at once brisk, practical, and comprehensive—framework for panel methods is developed by Bousquet (1990). The most important English-language compendium of varieties of panel methods is that of Katz and Plotkin (2001), though the FORTRAN code listings are somewhat cumbersome and unfortunately suffer from transcription errors. Good working demonstration panel programs in FORTRAN are presented by Kuethe and Chow (1998), Paraschivoiu (1998), and Moran (2003).

The complex velocity fields induced by real and imaginary uniform source strength segments along the real axis are derived by Paraschivoiu (1998).

The thickness formulae for NACA four-digit wing sections are given by Jacobs *et al.* (1933) and Abbott and von Doenhoff (1959).

References

Abbott, I.H. and von Doenhoff, A.E. (1959) *Theory of Wing Sections*. New York: Dover.

Abbott, I.H., von Doenhoff, A.E. and Stivers, L.S. (1945) Summary of airfoil data. Report 824, NACA.

Bousquet, J. (1990) *Aérodynamique: Méthode des singularités*. Toulouse: Cépaduès.

Hess, J.L. and Smith, A.M.O. (1967) Calculation of potential flow about arbitrary bodies. *Progress in Aerospace Sciences* **8**.

Hoerner, S.F. and Borst, H.V. (1985) *Fluid-Dynamic Lift*, 2nd edn. Bakersfield, CA: Hoerner Fluid Dynamics.

Jacobs, E.N., Ward, K.E. and Pinkerton, R.M. (1933) The Characteristics of 78 Related Airfoil Sections from Tests in the Variable-Density Wind Tunnel. Report 460, NACA.

Katz, J. and Plotkin, A. (2001) *Low-Speed Aerodynamics*, 2nd edn. Cambridge: Cambridge University Press.

Kuethe, A.M. and Chow, C.Y. (1998) *Foundations of Aerodynamics*, 5th edn. Chichester: John Wiley & Sons, Ltd.

Moran, J. (2003) *An Introduction to Theoretical and Computational Aerodynamics*. New York: Dover.

Paraschivoiu, I. (1998) *Aérodynamique Subsonique*. Éditions de l'École de Montréal.

Silverstein, A. (1935) Scale effect on Clark Y airfoil characteristics from N.A.C.A. full-scale wind-tunnel tests. Report 502, NACA.

UIUC (n.d.) UIUC Airfoil Coordinates Database .Accessed 3 Dec. 2011.

8.4   Conclusion to Part I: The Origin of Lift

This concludes the first part of this book, on plane ideal flow. Although, as will be dealt with the next part, all real flows are necessarily three-dimensional, and as will be touched on in the last part, none is ideal, there is a special significance to plane ideal flow in aerodynamics: it is able to explain the origin of lift, which is the phenomenon without which there would almost be no aeronautics—it would be reduced to such craft as can remain aloft by buoyancy or thrust, aerostats and rockets. It is fitting therefore at this point, particularly after invoking so much theoretical machinery and advanced mathematics, to pause and reflect on the physical origins of lift.

It's all about pressure, or rather the difference between the pressure below and above. This was met very early in the book, in Section 2.5.1.

So where does the pressure difference come from? From the Bernoulli Equation (2.14), which is an expression of the conservation of mechanical energy, we know it comes from a speed difference: the speed above has to exceed that below.

This implies that there will be positive circulation around the aerofoil; i.e. that there is such a circulation implies a relative excess of speed above over below, and therefore a pressure excess below over above, and therefore lift.

So where does circulation come from? In Chapters 6 and 8 it was posited as coming from vortices bound inside the wing section, but these were merely (very) convenient fictions. Of course aerodynamics happens only on the outside of the aerofoil, not inside it. In fact, the air-stream is divided in two by the aerofoil into streams passing above and below. These will be separated by semi-infinite streamlines terminating at stagnation points on the aerofoil. The location of these stagnation points will vary with the incidence of the free-stream, but it has been empirically learned that for reasonably thin aerofoils not inclined too much, a sharp trailing edge will fix the downstream stagnation point and therefore allows the circulation to be controlled monotonically by incidence, within a quite limited but usable range.

As Hoerner and Borst (1985) put it:

The airfoil is thus a device which forces certain stream tubes to go over the top.

Part Two

Three-Dimensional Ideal Aerodynamics

9

Finite Wings and Three-Dimensional Flow

9.1 Wings of Finite Span

9.1.1 Empirical Effect of Finite Span on Lift

Up to now we have only considered two-dimensional flow over two-dimensional wing sections. Thought of in three dimensions, this situation corresponds to a wing with an infinite span. Experimental data for the variation of the lift coefficient with wing span (or *aspect ratio*, $A\!R$, as defined in Section 1.2.2) show that the slope of the lift–incidence curve decreases as the aspect ratio decreases from infinity. The zero-lift incidence α_0, however, is approximately independent of aspect ratio. The most famous set of experimental results demonstrating this behaviour is that in figure 47 of Ludwig Prandtl's report 'Application of Modern Hydrodynamics to Aeronautics' (Prandtl 1921), resketched in Figure 11.7 below and in many other texts on aerodynamics.

9.1.2 Finite Wings and Three-dimensional Flow

The reason why three-dimensional lift coefficients differ from two-dimensional ones (i.e. why finite aspect ratio lift coefficients differ from infinite aspect ratio lift coefficients) is simple. As explained by Prandtl (1921), the generation of lift by a wing of finite aspect ratio is not compatible with a purely two-dimensional flow field.

As discussed in Section 8.4, the fact that a wing generates lift implies that the pressure beneath it exceeds, in the mean, the pressure above it. Consider points A, B, and T, in the air above, below, and near the tip of the wing, all in the same plane normal to the direction of flight (see Figure 9.1). If the wing is generating lift, the pressure at B should exceed that at A. Now, whether we proceed from B to T or from A to T, we must arrive at the same pressure at T, since the pressure at a point is single-valued. Therefore, there is a drop in pressure moving outwards along the lower surface and a rise in pressure moving outwards along the upper surface. Since air is forced in the direction of decreasing pressure, it must be expected that air passing over the left wing will be pushed inwards towards the plane of symmetry (in the direction from T to A). Similarly, air passing under the left wing will be pushed outwards away from B towards T.

Theory of Lift: Introductory Computational Aerodynamics in MATLAB®/Octave, First Edition. G. D. McBain.
© 2012 John Wiley & Sons, Ltd. Published 2012 by John Wiley & Sons, Ltd.

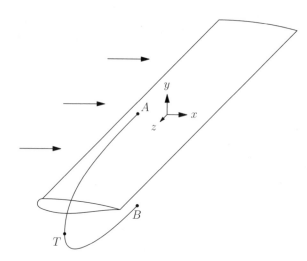

Figure 9.1 A wing of rectangular planform and spanwise-uniform thickness, viewed in oblique projection, showing the three-dimensional Cartesian axes of Section 9.2.1

For air particles passing over and under the right wing, the directions are reversed, being still towards and away from the symmetry plane, respectively, as illustrated in Figure 9.2. Stated otherwise, the air passing over the wing drifts inboard and that under the wing outboard.

It is clear that the flow is not two-dimensional: there is both spanwise variation to the flow and, except in the plane of symmetry ($z = 0$), a spanwise component of the velocity field. This is why the finite aspect ratio wind tunnel data differs from the infinite aspect ratio limit.

Quantifying this dependence on aspect ratio is important. It can be done theoretically by the Lanchester–Prandtl finite wing theory and the concept of the 'lifting line', but first we must digress to discuss some fundamentals of the three-dimensional flow of perfect fluids.

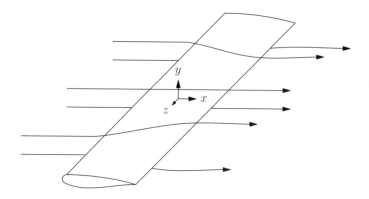

Figure 9.2 Schematic three-dimensional motion relative to the wing of Figure 9.1 of air induced by finite aspect ratio and the pressure difference associated with lift

9.2   Three-Dimensional Flow

9.2.1   Three-dimensional Cartesian Coordinate System

In our description of two-dimensional flow we used an x-axis parallel to the wing-section chord and directed positive aft. The y-axis was perpendicular to this, vaguely 'up'. In three dimensions we add to these a z-axis, with the direction chosen to form a *right-handed system*: i.e. z increases along the left wing. The axes are illustrated in Figures 9.1 and 9.2.

The unit vectors for the x, y, and z axes are **i**, **j**, and **k**. The corresponding components of velocity are u, v, and w.

9.2.2   Three-dimensional Governing Equations

As in two dimensions (Section 2.2.1), flow in three dimensions is governed by equations derived from two basic physical principles:

- the conservation of mass; and
- Newton's Second Law of motion.

For a perfect fluid (i.e. one with uniform density and free from shearing stresses), conservation of mass leads to the continuity equation and Newton's law leads to the Euler equation. The derivations in three dimensions are similar to those for two dimensions in Chapter 2, but before considering these equations in Section 9.4, we recall in Section 9.3 some vector notation and vector identities.

9.3   Vector Notation and Identities

Vectors are quantities with magnitude and direction and can be expressed by their Cartesian components: e.g.

$$\boldsymbol{v} = v_x\mathbf{i} + v_y\mathbf{j} + v_z\mathbf{k}.$$

9.3.1   Addition and Scalar Multiplication of Vectors

The sum and difference of two vectors are also vectors:

$$\boldsymbol{v} \pm \boldsymbol{w} = (v_x \pm w_x)\mathbf{i} + (v_y \pm w_y)\mathbf{j} + (v_z \pm w_z)\mathbf{k}.$$

A vector can be *scaled* by a real factor λ:

$$\lambda\boldsymbol{v} = \lambda v_x\mathbf{i} + \lambda v_y\mathbf{j} + \lambda v_z\mathbf{k}.$$

This is called *scalar multiplication*.

9.3.2   Products of Vectors

There are two products of two vectors: the *scalar product* and the *vector product*.
The scalar product is defined by

$$v \cdot w \equiv v_x w_x + v_y w_y + v_z w_z,$$

and is *commutative*:

$$w \cdot v = v \cdot w.$$

The scalar product of a vector with itself is the square of its magnitude

$$v \cdot v = v_x^2 + v_y^2 + v_z^2 \equiv |v|^2 \equiv v^2.$$

The scalar product can be used to obtain the (cosine of the) angle between two vectors:

$$\cos\theta = \frac{v \cdot w}{|v||w|}. \tag{9.1}$$

The *vector product* is defined by

$$v \times w \equiv (v_y w_z - v_z w_y)\mathbf{i} + (v_z w_x - v_x w_z)\mathbf{j} + (v_x w_y - v_y w_x)\mathbf{k}$$

$$\equiv \begin{vmatrix} \mathbf{i} & \mathbf{j} & \mathbf{k} \\ v_x & v_y & v_z \\ w_x & w_y & w_z \end{vmatrix},$$

and is *anticommutative*:

$$w \times v = -v \times w;$$

it can be used to get the sine of the angle between two vectors:

$$\sin\theta = \frac{|v \times w|}{|v||w|}.$$

There are also two products of three vectors.
The *triple scalar product* is

$$(v \times w) \cdot u$$

and satisfies

$$(v \times w) \cdot u = (u \times v) \cdot w = (w \times u) \cdot v$$
$$(v \times w) \cdot u = -(w \times v) \cdot u.$$

The *triple vector product* is

$$(v \times w) \times u$$

and satisfies

$$(v \times w) \times u \equiv (u \cdot v)w - (w \cdot u)v$$

and

$$(v \times w) \times u + (w \times u) \times v + (u \times v) \times w = 0.$$

9.3.3 Vector Derivatives

The *gradient* of a scalar s is a vector field defined by

$$\nabla s \equiv \frac{\partial s}{\partial x}\mathbf{i} + \frac{\partial s}{\partial y}\mathbf{j} + \frac{\partial s}{\partial z}\mathbf{k}.$$

The Taylor expansion of a three-dimensional scalar field is

$$s(\mathbf{r} + \delta\mathbf{r}) \sim s(\mathbf{r}) + \delta\mathbf{r} \cdot \nabla s + O(\delta r^2),$$

so that for an infinitesimal change of position,

$$\delta s = s(\mathbf{r} + \delta\mathbf{r}) - s(\mathbf{r}) = (\nabla s) \cdot \delta\mathbf{r}.$$

The chain rule leads directly to an expression for the gradient of a function of a scalar:

$$\nabla f(s) = f'(s)\nabla s. \tag{9.2}$$

The *divergence* of a vector field v is a scalar field defined by

$$\nabla \cdot v \equiv \frac{\partial v_x}{\partial x} + \frac{\partial v_y}{\partial y} + \frac{\partial v_z}{\partial z}.$$

The *curl* of a vector field is a vector field defined by

$$\nabla \times v \equiv \left(\frac{\partial v_z}{\partial y} - \frac{\partial v_y}{\partial z} \right)\mathbf{i} + \left(\frac{\partial v_x}{\partial z} - \frac{\partial v_z}{\partial x} \right)\mathbf{j} + \left(\frac{\partial v_y}{\partial x} - \frac{\partial v_x}{\partial y} \right)\mathbf{k}.$$

The *Laplacian* of a scalar field is a scalar field defined by

$$\nabla^2 s \equiv \nabla \cdot \nabla s = \frac{\partial^2 s}{\partial x^2} + \frac{\partial^2 s}{\partial y^2} + \frac{\partial^2 s}{\partial z^2}. \tag{9.3}$$

The Laplacian of a vector field is a vector field defined by

$$\nabla^2 v \equiv \nabla(\nabla \cdot v) - \nabla \times (\nabla \times v).$$

In terms of the Cartesian components this reduces to

$$\nabla^2 v = \mathbf{i}\nabla^2 v_x + \mathbf{j}\nabla^2 v_y + \mathbf{k}\nabla^2 v_z,$$

but this simple relation does not hold for components in non-Cartesian coordinate systems for which the unit vectors are not uniform (and so cannot be taken outside the differentiations).

Some identities relating vector derivatives are:

$$\nabla(st) = t\nabla s + (\nabla t)s$$
$$\nabla \cdot (s\boldsymbol{v}) = (\nabla s) \cdot \boldsymbol{v} + s\nabla \cdot \boldsymbol{v} \tag{9.4}$$
$$\nabla \cdot (\nabla \times \boldsymbol{v}) = 0 \tag{9.5}$$
$$\nabla \times (\nabla s) = 0$$
$$\nabla \times (s\boldsymbol{v}) = (\nabla s) \times \boldsymbol{v} + s\nabla \times \boldsymbol{v}$$
$$\nabla \times (\boldsymbol{v} \times \boldsymbol{w}) = \boldsymbol{v}(\nabla \cdot \boldsymbol{w}) - \boldsymbol{w}(\nabla \cdot \boldsymbol{v}) + (\boldsymbol{w} \cdot \nabla)\boldsymbol{v} - (\boldsymbol{v} \cdot \nabla)\boldsymbol{w}$$
$$\nabla(\boldsymbol{v} \cdot \boldsymbol{w}) = (\boldsymbol{v} \cdot \nabla)\boldsymbol{w} + (\boldsymbol{w} \cdot \nabla)\boldsymbol{v} + \boldsymbol{v} \times (\nabla \times \boldsymbol{w}) + \boldsymbol{w} \times (\nabla \times \boldsymbol{v})$$
$$\boldsymbol{v} \times (\nabla \times \boldsymbol{v}) = \nabla \left(\frac{|\boldsymbol{v}|^2}{2}\right) - (\boldsymbol{v} \cdot \nabla)\boldsymbol{v}. \tag{9.6}$$

9.3.4   Integral Theorems for Vector Derivatives

If the surface S encloses the three-dimensional region $\mathcal{R}$, the *divergence theorem* is

$$\iiint_{\mathcal{R}} \nabla \cdot \boldsymbol{v} \, dV = \iint_{S} \boldsymbol{v} \cdot d\boldsymbol{S}, \tag{9.7}$$

where the infinitesimal surface area vector $d\boldsymbol{S}$ has the direction of the outward normal to the surface.

The *gradient theorem* for scalars is

$$\iiint_{\mathcal{R}} \nabla s \, dV = \iint_{S} s \, d\boldsymbol{S}. \tag{9.8}$$

Stokes's theorem states

$$\iint_{S} (\nabla \times \boldsymbol{v}) \cdot d\boldsymbol{S} = 0;$$

however, if the surface S is not closed but is bounded by a simple closed curve C,

$$\iint_{S} (\nabla \times \boldsymbol{v}) \cdot d\boldsymbol{S} = \oint_{C} \boldsymbol{v} \cdot d\boldsymbol{s}. \tag{9.9}$$

For such a surface there is no 'outward' direction, but giving it an arbitrary definition we define the direction of C (and $d\boldsymbol{s}$) so that the surface and its normal are kept on the left during the circuit; e.g. if the surface is confined to the xy-plane and the normal is taken to be in the positive z-direction (in a right-handed coordinate system), the circuit should be traversed counterclockwise.

9.4 The Equations Governing Three-Dimensional Flow

The velocity of the fluid at a point (x, y, z) in space is denoted $\boldsymbol{q} = u\mathbf{i} + v\mathbf{j} + w\mathbf{k}$.

9.4.1 Conservation of Mass and the Continuity Equation

In three dimensions just as in two (Section 2.3), the conservation of mass means that the velocity must be divergence-free; i.e.

$$\nabla \cdot \boldsymbol{q} \equiv \frac{\partial u}{\partial x} + \frac{\partial v}{\partial y} + \frac{\partial w}{\partial z} = 0. \tag{9.10}$$

This can be derived by considering the rate of change of mass in an infinitesimal box fixed in space and aligned with the Cartesian axes, as shown for the x-component of velocity in Figure 9.3, or by applying the divergence theorem (Equation 9.7) to an arbitrary control volume.

9.4.2 Newton's Law and Euler's Equation

Again, just as in two dimensions (Section 2.4.1), the acceleration of a fluid particle is given by the substantial derivative of the velocity field, which in three dimensions also accounts for variations in the third (z) direction and for the w velocity component:

$$\frac{\mathrm{D}}{\mathrm{D}t} \equiv \frac{\partial}{\partial t} + u\frac{\partial}{\partial x} + v\frac{\partial}{\partial y} + w\frac{\partial}{\partial z};$$

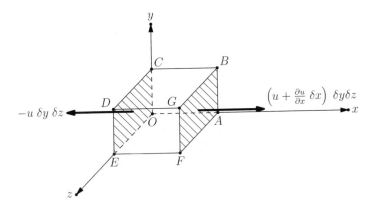

Figure 9.3 Contributions to the net outflow of an infinitesimal box fixed in space from the two faces normal to the x-axis

thus the acceleration is

$$
\begin{aligned}
\frac{Dq}{Dt} &= \frac{Du}{Dt}\mathbf{i} + \frac{Dv}{Dt}\mathbf{j} + \frac{Dw}{Dt}\mathbf{k} \\
&= \left(\frac{\partial u}{\partial t} + u\frac{\partial u}{\partial x} + v\frac{\partial u}{\partial y} + w\frac{\partial u}{\partial z} \right)\mathbf{i} \\
&\quad + \left(\frac{\partial v}{\partial t} + u\frac{\partial v}{\partial x} + v\frac{\partial v}{\partial y} + w\frac{\partial v}{\partial z} \right)\mathbf{j} \\
&\quad + \left(\frac{\partial w}{\partial t} + u\frac{\partial w}{\partial x} + v\frac{\partial w}{\partial y} + w\frac{\partial w}{\partial z} \right)\mathbf{k},
\end{aligned}
$$

Euler's Equation (2.6) asserts a balance between this and the net force per unit mass, and conservative forces such as gravity in a uniform density fluid can be accounted for by modifying the pressure (Section 2.4.4). Thus, the momentum balance is again given by the Euler equation

$$
\frac{Dq}{Dt} = -\frac{1}{\rho}\nabla p. \tag{9.11}
$$

The substantial derivative of the position is just the velocity:

$$
\frac{D}{Dt}(x\mathbf{i} + y\mathbf{j} + z\mathbf{k}) = q. \tag{9.12}
$$

9.5 Circulation

9.5.1 Definition of Circulation in Three Dimensions

As in Equation (2.15) for two-dimensional flow, the circulation along a curve in the fluid is defined as the integral of the tangential component of velocity along the curve:

$$
\Gamma \equiv \int_C q \cdot ds.
$$

Whereas in two dimensions in Section 2.5.3, the circuit was prescribed as clockwise (or rather the circulation was defined as minus the integral around the anticlockwise circuit), there is no absolute way of defining clockwise or anticlockwise in three dimensions; each circuit C must have its own sense defined in terms of its context. If the circuit bounds a surface, and that surface is pierced exactly once by a directed curve, a sense for the circuit can be derived by applying the right-hand rule to the directed curve. With such a bound surface but without reference to a piercing curve, a sense can be derived from the prescription that the surface be kept on the left as the circuit is traced.

9.5.2 The Persistence of Circulation

If the curve is allowed to drift with the fluid, an expression can be obtained for the time rate of change of the circulation along it in terms of quantities just at the two ends of the curve:

$$
\begin{aligned}
\frac{D\Gamma}{Dt} &= \int_C \frac{D}{Dt} (q \cdot ds) \\
&= \int_C \left(q \cdot \frac{D\,ds}{Dt} + \frac{Dq}{Dt} \cdot ds \right) \\
&= \int_C \left(q \cdot d\frac{Ds}{Dt} - \frac{1}{\rho} \{\nabla p\} \cdot ds \right) \\
&= \int_C \left(q \cdot dq - \frac{1}{\rho} dp \right) \\
&= \int_C d \left(\frac{q^2}{2} - \frac{p}{\rho} \right) \\
&= \left[\frac{q^2}{2} - \frac{p}{\rho} \right]_C .
\end{aligned}
$$

Here we have used Euler's Equation (9.11) to replace the substantial derivative of the velocity field with the pressure gradient, and Equation (9.12) to replace the substantial derivative of the position vector field.

If the curve is a closed circuit so that the two ends coincide, this difference vanishes and we have Kelvin's theorem: *the circulation around a circuit moving with an ideal fluid doesn't change*, $D\Gamma/Dt = 0$.

9.5.3 Circulation and Vorticity

Consider the circulation around the infinitesimal circuit $ABCDEFA$ in Figure 9.4. We have, for the six legs of the circuit,

$$
\Gamma_{AB} = + \left(v + \frac{\partial v}{\partial x} \delta x \right) \delta y
$$

$$
\Gamma_{BC} = - \left(u + \frac{\partial u}{\partial y} \delta y \right) \delta x
$$

$$
\Gamma_{CD} = + \left(w + \frac{\partial w}{\partial y} \delta y \right) \delta z
$$

$$
\Gamma_{DE} = - \left(v + \frac{\partial v}{\partial z} \delta z \right) \delta y
$$

$$
\Gamma_{EF} = + \left(u + \frac{\partial u}{\partial z} \delta z \right) \delta x
$$

$$
\Gamma_{FA} = - \left(w + \frac{\partial w}{\partial x} \delta x \right) \delta z,
$$

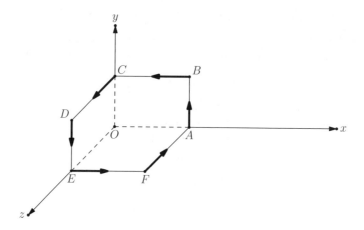

Figure 9.4   Diagram for circulation around an infinitesimal three-dimensional simple closed circuit

and the total circulation is

$$\Gamma = \left(\frac{\partial w}{\partial y} - \frac{\partial v}{\partial z}\right)\delta y\,\delta z + \left(\frac{\partial u}{\partial z} - \frac{\partial w}{\partial x}\right)\delta z\,\delta x + \left(\frac{\partial v}{\partial x} - \frac{\partial u}{\partial y}\right)\delta x\,\delta y$$
$$= \xi\,\delta A_x + \eta\,\delta A_y + \zeta\,\delta A_z, \tag{9.13}$$

where

$$\xi \mathbf{i} + \eta \mathbf{j} + \zeta \mathbf{k} \equiv \boldsymbol{\omega} \equiv \nabla \times \boldsymbol{q}$$

is the *vorticity* and

$$\delta A_x = \delta y\,\delta z$$
$$\delta A_y = \delta z\,\delta x$$
$$\delta A_z = \delta x\,\delta y,$$

are the areas of the projections of the surface normal to the x, y, and z axes, respectively. In fact, Equation (9.13) is a simple instance of Stokes's theorem, Equation (9.9).

Another way to calculate the circulation around this circuit is to note that it can be decomposed into the sum of the circulations around the three rectangles:

$$\Gamma = \Gamma_{ABCOA} + \Gamma_{OCDEO} + \Gamma_{EFAOE},$$

since the contributions of the legs involving the point O cancel when they are traced twice, once in each direction. See Figure 9.5. Here it is clear that the circulation around a circuit of an elementary plane surface is the product of the component of vorticity normal to the surface and the area of the surface. The generalization to arbitrarily oriented simple closed curves is obtained by putting q in Stokes's theorem, Equation (9.9):

$$\Gamma_C \equiv \oint_C \boldsymbol{q}\cdot\mathrm{d}\boldsymbol{s} = \iint_S \boldsymbol{\omega}\cdot\mathrm{d}\boldsymbol{S}.$$

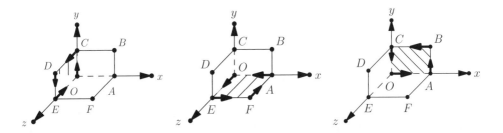

Figure 9.5 Decomposition of the surface in Figure 9.4 into plane parts normal to the coordinate axes

It is this difference, that the vorticity is a scalar in two dimensions but a vector in three dimensions, along with the essential connection between vorticity and lift, that is principally responsible for making three-dimensional aerodynamics so much more complicated than two-dimensional.

9.5.4 Rotational Form of Euler's Equation

Using the vector identity Equation (9.6), the acceleration term of Euler's Equation (9.11) can be rewritten

$$\frac{D\boldsymbol{q}}{Dt} \equiv \frac{\partial \boldsymbol{q}}{\partial t} + \boldsymbol{q} \cdot \nabla \boldsymbol{q} = \frac{\partial \boldsymbol{q}}{\partial t} + \nabla \left(\frac{q^2}{2} \right) + (\nabla \times \boldsymbol{q}) \times \boldsymbol{q},$$

so that Euler's equation may be written in the so-called *rotational form*

$$\frac{\partial \boldsymbol{q}}{\partial t} + \boldsymbol{\omega} \times \boldsymbol{q} = \frac{-1}{\rho} \nabla \left(p + \frac{\rho q^2}{2} \right). \tag{9.14}$$

9.5.5 Steady Irrotational Motion

A vector field with zero curl is called *irrotational*. If the velocity is steady

$$\frac{\partial \boldsymbol{q}}{\partial t} = 0$$

and irrotational

$$\nabla \times \boldsymbol{q} \equiv \boldsymbol{\omega} = \boldsymbol{0},$$

Euler's Equation (9.14) reduces to

$$\boldsymbol{0} = \frac{-1}{\rho} \nabla \left(p + \frac{\rho q^2}{2} \right);$$

i.e.

$$p + \frac{\rho q^2}{2} = \text{const.,} \qquad (9.15)$$

which is the form of *Bernoulli's equation* for irrotational motion. Thus, the equations of motion—the continuity Equation (9.10) and Euler's Equation (9.11)—are satisfied if the velocity is steady, divergence-free, and irrotational, and the pressure satisfies Bernoulli's equation.

Conversely, if the fluid in a flow system all originates from an irrotational upstream (as is usual in an aerodynamical problem, posed in the frame of reference of the aircraft), the circulation around each circuit in the far upstream will be zero, since the vorticity passing through any surface bounded by such a circuit will be zero (Stokes's theorem). These circuits continue to have zero circulation as they pass downstream over the aircraft (by Kelvin's theorem), and therefore continue to have zero vorticity through the surfaces closing them. Thus not only is an irrotational flow a solution of the governing equations, irrotationality is to be expected when the far upstream is irrotational.

Note that this does not imply that there can be no circulation around a loop encircling a wing: no closed circuit from the far upstream will form such a circuit.

9.6   Exercises

1. Verify Equation 9.12 by expanding it in Cartesian components.

2. Show that a two-dimensional velocity field q possessing a stream function ψ as in Section 3.6.2 can be derived from it by

$$q = \nabla \times (\psi \mathbf{k}).$$

   Hence (or otherwise) show that if ϕ and ψ are the potential and stream function of a plane ideal flow then

$$\nabla \phi = -\mathbf{k} \times \nabla \psi.$$

3. Show that the gradient theorem of Equation (9.8) follows from applying vector identities and the divergence theorem of Equation (9.7) to $s\mathbf{e}$, where $\mathbf{e}$ is any constant vector (Bousquet 1990).

4. Derive the two-dimensional divergence theorem from the three-dimensional one by applying the latter to a scalar field independent of z over a domain which is prismatic in the z-direction.

5. By expanding in Cartesian components, show that

$$\frac{D\mathbf{v}}{Dt} = \frac{\partial \mathbf{v}}{\partial t} + \nabla \left(\frac{v^2}{2} \right) + (\nabla \times \mathbf{v}) \times \mathbf{v}$$

   for any differentiable three-dimensional vector field $\mathbf{v}$ with magnitude $v \equiv \sqrt{\mathbf{v} \cdot \mathbf{v}}$.

9.7   Further Reading

Prandtl's argument on the essentially three-dimensional nature of the aerodynamics of wings has been repeated by Glauert (1926), Prandtl and Tietjens (1957), Milne-Thomson (1973), Kuethe and Chow (1976), Bertin (2002), and Anderson (2007). In particular, his figure 47 (referred to in Section 9.1.1 and the model for Figure 11.7 here) has been reproduced many times (Prandtl and Tietjens 1957, figure 173; Abbott and von Doenhoff 1959, figure 2; Kuethe and Chow 1976, figure 8a; Bertin 2002, figure 7.10; Moran 2003, figure 5.19a; Anderson 2007, figure 5.24).

The collection of vector identities given here is sufficient for most work in aerodynamics, and covers the union of the collections in Weatherburn (1924), Milne-Thomson (1973, chapter 21), Aris (1989, chapters 2–3), Bertin (2002, chapter 2), and Anderson (2007, chapter 2); for a truly extensive set, see Bousquet (1990). For an introduction to vector analysis, see Pope (2009, chapter 1) or for a complete treatment, Aris (1989).

More detailed derivations of the three-dimensional equations of motion have been given by Milne-Thomson (1973), Bertin (2002), and Anderson (2007).

Kelvin's theorem encapsulates much of the mechanics of three-dimensional ideal flow; it can derived in a few superficially different but essentially equivalent ways (Milne-Thomson 1973; Bertin 2002; Moran 2003).

References

Abbott, I.H. and von Doenhoff, A.E. (1959) *Theory of Wing Sections*. New York: Dover.

Anderson, J.D. (2007) *Fundamentals of Aerodynamics*, 4th edn. New York: McGraw-Hill.

Aris, R. (1989) *Vectors, Tensors, and the Basic Equations of Fluid Mechanics*. New York: Dover.

Bertin, J.J. (2002) *Aerodynamics for Engineers*, 4th edn. New Jersey: Prentice Hall.

Bousquet, J. (1990) *Aérodynamique: Méthode des singularités*. Toulouse: Cépaduès.

Glauert, H. (1926) *The Elements of Aerofoil and Airscrew Theory*. Cambridge: Cambridge University Press.

Kuethe, A.M. and Chow, C.Y. (1976) *Foundations of Aerodynamics*, 3rd edn. Chichester: John Wiley & Sons, Ltd.

Milne-Thomson, L.M. (1973) *Theoretical Aerodynamics*, 4th edn. New York: Dover.

Moran, J. (2003) *An Introduction to Theoretical and Computational Aerodynamics*. New York: Dover.

Pope, A. (2009) *Basic Wing and Airfoil Theory*. New York: Dover.

Prandtl, L. (1921) Applications of Modern Hydrodynamics to Aeronautics. Report 116, NACA.

Prandtl, L. and Tietjens, O.G. (1957) *Applied Hydro- and Aeromechanics*. New York: Dover.

Weatherburn, C.E. (1924) *Advanced Vector Analysis*. London: G. Bell & Sons Ltd.

10

Vorticity and Vortices

Vortices are as important in three-dimensional aerodynamics as in two-dimensional aerodynamics; however, whereas in two dimensions a vortex is merely a point, in three dimensions it is a line or curve. Some properties of three-dimensional vortices are considered here.

10.1 Streamlines, Stream Tubes, and Stream Filaments

We begin, however, with a discussion of a simpler concept in three-dimensional fluid kinematics: streamlines.

10.1.1 Streamlines

A *streamline* is a curve which is tangential to the velocity field along its length. Given a velocity field (steady or at an instant)

$$q(x, y, z) = u(x, y, z)\mathbf{i} + v(x, y, z)\mathbf{j} + w(x, y, z)\mathbf{k},$$

the streamline through a point $r_0 \equiv x_0\mathbf{i} + y_0\mathbf{j} + z_0\mathbf{k}$ can be traced by integrating the vector ordinary differential equation:

$$\frac{\mathrm{d}r}{\mathrm{d}s} = q(r) \tag{10.1}$$

with the initial condition $r = r_0$. The streamline is defined at an instant in time, and the independent variable s is a parameter increasing along the curve.

If the velocity field is steady, the streamlines coincide with the actual particle trajectories in time, since Equation (10.1) with t in place of s is the equation of motion of the particle.

If Equation (10.1) is written in the form

$$\mathrm{d}r = q\,\mathrm{d}s,$$

the condition that the streamline is tangential to the velocity field is obvious.

Theory of Lift: Introductory Computational Aerodynamics in MATLAB®/Octave, First Edition. G. D. McBain.
© 2012 John Wiley & Sons, Ltd. Published 2012 by John Wiley & Sons, Ltd.

10.1.2   Stream Tubes and Stream Filaments

Take a simple closed curve in the fluid, then the surface formed by all the streamlines passing through points on the curve is called a *stream tube*.

A *stream filament* is a stream tube of infinitesimal cross-section.

Consider a segment of a stream filament bounded by two cross-sections with areas δA_1 and δA_2. For conservation of mass (with uniform density), the volumetric flow-rate in at the area δA_1 has to match that out at the area δA_2. If the speed at the ends is q_1 and q_2, this gives

$$q_1 \, \delta A_1 = q_2 \, \delta A_2, \tag{10.2}$$

and in general the product of the speed and the area is constant along the stream filament. This product is called the *strength* of the stream filament; it is the volume rate of flow along the filament.

This implies that a stream filament cannot end within the fluid since that would mean the cross-sectional area vanishing and therefore the speed becoming infinite, which is impossible. Possible topologies for a streamline include a closed loop, or a curve with each end either at infinity or on a boundary. Streamlines cannot intersect each other or themselves except at a point of zero velocity, since there the direction of the velocity is moot. An impermeable boundary consists of streamlines since no flow ever crosses it; if a streamline starts or ends on a boundary, it must do so at a stagnation point.

A stream tube can be regarded as a bundle of stream filaments. Its strength is defined as the aggregate (i.e. sum or integral) of the strengths of its component filaments and again gives the volume rate of flow along the tube. It follows that:

- the strength of a stream tube or filament is constant along its length; and
- neither a stream tube nor filament can end at a point inside the fluid.

It also follows that a streamline cannot end in the fluid.

The first result here for a stream tube could also have been obtained by applying the divergence theorem (Equation 9.7) to the velocity field and a segment of the tube:

$$\iiint_V \nabla \cdot \mathbf{q} \, \mathrm{d}V = \iint_S \mathbf{q} \cdot \mathrm{d}\mathbf{S}.$$

Here the volume-integral vanishes by the continuity Equation (9.10). The bounding surface $\mathcal{S}$ consists of three parts: the two ends and the portion of the tube wall. The integral over the tube wall portion vanishes since the wall is everywhere parallel to the velocity field, by definition. Therefore, the divergence theorem reduces to

$$-\iint_{S_1} \mathbf{q} \cdot \mathrm{d}\mathbf{S} = \iint_{S_2} \mathbf{q} \cdot \mathrm{d}\mathbf{S}.$$

If the tube cross-section is taken infinitesimally small, this reduces to Equation (10.2); alternatively that equation could be arrived at by applying the mean value theorem of integral calculus.

10.2   Vortex Lines, Vortex Tubes, and Vortex Filaments

A *vortex line* is a curve which is tangential to the vorticity field along its length; thus a vortex line is to the vorticity as a streamline is to the velocity.

10.2.1   Strength of Vortex Tubes and Filaments

Continuing the analogy, the strength of a vortex filament is defined as the product of the magnitude of the vorticity and the cross-sectional area, and the strength of a vortex tube as the integral of the strengths of the filaments composing it.

If we apply Stokes's theorem to the velocity field and a cross-section of a vortex tube,

$$\iint_S (\nabla \times \boldsymbol{q}) \cdot \mathrm{d}S = \oint_C \boldsymbol{q} \cdot \mathrm{d}\boldsymbol{s}$$

or

$$\iint_S \boldsymbol{\omega} \cdot \mathrm{d}S = \Gamma_C,$$

which shows that the strength of a vortex tube is equal to the circulation around a circuit enclosing it.

10.2.2   Kinematic Properties of Vortex Tubes

The vorticity field is divergence-free, as follows from the vector identity Equation (9.5). Thus, all results about the velocity field that depend only on its being divergence-free carry over to analogous results about the vorticity field.

- The strength of a vortex tube or filament is constant along its length.
- Neither a vortex tube nor filament can end inside the fluid.

10.3   Helmholtz's Theorems

The preceding theorems all apply instantaneously, but two important theorems on the dynamics of vortex tubes in flows of perfect fluid were derived by Helmholtz, using Kelvin's theorem that the circulation around a fluid circuit remains constant during the flow.

10.3.1   'Vortex Tubes Move with the Flow'

Take a vortex tube and consider a piece S of its wall, bounded by a simple closed curve C. No vortex lines pierce S, so Stokes's theorem, Equation (9.9), implies that there is no circulation around C.

Now let S and C drift with the flow. As it drifts, the circulation around C remains zero, by Kelvin's theorem. Therefore, no vortex filaments pierce it during the flow and it remains part of the wall of the vortex tube; i.e.

vortex tubes move with the flow.

10.3.2 'The Strength of a Vortex Tube is Constant'

Now consider a simple closed curve around a vortex tube. The circulation around the curve is equal to the vortex tube strength. As the tube and the circuit move with the flow, the circuit remains on the tube (previous theorem) and its circulation remains constant (Kelvin's theorem). Therefore:

The strength of a vortex tube remains constant.

10.4 Line Vortices

10.4.1 The Two-dimensional Vortex

Every two-dimensional flow satisfying the continuity and Euler equations also satisfies the three-dimensional continuity and Euler equations. In particular, the two-dimensional vortex with strength Γ and centre $x = x'$, $y = y'$; i.e.

$$u - iv = \frac{i\Gamma}{2\pi \{(x + iy) - (x' + iy')\}}$$

or

$$u = \frac{+\Gamma}{2\pi} \frac{y - y'}{(x - x')^2 + (y - y')^2} \tag{10.3a}$$

$$v = \frac{-\Gamma}{2\pi} \frac{x - x'}{(x - x')^2 + (y - y')^2}; \tag{10.3b}$$

is a valid three-dimensional flow. It is a vortex filament of strength Γ along the line $x = x'$, $y = y'$, with the direction of the unit vector $-\mathbf{k}$.

10.4.2 Arbitrarily Oriented Rectilinear Vortex Filaments

Section 10.4.1 described the velocity field induced by a vortex filament parallel to the z-axis. This can be generalized for any rectilinear vortex filament by

$$\mathbf{q}(\mathbf{r}) = \frac{\Gamma \hat{\boldsymbol{\ell}} \times (\mathbf{r} - \mathbf{r}')}{2\pi |\hat{\boldsymbol{\ell}} \times (\mathbf{r} - \mathbf{r}')|^2}, \tag{10.4}$$

where $\hat{\boldsymbol{\ell}}$ is the unit vector in the direction of the filament, and $\boldsymbol{r}'$ is any point on the filament; the value of $\boldsymbol{q}(\boldsymbol{r})$ is independent of the choice of $\boldsymbol{r}'$. To see that $\hat{\boldsymbol{\ell}} \times (\boldsymbol{r} - \boldsymbol{r}')$ is independent of the choice of $\boldsymbol{r}$, consider any other point on the filament, $\boldsymbol{r}''$, since it and $\boldsymbol{r}'$ are both on the filament, we can write

$$\boldsymbol{r}'' = \boldsymbol{r}' + \lambda \hat{\boldsymbol{\ell}}$$

for some scalar λ, and

$$\begin{aligned} \hat{\boldsymbol{\ell}} \times (\boldsymbol{r} - \boldsymbol{r}'') &= \hat{\boldsymbol{\ell}} \times (\boldsymbol{r} - \{\boldsymbol{r}' + \lambda \hat{\boldsymbol{\ell}}\}) \\ &= \hat{\boldsymbol{\ell}} \times (\boldsymbol{r} - \boldsymbol{r}') - \hat{\boldsymbol{\ell}} \times (\lambda \hat{\boldsymbol{\ell}}) \\ &= \hat{\boldsymbol{\ell}} \times (\boldsymbol{r} - \boldsymbol{r}') - \mathbf{0}, \end{aligned}$$

as claimed.

The vector $\hat{\boldsymbol{\ell}} \times (\boldsymbol{r} - \boldsymbol{r}')$ is perpendicular to both the line vortex and the straight line from the filament to the point. Its magnitude is the (perpendicular) distance of the point from the filament. These two results follow from the basic properties of the vector cross product. Thus, in general, the velocity induced by a straight filament at a point is perpendicular to the plane containing the filament and the point and inversely proportional in magnitude to the distance of the point from the filament.

For example, in one of the vortices of a two-dimensional flow, we'll have $\hat{\boldsymbol{\ell}} = -\mathbf{k}$ (and of course $\boldsymbol{r} = x\mathbf{i} + y\mathbf{j} + z\mathbf{k}$) so that

$$\begin{aligned} \hat{\boldsymbol{\ell}} \times (\boldsymbol{r} - \boldsymbol{r}') &= -\mathbf{k} \times (\boldsymbol{r} - \boldsymbol{r}') \\ &= -\mathbf{k} \times \{(x - x')\mathbf{i} + (y - y')\mathbf{j} + (z - z')\mathbf{k}\} \\ &= +(y - y')\mathbf{i} - (x - x')\mathbf{j} \end{aligned}$$

and

$$\frac{\Gamma \hat{\boldsymbol{\ell}} \times (\boldsymbol{r} - \boldsymbol{r}')}{2\pi |\hat{\boldsymbol{\ell}} \times (\boldsymbol{r} - \boldsymbol{r}')|} = \frac{\Gamma \{(y - y')\mathbf{i} - (x - x')\mathbf{j}\}}{2\pi \{(x - x')^2 + (y - y')^2\}},$$

which matches Equation (10.3).

10.5   Segmented Vortex Filaments

10.5.1   The Biot–Savart Law

In three dimensions, a line vortex doesn't have to be a straight line along its whole length. It cannot end in the fluid, so must either be a loop, extend to infinity at either end, or perhaps have one or both ends on a boundary of the fluid.

Curved or bent line vortices can be modelled by building them up out of short segments of straight line vortices, and then the velocity field induced by the full line vortex obtained by integrating over these segments. In doing this, it must be remembered that a segment of a line vortex cannot exist in isolation, since vortex filaments cannot end in the fluid.

The velocity field induced at a point r by a segment of line vortex of vector length $d\ell$ and strength Γ is

$$dq = \frac{\Gamma}{4\pi} \frac{d\ell \times (r - r')}{|r - r'|^3} \tag{10.5}$$

where r' is now not any point on the vortex filament, but the location of the infinitesimal segment in question. The direction of the velocity is perpendicular to the plane containing the line of the line vortex and the point (a line and a point off the line define a plane). The sense of the velocity is such that:

1. the line vortex segment (directed by the vorticity);
2. the perpendicular from the line vortex to the point; and
3. the induced velocity.

form a right-handed triad.

We do not prove Equation (10.5), called the *Biot–Savart* law; however, we will verify that it is compatible with the known result for infinite rectilinear vortex filaments.

10.5.2 Rectilinear Vortex Filaments

Consider a finite segment straight line vortex at r' with directed length $d\ell = \hat{\ell}d\ell$, as in Figure 10.1, and compute the velocity at a point r off the line, i.e. an r for which $\hat{\ell} \times (r - r') \neq 0$, by integrating the Biot–Savart law, Equation (10.5).

Let ℓ be the vector from the beginning of the segment to the infinitesimal segment under consideration in the integrand, then

$$d\ell = \hat{\ell}\, d\ell$$
$$= -\hat{\ell}\, d(h \cot \beta)$$
$$= \hat{\ell}\, \frac{h}{\sin^2 \beta}\, d\beta.$$

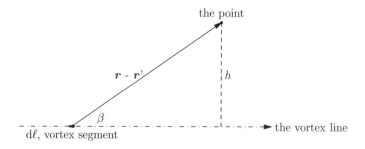

Figure 10.1   Velocity induced by an infinitesimal segment of a rectilinear vortex filament

Also,

$$|\hat{\boldsymbol{\ell}} \times (\boldsymbol{r} - \boldsymbol{r}')| = |\boldsymbol{r} - \boldsymbol{r}'| \sin \beta \equiv h,$$

the perpendicular distance of the point from the line.

Thus, the Biot–Savart law can be rewritten

$$
\begin{aligned}
\mathrm{d}\boldsymbol{q} &= \frac{\Gamma}{4\pi} \frac{\mathrm{d}\boldsymbol{\ell} \times (\boldsymbol{r} - \boldsymbol{r}')}{|\boldsymbol{r} - \boldsymbol{r}'|^3} \\
&= \frac{\Gamma}{4\pi} \frac{h}{\sin^2 \beta} \frac{\hat{\boldsymbol{\ell}} \times (\boldsymbol{r} - \boldsymbol{r}')}{|\boldsymbol{r} - \boldsymbol{r}'|^3} \, \mathrm{d}\beta \\
&= \frac{\Gamma}{4\pi} \frac{h}{\sin^2 \beta} \frac{\hat{\boldsymbol{\ell}} \times (\boldsymbol{r} - \boldsymbol{r}')}{(h/\sin \beta)^3} \, \mathrm{d}\beta \\
&= \frac{\Gamma}{4\pi} \frac{\sin \beta}{h^2} \hat{\boldsymbol{\ell}} \times (\boldsymbol{r} - \boldsymbol{r}') \, \mathrm{d}\beta \\
&= \frac{-\Gamma}{4\pi} \frac{\hat{\boldsymbol{\ell}} \times (\boldsymbol{r} - \boldsymbol{r}')}{|\hat{\boldsymbol{\ell}} \times (\boldsymbol{r} - \boldsymbol{r}')|^2} \, \mathrm{d}\cos \beta.
\end{aligned}
$$

For a straight line, h is independent of $\boldsymbol{\ell}$ (or β), and so is $\hat{\boldsymbol{\ell}} \times (\boldsymbol{r} - \boldsymbol{r}')$ as shown in Section 10.4.2. Hence,

$$\boldsymbol{q} = \frac{\Gamma}{4\pi} \frac{\hat{\boldsymbol{\ell}} \times (\boldsymbol{r} - \boldsymbol{r}')}{|\hat{\boldsymbol{\ell}} \times (\boldsymbol{r} - \boldsymbol{r}')|^2} (\cos \beta_1 - \cos \beta_2). \tag{10.6}$$

Here β_1 and β_2 are the biangular coordinates of the point at which the velocity is evaluated, with respect to the vortex segment, in the plane defined by the point and the vortex segment; i.e. the values of the angle β between the vortex filament and the vector from the filament to the point at which the velocity is evaluated, taken at the start and end of the filament; the angles are illustrated in Figure 10.2.

In this integrated formula and the ones to follow such as Equations (10.8) and (10.9), $\boldsymbol{r}'$ is any point lying on the same (infinite straight) line as the finite rectilinear segment of vortex filament. This arbitrariness can be exploited to simplify the vector algebra.

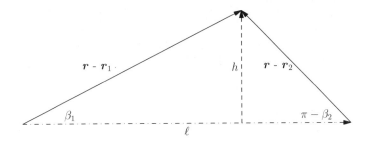

Figure 10.2   Velocity induced by a finite segment of a rectinlinear vortex filament

10.5.3   Finite Rectilinear Vortex Filaments

The cosines here can be expressed in terms of the position vectors of the start and end of the segment, r_1 and r_2 (illustrated in Figure 10.2), using the dot-product Equation (9.1) for the angle between two vectors:

$$\cos \beta_1 = \frac{\hat{\ell} \cdot (r - r_1)}{|r - r_1|}$$

$$\cos \beta_2 = \frac{\hat{\ell} \cdot (r - r_2)}{|r - r_2|}$$

$$\cos \beta_1 - \cos \beta_2 = \hat{\ell} \cdot \left(\frac{r - r_1}{|r - r_1|} - \frac{r - r_2}{|r - r_2|} \right). \tag{10.7}$$

10.5.4   Infinite Straight Line Vortices

If the line segment begins infinitely far away, $\beta_1 \sim 0$ and $\cos \beta_1 \sim 1$, whereas if it ends infinitely far away, $\beta_2 \sim \pi$ and $\cos \beta_2 \sim -1$, so Equation (10.7) becomes 2.

If the line vortex becomes infinitely long at both ends,

$$q \sim \frac{\Gamma}{4\pi} \frac{\hat{\ell} \times (r - r')}{|\hat{\ell} \times (r - r')|^2} (1 - \{-1\})$$

$$= \frac{\Gamma}{2\pi} \frac{\hat{\ell} \times (r - r')}{|\hat{\ell} \times (r - r')|^2}, \tag{10.8}$$

which is exactly the Equation (10.4) for a two-dimensional vortex.

10.5.5   Semi-infinite Straight Line Vortex

A case of practical interest later in lifting line theory is the velocity induced by a semi-infinite straight line vortex at a point in the plane perpendicular to the line vortex and containing the end of the semi-infinite segment. In this case, $\beta_1 \sim 0$ but $\beta_2 = \pi/2$, so that $\cos \beta_2 = 0$, Equation (10.7) becomes 1 and

$$q \sim \frac{\Gamma}{4\pi} \frac{\hat{\ell} \times (r - r')}{|\hat{\ell} \times (r - r')|^2} (1 - 0)$$

$$= \frac{\Gamma}{4\pi} \frac{\hat{\ell} \times (r - r')}{|\hat{\ell} \times (r - r')|^2}, \tag{10.9}$$

This result could also have been deduced by a symmetry argument, since it is half the Equation (10.8) for an infinite line vortex and that result could be considered as being due to the sum of the contributions from the two semi-infinite line vortices on either side.

10.5.6   *Truncating Infinite Vortex Segments*

Infinite quantities are easy enough to handle in manual calculations, indeed are often easier since the neglect of accompanying finite quantities leads to simplifications; however, they can be a nuisance in computer programs, necessitating checks against floating point overflow and then special cases. Since the influence of an element of vortex segment diminishes, according to Equation (10.5), like the square of distance, the more remote elements are less important and a practical alternative approach is simply to truncate the infinite leg at some great distance from the region of interest.

If the vortex line segment begins infinitely far away, in the $-\hat{\ell}$ direction, one could say $r_1 \sim r - \Lambda\hat{\ell}$ as the length $\Lambda \to \infty$; thus

$$r - r_1 \sim \Lambda\hat{\ell}$$
$$|r - r_1| \sim \Lambda$$
$$\frac{r - r_1}{|r - r_1|} \sim \hat{\ell}$$
$$\hat{\ell} \cdot \frac{r - r_1}{|r - r_1|} \sim 1,$$

which is the exact answer without truncation used in Equations (10.8) and (10.9).

Similarly, if the distant end r_2 is approximated by $r + \Lambda\hat{\ell}$,

$$\lim_{\Lambda \to \infty} \hat{\ell} \cdot \frac{r - r_2}{|r - r_2|} = -1,$$

which too was used in Equation (10.8).

The question of how large Λ needs to be taken is a practical one. It depends a little on the problem and the accuracy sought. Katz and Plotkin (2001) suggest 'at least twenty wing spans' but use 100 to be safe in their FORTRAN code (p. 580). Unlike adding panels, increasing Λ to more closely approach infinity does not increase the computational cost of the model, so there is little incentive to economize. A good idea, if time permits, is to experiment with changing this value and see how much it influences the flow near the wing and particularly the predictions of quantities of interest.

On the other hand, having distinct formulae for the trailing vortices does not unduly complicate a computer program; Moran (2003) lists an equivalent FORTRAN code not exploiting this approximation.

10.5.7   *Implementing Line Vortices in Octave*

Although the preceding discussion has edged towards issues of implementation, the details are deferred until Section 13.4, since it is only for multiple vortices that a computer program is required; nevertheless, this forward-notice is placed here as those programs can be useful in checking and experimenting with the exercises of the present chapter.

10.6 Exercises

1. Find the two points in the two-dimensional velocity field outside the unit circle with complex velocity given by Equation (3.6) at which streamlines intersect.

2. Show that the velocity q induced at a point r by a rectilinear vortex segment of strength Γ running from $-b\mathbf{k}/2$ to $+b\mathbf{k}/2$ is

$$q(r) = \frac{\Gamma}{4\pi} \left(\frac{z + \frac{b}{2}}{|r + \frac{b}{2}\mathbf{k}|} - \frac{z - \frac{b}{2}}{|r - \frac{b}{2}\mathbf{k}|} \right) \frac{r \times \mathbf{k}}{|r \times \mathbf{k}|^2}.$$

3. Show that the velocity induced by a semi-infinite rectilinear vortex segment of strength Γ running in the negative x-direction to $b\mathbf{k}/2$ is

$$q(r) = \frac{\Gamma}{4\pi} \left(1 + \frac{x}{|r - \frac{b}{2}\mathbf{k}|} \right) \frac{r \times \mathbf{i} - \frac{b}{2}\mathbf{j}}{|r \times \mathbf{i} - \frac{b}{2}\mathbf{j}|^2}.$$

4. Show that the velocity field derived in the preceding exercise reduces
   (a) on the z-axis to

$$q(z\mathbf{k}) = \frac{\Gamma\mathbf{j}}{4\pi \left(z - \frac{b}{2} \right)},$$

   and
   (b) in particular at the origin to

$$q(0) = \frac{-\Gamma\mathbf{j}}{2\pi b}.$$

5. Since

$$|\hat{\boldsymbol{\ell}} \times (r - r')| = h$$

   and

$$|(r - r_1) \times (r - r_2)| = h\,|r_2 - r_1|,$$

   and both vectors point in the same direction (out of the page in Figure 10.2), show that Equations (10.6), (10.8), and (10.9) for the velocity induced by the segments finite, infinite, and semi-infinite, can be rewritten using

$$\frac{\hat{\boldsymbol{\ell}} \times (r - r')}{|\hat{\boldsymbol{\ell}} \times (r - r')|^2} = \frac{(r - r_1) \times (r - r_2)}{|r_2 - r_1|^2|(r - r_1) \times (r - r_2)|^2},$$

   if it is more convenient. Bertin (2002) recommended this form.

10.7   Further Reading

For other references to line vortices in three dimensions, see Glauert (1926), Karamcheti (1966), Batchelor (1967), Milne-Thomson (1973), Bousquet (1990), Kuethe and Chow (1998), Bertin (2002), or Anderson (2007).

Milne-Thomson (1973) warns that the Biot–Savart law can be misleading since it refers to an infinitesimal segment of a vortex line, which is something that cannot exist independently but only as part of a vortex line not ending in the fluid. Bousquet (1990) avoids this awkwardness by dealing only with an integral form of the law; this is perhaps to be preferred, though it requires some facility with the theory of multidimensional *distributions*—a distribution being a generalized function (Lighthill 1958) which is defined not in terms of its values at points but rather by the values of line (or surface or volume) integrals of products of it with other functions, the classic example being Dirac's delta $\delta(x)$, defined by $\int_{-\infty}^{\infty} \delta(x) f(x)\, \mathrm{d}x = f(0)$ for any function f smooth at the origin.

References

Anderson, J.D. (2007) *Fundamentals of Aerodynamics*, 4th edn. New York: McGraw-Hill.

Batchelor, G.K. (1967) *An Introduction to Fluid Dynamics*. Cambridge: Cambridge University Press.

Bertin, J.J. (2002) *Aerodynamics for Engineers*, 4th edn. New Jersey: Prentice Hall.

Bousquet, J. (1990) *Aérodynamique: Méthode des singularités*. Toulouse: Cépaduès.

Glauert, H. (1926) *The Elements of Aerofoil and Airscrew Theory*. Cambridge: Cambridge University Press.

Karamcheti, K. (1966) *Principles of Ideal-Fluid Aerodynamics*. New York: John Wiley & Sons, Ltd.

Katz, J. and Plotkin, A. (2001) *Low-Speed Aerodynamics* 2nd edn. Cambridge: Cambridge University Press.

Kuethe, A.M. and Chow, C.Y. (1998) *Foundations of Aerodynamics*, 5th edn. New York: John Wiley & Sons, Ltd.

Lighthill, M.J. (1958) *Introduction to Fourier Analysis and Generalised Functions*. Cambridge: Cambridge University Press.

Milne-Thomson, L.M. (1973) *Theoretical Aerodynamics*, 4th edn. New York: Dover.

Moran, J. (2003) *An Introduction to Theoretical and Computational Aerodynamics*. New York: Dover.

11

Lifting Line Theory

The first successful quantitative model used for predicting the aerodynamic force on a wing of finite span was the *lifting line theory*.

11.1 Basic Assumptions of Lifting Line Theory

Lifting line theory in its simplest form assumes that:

- the thickness and chord are much shorter than the span;
- the wing is unswept; and
- the flight is steady and perpendicular to the span.

With the above assumptions, the wing is modelled (on the length scale of the span) as the line segment

$$x = y = 0, \qquad \frac{-b}{2} \leqslant z \leqslant \frac{b}{2}. \tag{11.1}$$

This line segment is called the *lifting line*.

11.2 The Lifting Line, Horseshoe Vortices, and the Wake

If the wing is to be producing lift, the Kutta–Joukowsky theorem (3.15a) leads us to expect that there will be circulation around it in circuits lying in planes parallel to the plane of symmetry (i.e. planes of constant z) and encircling the wing.

11.2.1 Deductions from Vortex Theorems

Stokes's theorem (Section 10.2.1) then requires that there are vortex filaments running along the wing. In the lifting line theory, these filaments are assumed to run along the lifting line, given by Equation (11.1).

Theory of Lift: Introductory Computational Aerodynamics in MATLAB®/Octave, First Edition. G. D. McBain.
© 2012 John Wiley & Sons, Ltd. Published 2012 by John Wiley & Sons, Ltd.

However, vortex filaments cannot end in the fluid (Section 10.2.2) so this description is physically incomplete. Since, by Helmholtz's theorems, vortex filaments move with the fluid, the simplest consistent model is that wherever each filament would end on the lifting line, it instead trails behind the wing (back to a spanwise starting vortex at the take-off strip, or, essentially, to infinity).

11.2.2 Deductions from the Wing Pressure Distribution

This picture of vortex filaments trailing behind the wing is consistent with the difference in the spanwise flow components over and under the wing induced by the pressure differences associated with the generation of lift (Section 9.1.2). When the upper and lower streams reunite at the trailing edge, they have the same speed (by Bernoulli's equation, as the pressure is single-valued) and also the same u and v components (by the Kutta–Joukowsky condition); however, their spanwise components are different: inboard for the upper stream and outboard for the lower. Thus there forms a surface in the air behind the trailing edge across which the tangential (specifically spanwise) component of velocity is discontinuous. This is a *vortex sheet*, composed of vortex filaments, being the trailing legs of the vortex filaments inferred above from the vortex theorems.

11.2.3 The Lifting Line Model of Air Flow

The model of the wing therefore consists of a collection of *horseshoe vortices*, each of which consists of a segment on the lifting line called a *bound vortex* (since it is constrained to move with the wing rather than allowed to drift in the flow) and two semi-infinite vortex filaments behind the wing called the *trailing vortices*. Together, the bound vortices of all the horseshoe vortices constitute the lifting line and represent the wing, and the collection of trailing vortices represent the *wake*.

The flow around the wing is then taken as the sum of the contributions from the free stream

$$\boldsymbol{q}_\infty = q_\infty(\mathbf{i}\cos\alpha + \mathbf{j}\sin\alpha) \tag{11.2}$$

and the horseshoe vortices constituting the lifting line and the wake.

11.2.4 Horseshoe Vortex

The horseshoe vortices are significant insofar as they affect the flow near the wing (that is, significant in the theory of lift—they can also be significant to things behind the wing, and so are an important consideration in air traffic control).

The bound vortex part of the horseshoe doesn't induce any velocity on the lifting line on which it lies, but the trailing vortices do.

Consider as an example, a horseshoe vortex filament of strength Γ running from $\infty\mathbf{i} + z'\mathbf{k}$ to $z'\mathbf{k}$ to $-z'\mathbf{k}$ to $\infty\mathbf{i} - z'\mathbf{k}$, as illustrated in Figure 11.1. The unit vectors of the three legs are $-\mathbf{i}$, $-\mathbf{k}$, and $+\mathbf{i}$, respectively. The strength, Γ, must be common to the three legs, by the vortex laws (Section 10.2.2).

Figure 11.1   A rectangular horseshoe vortex, lying in the zx-plane with vertices at $\pm z'\mathbf{k}$

The velocity induced on the line of the bound vortex ($x = y = 0$) can be calculated from two applications of the Equation (10.9) for points in the perpendicular plane of the end of a semi-infinite rectilinear vortex filament

$$q = \frac{\Gamma}{4\pi} \frac{\hat{\boldsymbol{\ell}} \times (\mathbf{r} - \mathbf{r}')}{|\hat{\boldsymbol{\ell}} \times (\mathbf{r} - \mathbf{r}')|^2}.$$

Here, we can take $\hat{\boldsymbol{\ell}} = \mp\mathbf{i}$ and $\mathbf{r}' = \pm z'\mathbf{k}$ for the two trailing vortices so that, for some point on the lifting line $\mathbf{r} = z\mathbf{k}$,

$$\hat{\boldsymbol{\ell}} \times (\mathbf{r} - \mathbf{r}') = \mp\mathbf{i} \times (z \mp z')\mathbf{k} = \pm(z \mp z')\mathbf{j}.$$

Therefore,

$$
\begin{aligned}
q(z\mathbf{k}) &= \frac{\Gamma}{4\pi} \left\{ \frac{z - z'}{|z - z'|^2} - \frac{z + z'}{|z + z'|^2} \right\} \mathbf{j} \\
&= \frac{\Gamma}{4\pi} \left\{ \frac{(z + z')^2(z - z') - (z - z')^2(z + z')}{(z + z')^2(z - z')^2} \right\} \mathbf{j} \\
&= \frac{\Gamma}{4\pi} \left\{ \frac{(z + z') - (z - z')}{(z + z')(z - z')} \right\} \mathbf{j} \\
&= \frac{-\Gamma z'\mathbf{j}}{2\pi\left(z'^2 - z^2\right)}.
\end{aligned}
\tag{11.3}
$$

Notice that, having assumed that the trailing vortices lie in the zx-plane (the $y = 0$ plane), the velocity induced at the bound vortex by the trailing vortices is purely perpendicular to this plane; i.e. vertical. For $|z| < z'$ (i.e. on the bound vortex), $v \equiv \mathbf{j} \cdot \mathbf{q} < 0$, so that this velocity is called *downwash* and denoted v_{w}. The downwash is plotted in Figure 11.2.

11.2.5   Continuous Trailing Vortex Sheet

A difficulty with using a single horseshoe vortex to model a wing is that the downwash given by Equation (11.3) is infinite at $z = \pm z'$. This difficulty can be obviated by using a continuous

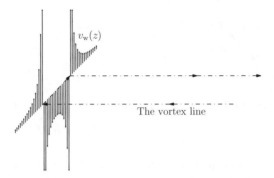

Figure 11.2   The downwash induced along the lifting line by a horseshoe vortex (like the one in Figure 11.1), according to Equation (11.3)

trailing sheet of vortex filaments. If the trailing vortex filament from $z'\mathbf{k}$ to $\infty\mathbf{i} + z'\mathbf{k}$ has strength $\gamma(z')\delta z'$, it contributes

$$\delta q(z\mathbf{k}) = \frac{\gamma(z')\delta z'\mathbf{i} \times (z - z')\mathbf{k}}{4\pi(z - z')^2}$$

$$= \frac{-\gamma(z')\mathbf{j}}{4\pi(z - z')}\,\delta z'$$

to the downwash at $\mathbf{r} = z\mathbf{k}$. The total downwash v_w is

$$v_w(z) \equiv \mathbf{j} \cdot q(z\mathbf{k}) = -\frac{1}{4\pi} \int_{-b/2}^{b/2} \frac{\gamma(z')\,dz'}{z - z'}. \tag{11.4}$$

Since vortex filaments cannot end in the fluid, and since it is the strength of the bound vortex filaments that makes up the circulation around a section of the wing, the strength of the trailing vortex filament arriving at the lifting line at $z = z'$ must be related to the circulation there by

$$\Gamma(z') + \gamma(z')\,\delta z' = \Gamma(z' + \delta z') \sim \Gamma(z') + \frac{d\Gamma}{dz'}\,\delta z' \tag{11.5a}$$

$$\gamma(z') \sim \frac{d\Gamma}{dz'}; \tag{11.5b}$$

as in Figure 11.3. Thus, in terms of the circulation distribution, the downwash is

$$v_w(z) = \frac{-1}{4\pi} \int_{-b/2}^{b/2} \frac{d\Gamma}{dz'} \frac{dz'}{z - z'}. \tag{11.6}$$

11.2.6   The Form of the Wake

In lifting line theory, the wake is assumed to be confined to a plane strip behind the wing. In fact, the situation is not so simple. We have seen (Section 11.2.4) that the horseshoe vortex filaments constituting the wake generate a downwash inside the horseshoe. Outside the horseshoe, they generate an upwash. Since vortex filaments move with the fluid, the effect of this is that the

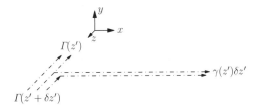

Figure 11.3 Relation between the local combined strength of the trailing vortex filaments in a strip of spanwise width $\delta z'$ and the distribution of circulation along the lifting line, as given by Equation (11.5)

outer filaments of the sheet drift upwards and the edges of the sheet curl inwards. These curled edges form what are called the two *wing-tip vortices*, and are clearly visible when there is dust in the air; e.g. in crop-spraying (Kuethe and Chow 1998, figure 6.5). They are also made visible very commonly because, being regions of concentrated vorticity and therefore high speed and, via Bernoulli's equation, low pressure, they cause atmospheric moisture to condense into liquid droplets (Bertin 2002, figure 7.3).

In spite of this very visible phenomenon, the model of a flat wake is still appropriate for computing the downwash on the wing because the curling occurs while the air is passing downstream behind the wing, so that immediately behind the wing the wake is flatter than implied by the photographs. From the Biot–Savart law (Section 10.5.1), the importance of a piece of a vortex filament decreases like the inverse square of the distance; therefore, it is the nearer part of the wake that is more important.

11.3 The Effect of Downwash

We have established that the system of trailing vortex filaments in the wake causes a downwash at the wing. What effect does this have on the aerodynamics?

Basically, instead of experiencing the free-stream velocity given by Equation (11.2), the section at $r = z\mathbf{k}$ experiences that plus the downwash; i.e.

$$q(z) = (q_\infty \cos \alpha)\mathbf{i} + \{q_\infty \sin \alpha + v_w(z)\}\,\mathbf{j}.$$

11.3.1 Effect on the Angle of Incidence: Induced Incidence

One effect is that instead of experiencing the geometric angle of incidence

$$\alpha \equiv \arctan \frac{q_\infty \sin \alpha}{q_\infty \cos \alpha} = \alpha,$$

the section experiences

$$\alpha'(z) = \arctan \frac{q_\infty \sin \alpha + v_w(z)}{q_\infty \cos \alpha},$$

which is called the *effective incidence*. For small angles ($\tan\alpha \sim \alpha$) and small downwash ($v_{\mathrm{w}} \ll q_\infty$) this reduces to

$$\alpha'(z) \sim \alpha + \frac{v_{\mathrm{w}}(z)}{q_\infty}.$$

The angle

$$\alpha_{\mathrm{i}}(z) \equiv \alpha - \alpha'(z) = \frac{-v_{\mathrm{w}}(z)}{q_\infty} \tag{11.7}$$

is called the *induced incidence*.

Equation (11.7) may be combined with Equation (11.6) to give

$$\alpha_{\mathrm{i}}(z) = \frac{-1}{4\pi q_\infty} \int_{-b/2}^{b/2} \frac{\mathrm{d}\Gamma}{\mathrm{d}z'} \frac{\mathrm{d}z'}{z' - z}. \tag{11.8}$$

11.3.2 Effect on the Aerodynamic Force: Induced Drag

In lifting line theory, we assume that the aerodynamics of each two-dimensional section is basically the same as in two dimensions, except that the section experiences the free stream modified by the addition of the downwash.

Since the free stream is rotated by the induced incidence α_{i}, the lift and drag components of the aerodynamic force per unit span are altered in accordance with Equations (1.1a)–(1.1b) to

$$C'_\ell = C_\ell \cos\alpha_{\mathrm{i}} - C_d \sin\alpha_{\mathrm{i}}$$
$$C'_d = C_\ell \sin\alpha_{\mathrm{i}} + C_d \cos\alpha_{\mathrm{i}},$$

where the unprimed coefficients refer to two-dimensional conditions and the primed to three-dimensional. For perfect fluid flow $C_d = 0$, and even when viscous effects are accounted for, $C_d \ll C_l$ for a functioning wing section below stall incidence. With the small angle approximation for α_{i},

$$C'_\ell = C_\ell \tag{11.9a}$$
$$C'_d = C_\ell \alpha_{\mathrm{i}}. \tag{11.9b}$$

Essentially, since the effective free stream has been rotated downwards and the aerodynamic force acts, in accordance with the Kutta–Joukowsky theorem, at right-angles to this, some of the two-dimensional lift force has been rotated backwards and acts as a drag; this drag is called the *induced drag*.

11.4 The Lifting Line Equation

In Equation (11.8), we can substitute for the local circulation using the Kutta–Joukowsky Equation (3.15a):

$$\Gamma = \frac{\ell(z)}{\rho q_\infty} = \frac{1}{2} q_\infty c(z) C_\ell(z)$$

to get Prandtl's lifting line equation

$$\alpha_i(z) = -\frac{1}{8\pi} \int_{-b/2}^{b/2} \frac{d(cC_\ell)}{dz'} \frac{dz'}{z' - z}. \tag{11.10}$$

11.4.1 Glauert's Solution of the Lifting Line Equation

The lifting line Equation (11.10) is more easily solved by changing the variable integration from the spanwise coordinate z to the *eccentric angle* θ via

$$\theta \equiv \arccos \frac{-2z}{b} \tag{11.11}$$

$$z \equiv \frac{-b}{2} \cos \theta,$$

as illustrated in Figure 11.4, the lifting line Equation (11.10) in terms of the eccentric angle is

$$\alpha_i = \frac{-1}{\pi} \int_0^\pi \frac{d\left(\frac{cC_\ell}{4b}\right)}{d\theta'} \frac{d\theta'}{\cos \theta - \cos \theta'}. \tag{11.12}$$

Comparing this with Glauert's integral (Equation 5.17)

$$\int_0^\pi \frac{\cos n\theta' \, d\theta'}{\cos \theta - \cos \theta'} = \frac{-\pi \sin n\theta}{\sin \theta},$$

we see that if we can expand the spanwise lift loading as

$$\frac{cC_\ell}{4b} = \sum_{j=1}^\infty A_j \sin j\theta \tag{11.13}$$

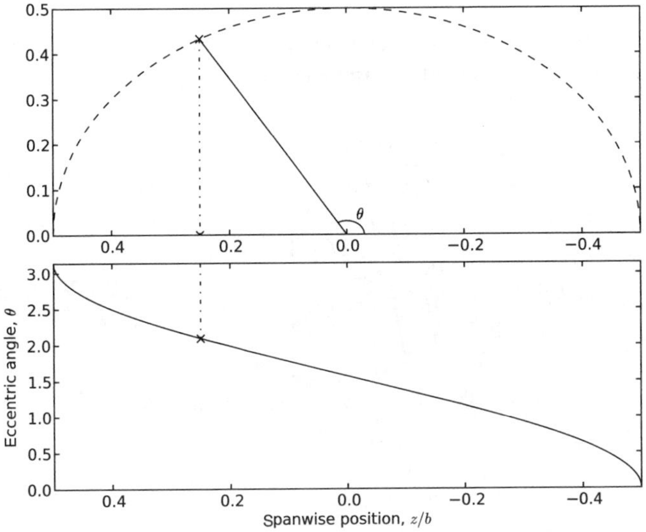

Figure 11.4 The eccentric angle defined by Equation (11.11) and used to describe spanwise lift loadings

so that

$$\frac{d}{d\theta}\frac{cC_\ell}{4b} = \sum_{j=1}^{\infty} jA_j \cos j\theta, \tag{11.14}$$

then the lifting line Equation (11.12) gives the induced incidence as

$$
\begin{aligned}
\alpha_i &= \frac{-1}{\pi} \int_0^\pi \sum_{j=1}^{\infty} jA_j \cos j\theta' \frac{d\theta'}{\cos\theta - \cos\theta'} \\
&= \sum_{j=1}^{\infty} jA_j \left\{ \frac{-1}{\pi} \int_0^\pi \frac{\cos j\theta' d\theta'}{\cos\theta - \cos\theta'} \right\} \\
&= \sum_{j=1}^{\infty} \frac{jA_j \sin j\theta}{\sin\theta}.
\end{aligned}
\tag{11.15}
$$

Note that the lift loading Equation (11.13) goes to zero at the wing tips $z = \pm\frac{b}{2}$ (i.e. $\theta = 0$ and π), which makes sense since the pressure difference has to vanish there. Any function over $-\frac{b}{2} < z < +\frac{b}{2}$ could be represented by a trigonometric series like Equation (11.13) but with cosines as well as sines, but the cosines wouldn't vanish at the tips and so the coefficients of the cosine terms would have to be zero for a function vanishing at the tips.

Note also that if the lift loading is symmetric, A_j must be zero for all even j.

11.4.2   Wing Properties in Terms of Glauert's Expansion

Lift

Assuming we were able to determine Glauert's expansion coefficients A_j in Equation (11.13), we could then compute the wing's lift coefficient from

$$
\begin{aligned}
C_L &\equiv \frac{\int_{-b/2}^{b/2} \ell \, dz}{\frac{1}{2}\rho q_\infty^2 b\bar{c}} \\
&= \frac{1}{b\bar{c}} \int_{-b/2}^{b/2} cC_\ell \, dz \\
&= \frac{4b}{2\bar{c}} \int_0^\pi \frac{cC_\ell}{4b} \, d\theta \\
&= 2\!R \sum_{j=1}^{\infty} A_j \int_0^\pi \sin j\theta \sin\theta \, d\theta,
\end{aligned}
$$

where R is the aspect ratio from Section 1.2.2

$$R \equiv \frac{b}{\bar{c}}. \tag{11.16}$$

Using the trigonometric integrals (for integer m and n)

$$\int_0^\pi \sin m\theta \sin n\theta \, d\theta = \begin{cases} \frac{\pi}{2}, & m = n; \\ 0, & m \neq n, \end{cases}$$

the lift coefficient is simply

$$C_L = \pi \mathcal{R} A_1. \tag{11.17}$$

Rolling Moment

Similarly, the lift distribution Equation (11.13) could be used to compute the *rolling moment* (the component of the moment about the x-axis); of course, this vanishes when the lift distribution is symmetric:

$$c(-z)C_\ell(-z) = c(z)C_\ell(z).$$

Induced Drag Coefficient

Using Equation (11.9b) for the sectional induced drag coefficient, the wing's induced drag coefficient is

$$
\begin{aligned}
C_D &= \frac{D}{\frac{1}{2}\rho q_\infty^2 b\bar{c}} \\
&= \frac{\int_{-b/2}^{b/2} d \, dz}{\frac{1}{2}\rho q_\infty^2 b\bar{c}} \\
&= \frac{\int_{-b/2}^{b/2} \frac{1}{2}\rho q_\infty^2 cC_d \, dz}{\frac{1}{2}\rho q_\infty^2 b\bar{c}} \\
&= \frac{\int_{-b/2}^{b/2} cC_d \, dz}{b\bar{c}} \\
&= \frac{1}{b\bar{c}} \int_{-b/2}^{b/2} cC_l \alpha_i \, dz \\
&= \frac{4}{\bar{c}} \int_{-b/2}^{b/2} \frac{cC_l}{4b} \alpha_i \, dz \\
&= 2\mathcal{R} \int_0^\pi \frac{cC_l}{4b} \alpha_i \sin\theta \, d\theta \\
&= 2\mathcal{R} \int_0^\pi \left(\sum_n A_n \sin n\theta \right) \left(\sum_m m A_m \frac{\sin m\theta}{\sin\theta} \right) \sin\theta \, d\theta \\
&= 2\mathcal{R} \sum_m \sum_n m A_m A_n \int_0^\pi \sin m\theta \sin n\theta \, d\theta \\
&= 2\mathcal{R} \sum_n n A_n^2 \frac{\pi}{2};
\end{aligned}
$$

i.e.

$$C_D = \pi\!R \sum_n n A_n^2. \tag{11.18}$$

Thus while only the first term of Glauert's expansion (Equation 11.13) contributes to the lift, all terms induce drag; moreover, each term induces a positive amount of drag since squares are nonnegative.

11.5   The Elliptic Lift Loading

Solving the lifting line equation for the lift distribution is quite difficult, but an easier problem is the investigation of a given lift distribution. Since only the first term in Glauert's expansion as given by Equation (11.13) contributes to the lift, let's examine first the lift distribution just consisting of this term:

$$\frac{cC_\ell}{4b} = A_1 \sin\theta = \frac{C_L}{\pi\!R}\sin\theta = \frac{C_L}{\pi\!R}\sqrt{1 - \left(\frac{2z}{b}\right)^2},$$

or in dimensional terms

$$\ell = \frac{4L}{\pi b}\sqrt{1 - \left(\frac{2z}{b}\right)^2}, \tag{11.19}$$

as plotted in Figure 11.5. It is called the *elliptic lift loading*.

The corresponding induced incidence is given by the lifting line Equation (11.15) as

$$\alpha_i = \frac{C_L}{\pi\!R}, \tag{11.20}$$

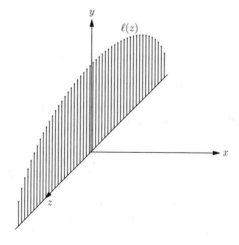

Figure 11.5   Elliptic lift loading, as given by Equation (11.19)

which is uniform across the span. The elliptic lift loading is therefore the only lift loading with this property.

11.5.1 Properties of the Elliptic Lift Loading

By Equation (11.18), an elliptically loaded wing's induced drag coefficient is

$$C_D = \pi \mathcal{R} A_1^2 = \pi \mathcal{R} \left(\frac{C_L}{\pi \mathcal{R}} \right)^2 = \frac{C_L^2}{\pi \mathcal{R}}, \tag{11.21}$$

Drag polars of Equation (11.21) are plotted in Figure 11.6.

Same Lift Coefficient, Different Aspect Ratio

If we take two elliptically loaded wings with different aspect ratios ($\mathcal{R}_1$ and $\mathcal{R}_2$) and vary their geometric incidences (α_1 and α_2) so that they have the same lift coefficients C_L, the induced incidences $\alpha_{i,1}$ and $\alpha_{i,2}$ must be related by

$$\alpha_{i,1} - \alpha_{i,2} = \frac{C_L}{\pi} \left(\frac{1}{\mathcal{R}_1} - \frac{1}{\mathcal{R}_2} \right).$$

This follows by forming Equation (11.20) for each wing and subtracting the two equations. Similarly, on forming Equation (11.21) and subtracting, the induced drag coefficients must be

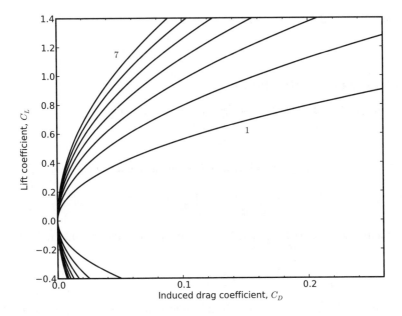

Figure 11.6 Theoretical drag–lift polars for elliptic lift loading and $\mathcal{R} = 1, 2, 3, \ldots, 7$. N.B.: drag includes only induced drag, not profile drag (skin friction and form drag); cf. Prandtl (1921, figure 46)

related by

$$C_{D,1} - C_{D,2} = \frac{C_L^2}{\pi}\left(\frac{1}{\mathcal{R}_1} - \frac{1}{\mathcal{R}_2}\right).$$

These two equations, given by Prandtl (1921), are extremely useful: they allow the prediction of the properties of a wing with one aspect ratio from measurements or computations of the properties at another aspect ratio. For example, (slightly generalized as discussed in Section 11.7.1) versions of these were used by Silverstein (1935) to transform experimental results at $\mathcal{R} = 6$ to an equivalent infinite aspect ratio basis.

Elliptic Lift Loading Minimizes Induced Drag

The induced drag coefficient for the general lift loading Equation (11.18) can be expressed as

$$C_D = \pi\mathcal{R}\sum_{n=1}^{\infty} nA_n^2 = \pi\mathcal{R}A_1^2\sum_{n=1}^{\infty} n\left(\frac{A_n}{A_1}\right)^2 = C_{D,\text{ell}}\left\{1 + \sum_{n=2}^{\infty} n\left(\frac{A_n}{A_1}\right)^2\right\},$$

where

$$C_{D,\text{ell}} = \pi\mathcal{R}A_1^2$$

is the induced drag coefficient of a wing with the same lift coefficient but with elliptic lift loading. Since the sum contains only nonnegative terms and vanishes for elliptic loading (for which A_1 is the only nonzero sine coefficient), a most important result follows:

The induced drag coefficient is a minimum for elliptic loading.

11.6   Lift–Incidence Relation

Say we know the lift–incidence relation for infinite aspect ratio:

$$C_{L,\infty} = f(\alpha_\infty).$$

Then for a finite aspect ratio, $\mathcal{R}$, with elliptic loading the induced incidence is

$$\alpha_i = \frac{C_L}{\pi\mathcal{R}},$$

and at geometric incidence α the lift coefficient is that for the infinite aspect ratio at geometric incidence

$$\alpha_\infty = \alpha - \alpha_i.$$

Since, for elliptic loading, the downwash and induced incidence are uniform along the span, if the wing is untwisted, the sectional lift coefficient C_ℓ will be too and the total lift coefficient C_L will change proportionally.

11.6.1   Linear Lift–Incidence Relation

If the infinite aspect ratio (two-dimensional) lift–incidence relation is linear,

$$C_L = m(\alpha - \alpha_i - \alpha_0),$$

but if the induced incidence is given by Equation (11.20) then

$$C_L = \frac{m}{1 + \dfrac{m}{\pi \mathcal{R}}}(\alpha - \alpha_0); \qquad (11.22)$$

thus, if the two-dimensional lift–incidence slope is m, the finite aspect ratio slope is

$$\frac{dC_L}{d\alpha} = \frac{m}{1 + \dfrac{m}{\pi \mathcal{R}}}. \qquad (11.23)$$

For $0 \leqslant \mathcal{R} \leqslant \infty$, this is less than m. Note also in Equation (11.22) that the zero-lift incidence α_0 is independent of aspect ratio. The result of Equation (11.22) for a thin aerofoil ($m = 2\pi$) is illustrated in Figure 11.7.

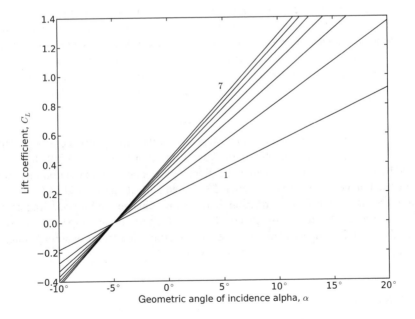

Figure 11.7   Theoretical lift–incidence relations for thin wings with zero-lift incidence $\alpha_0 = -5°$ and aspect ratio $\mathcal{R} = 1, 2, 3, \ldots, 7$, as indicated; cf. Prandtl (1921, figure 47)

11.7 Realizing the Elliptic Lift Loading

If the effective wing section lift coefficient is the same function of the effective incidence across the span (i.e. no twist), C_ℓ is uniform too:

$$C_\ell = C_L.$$

For this to be compatible with the elliptical lift loading, since

$$c(z) \equiv \frac{\ell(z)}{\frac{1}{2}\rho q_\infty^2 C_\ell},$$

the chord $c(z)$ must vary like Equation (11.19)

$$\ell(z) = \frac{2}{\pi}\rho q_\infty^2 \bar{c} C_L \sqrt{1 - \left(\frac{2z}{b}\right)^2};$$

i.e. elliptically

$$\frac{c(z)}{\bar{c}} = \frac{4}{\pi}\sqrt{1 - \left(\frac{2z}{b}\right)^2}.$$

11.7.1 Corrections to the Elliptic Loading Approximation

Practical wings are rarely constructed with an elliptic variation of chord length, since this is more expensive to manufacture than rectangular or trapezoidal planforms. Nevertheless, the simple results for elliptic loading are appealing and useful at least as a first approximation. They are frequently used in a generalized form, with correction factors. For example, in place of the elliptic loading induced drag coefficient Equation (11.21), Abbott and von Doenhoff (1959) recommend a corrected formula which for untwisted unswept wings reduces to

$$C_D = \frac{C_L^2}{\pi \mathcal{R} u}$$

where u is an 'induced-drag factor' which depends on the taper ratio and aspect ratio. The values of u may be read off the charts of Abbott and von Doenhoff (1959, figure 10), or computed by the methods in the Chapter 12. Rectangular-planform correction factors of this type were used in the report of Silverstein (1935) cited in Sections 1.6 and 11.5.1.

It should be noted that these correction factors don't change the induced drag coefficient by more than about a tenth over the practical range of taper ratio and aspect ratio (Anderson 2001), so that even the uncorrected formulae are sometimes useful for obtaining rough initial estimates.

11.8 Exercises

1. Derive an expression for the rolling moment, analogous to Equation (11.18) for the induced drag. Show that it does vanish when the spanwise distribution of lift is symmetric.

11.9   Further Reading

Lifting line theory is covered in most texts on aerodynamics; e.g. Glauert (1926), Jones (1942), Prandtl and Tietjens (1957), Abbott and von Doenhoff (1959), Karamcheti (1966), Batchelor (1967), Milne-Thomson (1973), Ashley and Landahl (1985), Thwaites (1987), Kuethe and Chow (1998), Paraschivoiu (1998), Bertin (2002), Moran (2003), and Anderson (2007).

References

Abbott, I.H. and von Doenhoff, A.E. (1959) *Theory of Wing Sections*. New York: Dover.

Anderson, J.D. (2001) *Fundamentals of Aerodynamics*, 3rd edn. New York: McGraw-Hill.

Anderson, J.D. (2007) *Fundamentals of Aerodynamics*, 4th edn. New York: McGraw-Hill.

Ashley, H. and Landahl, M. (1985) *Aerodynamics of Wings and Bodies*. New York: Dover.

Batchelor, G.K. (1967) *An Introduction to Fluid Dynamics*. Cambridge: Cambridge University Press.

Bertin, J.J. (2002) *Aerodynamics for Engineers*, 4th edn. Upper Saddle River, New Jersey: Prentice Hall.

Glauert, H. (1926) *The Elements of Aerofoil and Airscrew Theory*. Cambridge: Cambridge University Press.

Jones, B. (1942) *Elements of Practical Aerodynamics*, 3rd edn. New York: John Wiley & Sons, Ltd.

Karamcheti, K. (1966) *Principles of Ideal-Fluid Aerodynamics*. New York: John Wiley & Sons, Ltd.

Kuethe, A.M. and Chow, C.Y. (1998) *Foundations of Aerodynamics*, 5th edn. New York: John Wiley & Sons, Ltd.

Milne-Thomson, L.M. (1973) *Theoretical Aerodynamics*, 4th edn. New York: Dover.

Moran, J. (2003) *An Introduction to Theoretical and Computational Aerodynamics*. New York: Dover.

Paraschivoiu, I. (1998) *Aérodynamique Subsonique*. Éditions de l'École de Montréal.

Prandtl, L. (1921) Applications of Modern Hydrodynamics to Aeronautics. Report 116, NACA.

Prandtl, L. and Tietjens, O.G. (1957) *Applied Hydro- and Aeromechanics*. New York: Dover.

Silverstein, A. (1935) Scale effect on Clark Y airfoil characteristics from N.A.C.A. full-scale wind-tunnel tests. Report 502, NACA.

Thwaites, B. (ed.) (1987) *Incompressible Aerodynamics*. New York: Dover.

12

Nonelliptic Lift Loading

12.1 Solving the Lifting Line Equation

12.1.1 The Sectional Lift–Incidence Relation

The pair of Equations (11.15) and (11.13) giving the trigonometric expansions for the induced incidence and the spanwise lift loading can be closed by introducing a relation between C_ℓ and α_i, typically via the effective incidence $\alpha' = \alpha - \alpha_i$. This could be nonlinear, coming from experimental data, such as the graphs of Abbott *et al.* (1945) or Abbott and von Doenhoff (1959), or from theoretical models or numerical simulations, such as those developed in Chapters 6–8.

12.1.2 Linear Sectional Lift–Incidence Relation

Assume a linear sectional lift–incidence relation

$$C_\ell = m(\alpha' - \alpha_0) = m(\alpha - \alpha_i - \alpha_0),$$

rewrite this as

$$\frac{4b}{mc}\frac{cC_\ell}{4b} + \alpha_i = \alpha - \alpha_0$$

and the system becomes

$$\sum_{j=1}^{\infty} \left\{ \left(\frac{4b}{mc} + \frac{j}{\sin\theta} \right) \sin j\theta \right\} A_j = \alpha - \alpha_0. \qquad (12.1)$$

12.1.3 Finite Approximation: Truncation and Collocation

Now truncate the summation at $r - 1$ terms

$$\sum_{j=1}^{r-1} \left\{ \left(\frac{4b}{mc} + \frac{j}{\sin\theta} \right) \sin j\theta \right\} A_j = \alpha - \alpha_0. \qquad (12.2)$$

Theory of Lift: Introductory Computational Aerodynamics in MATLAB®/Octave, First Edition. G. D. McBain.
© 2012 John Wiley & Sons, Ltd. Published 2012 by John Wiley & Sons, Ltd.

and enforce the equation at $r - 1$ points, equally spaced in eccentric angle along the span,

$$\theta_i = \frac{i\pi}{r} \qquad (i = 1, 2, \ldots, r - 1), \tag{12.3}$$

to get the linear system

$$\sum_{j=1}^{r-1} \beta_{ij} A_j = (\alpha - \alpha_0)_i \qquad (i = 1, 2, \ldots, r - 1) \tag{12.4}$$

with coefficient matrix

$$\beta_{ij} = \left(\frac{4b}{m_i c_i} + \frac{j}{\sin \frac{i\pi}{r}} \right) \sin \frac{ij\pi}{r},$$

where m_i, c_i, α_i, and $(\alpha_0)_i$ are the values of the lift–incidence slope, chord, geometric incidence, and zero-lift incidence, respectively, at the spanwise station $\theta = \theta_i = i\pi/r$. These quantities are assumed to be known.

Just as in the lumped vortex method of Chapter 7, the idea of obtaining a numerical solution to an equation like Equation (12.2) by requiring it to be satisfied exactly at a finite number of discrete points is called *collocation*. Notice that the points given by Equation (12.3), illustrated in Figure 12.1 for $r = 8$, are the zeros of the first neglected mode:

$$\sin r\theta_i = \sin i\pi = 0 \qquad (i = 1, 2, \ldots, r - 1).$$

If the wing is untwisted, so that α is constant (no geometric twist) and α_0 is constant (no aerodynamic twist), the equation can be solved for unit aerodynamic incidence $\alpha - \alpha_0 = 1$, and results for other incidences obtained by scaling. This is equivalent to dividing each of

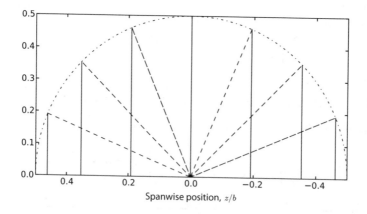

Figure 12.1 Construction of the collocation points by Equation (12.3) for $r = 8$

Listing 12.1 `lline.m`: Octave implementation of Glauert's method for lifting line equation. The chord lengths are normalized by span, and the loading sine coefficients by the aerodynamic incidence (which is assumed uniform).

```
function A = lline (c)
   r = length (c) + 1;
   J = 1:(r - 1);
   S = sin (J' * J * pi / r);
   B = bsxfun (@plus, 2/pi ./ c, 1 ./ S(:,1) * J) .* S;
   A = B \ ones (r - 1, 1);
```

the sine coefficients A_n by $\alpha - \alpha_0$. That is, for constant $\alpha - \alpha_0$, Equation (12.4) can be rewritten as

$$\sum_{j=1}^{r-1} \beta_{ij} \left(\frac{A_j}{\alpha - \alpha_0} \right) = 1 \qquad (i = 1, 2, \ldots, r - 1).$$

12.1.4 Computer Implementation

The system can be coded and solved in Octave. The special case of a thin wing, for which $m = 2\pi$ is given in Listing 12.1.

Once the expansion coefficients A_n have been obtained, the lift and induced drag coefficients can be computed using Equations (11.17) and (11.18), either for arbitrary incidence, or for a particular incidence. These are concisely implemented in Octave in Listing 12.2 and demonstrated in the following example.

Listing 12.2 `lline_CLCD`: Octave implementation of the lifting line theory Equations (11.17) and (11.18) for computing the lift and drag coefficients.

```
function [CL, CD] = lline_CLCD (AR, A)
   CL = pi * AR * A(1);
   CD = pi * AR * (1:length (A)) * A.^2;
```

12.1.5 Example: a Rectangular Wing

Kuethe and Chow (1998) computed the lift and induced drag coefficients for a thin ($m = 2\pi$) flat ($\alpha_0 = 0$) rectangular wing with $\mathcal{R} = 6$ using $r = 8$ as $C_L = 4.5273\alpha$ and

Listing 12.3 `rectlline.m`: Simple Octave function to specialize `lline.m` from Listing 12.1 to rectangular wings. The chord lengths are normalized by span, and the loading sine coefficients by the aerodynamic incidence (which is assumed uniform).

```
function [CL, CD] = rectlline (AR, r)
   [CL, CD] = lline_CLCD (AR, lline (ones (r-1, 1) / AR));
```

Listing 12.4 Octave code to calculate C_L and C_D for a thin untwisted tapered wing.

```
AR = 9; lambda = 0.4;
aalpha = (4 - (-1.2)) * (pi/180);
r = 8; theta = (1:(r-1))' * pi/r;
cr = 2 / AR / (1 + lambda);
c = cr * (1 - (1 - lambda) * (abs (cos (theta))));
A = lline (c);
[CL, CD] = lline_CLCD (AR, A);
CL * aalpha, CD * aalpha^2
```

$C_D = 1.1378\alpha^2$. These results can be reproduced using `rectlline.m` from Listing 12.3, with which `[CL, CD] = rectlline (6, 8)` returns `CL=4.5273` and `CD=1.1378`.

In comparison, the elliptic loading approximation of Equation (11.23) would predict a lift–incidence slope of

$$\frac{m}{1 + \frac{m}{\pi R}} = \frac{2\pi}{1 + \frac{2}{6}} = \frac{3\pi}{2} \doteq 4.712\dots.$$

Thus for the same absolute incidence $\alpha - \alpha_0$, the rectangular wing would generate nearly 4% less lift than an elliptically loaded wing with the same planform area and aspect ratio.

Bertin (2002) computed the lift and induced drag for a thin cambered ($\alpha_0 = -1.2°$) untwisted tapered wing with $R = 9$ and taper ratio $\lambda \equiv c_t/c_r = 0.4$. Using $r = 8$, he found that at geometric incidence $\alpha = 4°$ the lift and induced drag coefficients are $C_L = 0.4654$ and $C_D = 0.00776$.

The chord length distribution for a tapered wing can be calculated from:

$$\frac{c}{b} = \frac{2\left\{1 - (1 - \lambda)\left|\frac{2z}{b}\right|\right\}}{R(1 + \lambda)}.$$

Using this and `lline.m` (Listing 12.1), the results ($C_L \doteq 0.46538$ and $C_D \doteq 0.0077661$) are reproduced with the code in Listing 12.4.

12.2 Numerical Convergence

For quick hand calculations, $r = 8$ is usually regarded as sufficient (Glauert 1926; Milne-Thomson 1973; Kuethe and Chow 1998) though $r = 20$ was conventional in NACA work (Sivells and Neely 1947; Abbott and von Doenhoff 1959).

The method was applied to a thin untwisted rectangular wing of aspect ratio 6 for increasing r and the absolute relative error in the aerodynamic force coefficients plotted in Figure 12.2. The exact answer is not known, so the error is estimated by taking the values at $r = 200$ as reference; i.e. $C_L/\alpha \doteq 4.530424981$ and $C_D/\alpha^2 = 1.141453135$.

Note that the drag coefficient converges slightly more slowly than the lift coefficient, since it depends on all of the sine coefficients and moreover gives an increasing weighting to sine coefficients of higher order.

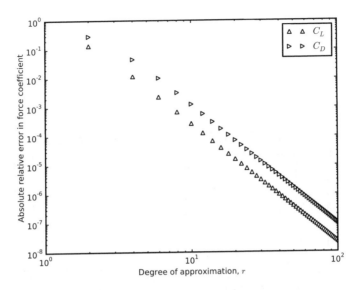

Figure 12.2 Convergence plot for the lift coefficient predicted by Glauert's method for a thin untwisted rectangular wing of aspect ratio 6

12.3 Symmetric Spanwise Loading

The above method works for an arbitrary spanwise loading, but very often (as in the two examples of Section 12.1.5) we are most interested in spanwise symmetric cases. As may be inferred from the graphs of the first few shape functions from Glauert's expansion Equation (11.13) in Figure 12.3, in these cases, only the odd terms in Glauert's expansion are required:

$$\frac{cC_\ell}{4b} = \sum_{j=1}^{\infty} A_{2j-1} \sin(2j-1)\theta.$$

Exploiting this leads to a considerable reduction in work. The number of arithmetic operations required to solve a linear system is proportional to the cube of the number of equations; therefore, halving the number means an eighth the work. This is particularly important for hand calculations.

Analogous to Equations (11.14), (11.15), and (12.1), we have

$$\frac{d}{d\theta} \frac{cC_\ell}{4b} = \sum_{j=1}^{\infty} (2j-1)A_{2j-1} \cos(2j-1)\theta$$

$$\alpha_i = \sum_{j=1}^{\infty} \frac{(2j-1)A_{2j-1} \sin(2j-1)\theta}{\sin \theta}$$

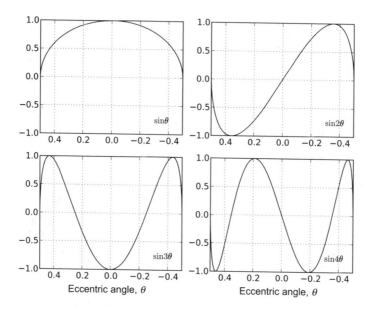

Figure 12.3 The first few modes of Glauert's expansion as given by Equation (11.13) for the spanwise lift loading on the span $\frac{b}{2} > z > \frac{-b}{2}$, showing the even and odd symmetry of the odd and even modes respectively

and

$$\sum_{j=1}^{\infty} \left\{ \left(\frac{4b}{mc} + \frac{2j-1}{\sin\theta} \right) \sin(2j-1)\theta \right\} A_{2j-1} = \alpha - \alpha_0,$$

respectively.

For even r, we truncate the series at $j = r/2$ (i.e. $2j - 1 = r - 1$) and enforce the equation at

$$\theta_i = \frac{i\pi}{r}, \qquad (i = 1, 2, \ldots, r/2).$$

These collocation points lie on the right wing (with the last on the plane of symmetry, $z = 0$). The discrete linear system is then

$$\sum_{j=1}^{r/2} \beta_{ij}^{(s)} A_{2j-1} = (\alpha - \alpha_0)_i$$

where the coefficients are now

$$\beta_{ij}^{(s)} = \left(\frac{4b}{m_i c_i} + \frac{2j-1}{\sin\frac{i\pi}{r}} \right) \sin\frac{(2j-1)i\pi}{r}.$$

The Octave function lline.m (Listing 12.1) is easily modified for this special case, as in Listing 12.5.

Listing 12.5 `lline_symmetric`: modification of `lline.m` (Listing 12.1) for the special case of symmetric spanwise lift loading.

```
function [A, B] = lline_symmetric (c)
  n = length (c);
  I = 1:n;
  J = 2 * I - 1;
  S = sin (I' * J * pi / 2 / n);
  B = bsxfun (@plus, 2/pi ./ c, 1 ./ S(:,1) * J) .* S;
  A = [B \ ones(n, 1), zeros(n, 1)]';
  A = A(:);
```

12.3.1 Example: Exploiting Symmetry

For the thin untwisted rectangular wing of aspect ratio 6 from Section 12.1.5, with $r = 8$, the coefficient matrix is

$$
B^{(s)} = \begin{bmatrix}
+2.4617 & +10.7716 & +15.6000 & +8.4617 \\
+3.7009 & +5.7009 & -7.7009 & -9.7009 \\
+4.5290 & -2.7044 & -3.5328 & +10.5290 \\
+4.8197 & -6.8197 & +8.8197 & -10.8197
\end{bmatrix},
$$

so that the linear system for arbitrary absolute incidence $\alpha - \alpha_0$ is

$$
\begin{bmatrix}
+2.4617 & +10.7716 & +15.6000 & +8.4617 \\
+3.7009 & +5.7009 & -7.7009 & -9.7009 \\
+4.5290 & -2.7044 & -3.5328 & +10.5290 \\
+4.8197 & -6.8197 & +8.8197 & -10.8197
\end{bmatrix}
\begin{Bmatrix} A_1 \\ A_3 \\ A_5 \\ A_7 \end{Bmatrix}
= (\alpha - \alpha_0) \begin{Bmatrix} 1 \\ 1 \\ 1 \\ 1 \end{Bmatrix}
$$

which has the solution

$$
\begin{Bmatrix} A_1 \\ A_3 \\ A_5 \\ A_7 \end{Bmatrix}
= (\alpha - \alpha_0) \begin{Bmatrix} 0.2401797 \\ 0.0288983 \\ 0.0057044 \\ 0.0010011 \end{Bmatrix}.
$$

The predicted force coefficients are then

$$
C_L = \pi \mathcal{R} A_1 \doteq 4.527(\alpha - \alpha_0)
$$
$$
C_D = \pi \mathcal{R} \left(A_1^2 + 3A_3^2 + 5A_5^2 + 7A_7^2 + \cdots \right)(\alpha - \alpha_0)^2 \doteq 1.138(\alpha - \alpha_0)^2.
$$

Both coefficients are correct to three significant figures.

In this case, the magnitude of the sine coefficients A_n decreases rapidly with increasing n. This means that the spanwise loading is mostly elliptical (the contribution from A_1), even though the wing planform is rectangular. Also, it suggests that the truncation at $r = 8$ is reasonable (the omitted sine coefficients are probably smaller still).

For $r = 4$,

$$\mathbf{B}^{(s)} = \begin{bmatrix} 3.7009 & 5.7009 \\ 4.8197 & -6.8197 \end{bmatrix},$$

$$\begin{Bmatrix} A_1 \\ A_3 \end{Bmatrix} = (\alpha - \alpha_0) \begin{Bmatrix} 0.237510 \\ 0.021222 \end{Bmatrix},$$

and $C_L \doteq 4.477(\alpha - \alpha_0)$ and $C_D \doteq 1.089(\alpha - \alpha_0)^2$; so the coefficients are still correct to two significant figures.

12.4   Exercises

1. Show that the numerical results of Section 12.3.1 can be obtained with the Octave code in Listing 12.6.

Listing 12.6   Octave code for Exercise 12.4.1.

```
AR = 6;
for r = [8, 4]
   [A, B] = lline_symmetric (ones (r/2, 1) / AR)
   [CL, CD] = lline_CLCD (AR, A)
end%for
```

References

Abbott, I.H. and von Doenhoff, A.E. (1959) *Theory of Wing Sections*. New York: Dover.
Abbott, I.H., von Doenhoff, A.E. and Stivers, L.S. (1945) Summary of airfoil data. Report 824, NACA.
Bertin, J.J. (2002) *Aerodynamics for Engineers*, 4th edn. Upper Saddle River, New Jersey: Prentice Hall.
Glauert, H. (1926) *The Elements of Aerofoil and Airscrew Theory*. Cambridge: Cambridge University Press.
Kuethe, A.M. and Chow, C.Y. (1998) *Foundations of Aerodynamics*, 5th edn. New York: John Wiley & Sons, Ltd.
Milne-Thomson, L.M. (1973) *Theoretical Aerodynamics*, 4th edn. New York: Dover.
Sivells, J.C. and Neely, R.H. (1947) Method for Calculating Wing Characteristics by Lifting-Line Theory using Nonlinear Section Lift Data. Report 865, NACA.

13

Lumped Horseshoe Elements

Chapter 7 on lumped vortex elements stepped back from the infinite trigonometric series expansions of thin aerofoil theory in Chapter 6 to see how much of the mechanics of a wing section could be mimicked with a single lumped vortex of the type previously encountered in the preliminary fundamental work on plane ideal flow in Chapter 3, and then extended this by spacing two and then several lumped vortices along the camber line. In this chapter, an analogous approach is taken to the three-dimensional aerodynamics of wings of finite span; that is, a step back from the infinite series expansions of lifting line theory in Chapters 11 and 12 to see what can be done with one or a few discrete horseshoe vortices of the type discussed in the preliminary fundamental work on three-dimensional ideal flow in Chapter 10.

13.1   A Single Horseshoe Vortex

Recall from Equation (11.3) that a single horseshoe vortex of strength Γ with bound segment

$$\frac{-b_h}{2} < z < \frac{b_h}{2}$$

and parallel free vortices running from and to $+\infty \mathbf{i}$, as drawn in Figure 13.1, induces the velocity at the origin, the midpoint of its bound vortex,

$$\boldsymbol{q}_h(0) = \frac{-\Gamma \mathbf{j}}{\pi b_h}. \tag{13.1}$$

Note that this discrete horseshoe vortex has been given its own breadth b_h, which may not necessarily coincide with that of the wing it represents.

Let this single discrete horseshoe vortex be placed in a uniform stream at incidence α,

$$\boldsymbol{q}_\infty = q_\infty \cos \alpha \mathbf{i} + q_\infty \sin \alpha \mathbf{j},$$

then the total velocity at the origin is

$$\boldsymbol{q}(0) = q_\infty \cos \alpha \mathbf{i} + \left(q_\infty \sin \alpha - \frac{\Gamma}{\pi b_h} \right) \mathbf{j}.$$

Theory of Lift: Introductory Computational Aerodynamics in MATLAB®/Octave, First Edition. G. D. McBain.
© 2012 John Wiley & Sons, Ltd. Published 2012 by John Wiley & Sons, Ltd.

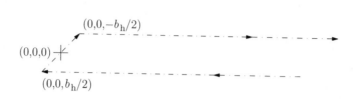

$(0,0,-b_h/2)$

$(0,0,0)$

$(0,0,b_h/2)$

Figure 13.1 Single horseshoe vortex of span b_h and semi-infinite legs in the x-direction. A collocation point is marked at the midpoint of the central finite segment

If the wing, as represented by this lifting line lumped as a single horseshoe vortex, is to be impermeable, the y-component of this must vanish, so we have one relation for the strength of the horseshoe vortex:

$$\Gamma = \pi b_h q_\infty \sin \alpha.$$

In the context of lifting line theory, only small angles of incidence are entertained so again using Equation (5.8) for the sine,

$$\Gamma \sim \pi b_h q_\infty \alpha. \tag{13.2}$$

The lift per unit spanwise section of the horseshoe is given by the Kutta–Joukowsky Theorem of Equation (3.15a) as $\ell = \rho q_\infty \Gamma$. Integrating this along the breadth of the bound vortex gives the total lift as

$$L = \rho b_h q_\infty \Gamma = \pi \rho b_h^2 q_\infty^2 \sin \alpha. \tag{13.3}$$

Now let's try and relate this to a wing with elliptic spanwise loading; from Equation (11.22),

$$C_L \equiv \frac{2L}{\rho q_\infty^2 b \bar{c}} = \frac{m}{1 + \frac{m}{\pi \mathcal{R}}}(\alpha - \alpha_0)$$

so

$$L = \frac{\rho q_\infty^2 b \bar{c} m (\alpha - \alpha_0)}{2\left(1 + \frac{m}{\pi \mathcal{R}}\right)} \tag{13.4}$$

is the lift for the wing to be modelled. Equating Equation (13.4) to the Equation (13.3) of the lift of the horseshoe, for:

- $\alpha_0 = 0$, since the horseshoe can only model wings without camber, and
- $m = 2\pi$, as predicted by thin aerofoil theory in Equation (6.11)

yields

$$b_h = \sqrt{\frac{b^2 \bar{c}}{b + 2\bar{c}}}. \tag{13.5}$$

Notice that the breadth of the bound vortex segment has to take into account the chord as well as the span; the former was hidden in the lifting line theory, but its influence remained in the normalization used for the lift coefficient and so explicitly reappeared once we reverted from lift coefficient to lift.

13.1.1   Induced Incidence of the Lumped Horseshoe Element

The induced incidence for this lumped horseshoe element is given as usual by Equation (11.7), so at the collocation point, using Equation (13.1) and Equation (13.2),

$$\alpha_i = \frac{-v_w}{q_\infty} = \frac{\Gamma}{\pi b_h q_\infty}$$
$$= \frac{\pi b_h q_\infty \alpha}{\pi b_h q_\infty}$$
$$= \alpha.$$

The simplicity of this result is not a coincidence, it is necessary; that is, for a single lumped horseshoe to provide sufficient circulation to stop the free-stream passing through the wing, it must provide an induced incidence exactly equal to the geometric incidence. This is just as in Section 7.1.3 where the two-dimensional lumped vortex element repredicted the correct circulation after it had been calibrated by choosing its collocation point.

13.2   Multiple Horseshoes along the Span

If two horseshoe vortices of equal strength were placed together, as shown in Figure 13.2, so that their two bound vortex segments were collinear, their four semi-infinite rectilinear trailing

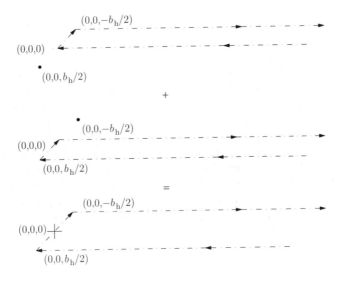

Figure 13.2   The inboard trailing vortices of two identical discrete horseshoe vortices abreast mutually cancel, leaving a single discrete horseshoe

vortices parallel and coplanar, and the right side of the left one coincided with the left side of the right, then the two inside trailing vortex legs would mutually cancel and the two bound vortex segments would coalesce into one of twice the span; the two outside vortices legs would be unaffected. That is to say, if the first horseshoe had corners $\left(0, 0, \frac{b}{2}\right)$ and $(0, 0, 0)$ and the second $(0, 0, 0)$ and $\left(0, 0, \frac{-b}{2}\right)$, while the four legs either start from or tend towards $(\infty, 0, 0)$, then their effective sum would have corners $\left(0, 0, \frac{b}{2}\right)$ and $\left(0, 0, \frac{-b}{2}\right)$ and the same infinite origin and terminus.

This cancellation means that the model has not gained any degrees of freedom and does not really refine the resolution of the spanwise lift loading; however, we might ask ourselves to what extent it models the flight of two planes abreast, tip-to-tip.

To avoid this cancellation, let's try adding abreast two of the single horseshoe vortex models from Section 13.1. Since these try to model elliptic loading, which is a loading that vanishes at the wing-tips (since the pressure above and below must agree there where the wing terminates), their bound vortex segments don't extend all the way across the span; therefore, their two inside trailing vortices will not exactly cancel. To be specific, say the left horseshoe represents a wing spanning $b > z > 0$ along the z-axis and the right $0 > z > -b$, then their bound segments each have breadth b as given by Equation (13.5) (and depending also on the mean chord or aspect ratio). This is perhaps a better model of flight abreast, but not a better model of a broader wing; i.e. does not provide a more realistic picture of the spanwise loading.

What if we placed side-by-side along the z-axis three or four horseshoes, symmetrically about the x-axis, but possibly permitting the strength of the inner one or two to differ from that of the outer two? If two parallel coplanar horseshoes of different strength are placed exactly side by side, the coincident trailing vortices will only partially cancel, and the resulting combined bound vortex segment will have a finite step in its strength. This is starting to get interesting: this begins to have degrees of freedom; this begins to permit modelling of spanwise loading—this is, of course, the beginnings of a finite analogue of the continuous lifting line of Chapter 11.

For a symmetric spanwise loading, the use of three and four horseshoes is really equivalent, since if four, the middle pair will have equal strengths and so are indistinguishable from a single horseshoe of twice the span. Given the equivalence, we henceforth consider only even numbers of horseshoes, because this leads more simply to a treatment of such symmetrical cases: we can then consider only those horseshoes of one half (with influences augmented by their missing mirror images), but do not have to split a central horseshoe straddling the plane of symmetry.

So, consider then a row of four horseshoe vortices with collinear bound vortex segments end to end along the z-axis representing a wing with spanwise symmetric loading. How many degrees of freedom does this system have? There are the two vortex strengths (inboard and outboard), and the two breadths (again inboard and outboard). If the total breadth of the four is fixed as the span, that reduces the number of parameters to three. As an aside, note that this four-horseshoe model also includes as a special case the single-horseshoe model of Section 13.1, since it reduces to it if the strengths of the two outboard vortices are set to zero. Actually though, the way the vortex strengths should be set is so that the downwash at the collocation points balances the oncoming free-stream. That supplies exactly as many conditions as there are horseshoes, and so only the choice of horseshoe breadths remains indeterminate.

A simple choice is not to exploit this freedom to improve the model but just choose something convenient, like equal breadths, or perhaps panels with equal breadth in terms not of the spanwise Cartesian coordinate z but the eccentric angle θ of Equation (11.11) and Figure 11.4.

13.2.1 A Finite-step Lifting Line in Octave

To get started with implementing horseshoe vortices in Octave, let's consider just this problem of a row of equal-breadth horseshoes as a finite-step representation of a lifting line, as shown in Figure 13.3. Say we have n horseshoes, with the jth running from $z_j^{(v)}$ to $z_{j+1}^{(v)}$ where

$$\frac{z_j^{(v)}}{b} \equiv \frac{1}{2} - \frac{j-1}{n} \qquad (j = 1, 2, \ldots, n+1) \tag{13.6}$$

so that $z_1^{(v)}$ and $z_{n+1}^{(v)}$ correspond to the left and right wing-tips, respectively. The n collocation points at the midpoints of the bound vortex segments bisect these:

$$\frac{z_i^{(c)}}{b} \equiv z_i^{(v)} - \frac{1}{2n} \qquad (i = 1, 2, \ldots, n). \tag{13.7}$$

The downwash induced by the semi-infinite vortex segment running back from the jth point of Equation (13.6) at the ith collocation point of Equation (13.7) can be deduced from Equation (10.9) as

$$\frac{\Gamma_j - \Gamma_{j-1}}{4\pi \left\{ z_i^{(c)} - z_j^{(v)} \right\}}. \tag{13.8}$$

The difference in the numerator here is to be understood as simply Γ_1 for $j = 1$ and $-\Gamma_n$ for $j = n+1$, i.e. as if $\Gamma_0 = \Gamma_{n+1} = 0$. This vector of $n + 1$ unknown trailing vortex strengths

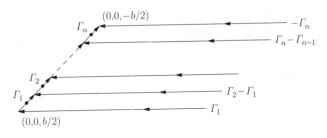

Figure 13.3   Finite-step lifting-line model of Section 13.2.1, using n lumped horseshoe vortex elements of equal breadth but different strengths to be determined by collocation at the midpoints of their bound vortex segments on the z-axis

Listing 13.1 Demonstration of the use Octave's **function toeplitz** to generate the bidiagonal matrix converting n horseshoe strengths to $n + 1$ trailing vortex strengths, as in Equation (13.9). See Listing 13.2 for the output.

```
n = 4; toeplitz ([1; -1; zeros(n-1, 1)], [1, zeros(1, n-1)])
```

$\{\Gamma_j - \Gamma_{j-1}\}_{j=1}^{n+1}$ can be constructed from the vector of n unknown horseshoe vortex strengths $\{\Gamma_j\}_{j=1}^{n}$ by a matrix–vector product:

$$
\begin{bmatrix}
1 & 0 & \cdots & & & \\
-1 & 1 & 0 & \cdots & & \\
0 & -1 & 1 & 0 & & \\
 & & & & \cdots & \\
\vdots & \ddots & \ddots & \ddots & & \\
 & & & 0 & -1 & 1 \\
 & & & & 0 & -1
\end{bmatrix}
\left\{
\begin{array}{c}
\Gamma_1 \\ \Gamma_2 \\ \vdots \\ \Gamma_n
\end{array}
\right\}
=
\left\{
\begin{array}{c}
\Gamma_1 \\ \Gamma_2 - \Gamma_1 \\ \vdots \\ -\Gamma_n
\end{array}
\right\}.
\tag{13.9}
$$

The bidiagonal $(n + 1) \times n$ matrix here, with its two constant nonzero diagonals, is a *Toeplitz* matrix, which Octave can generate from its first column and row with the **zeros** and **toeplitz** functions; e.g. as in Listing 13.1 for $n = 4$, which produces the output shown in Listing 13.2.

The rest of the influence coefficient is very similar to that appearing in the matrix–vector Equation (7.3) for the lumped vortex element method and so can be constructed in Octave in the same way.

The condition to be enforced is that the combined downwash at the collocation points balances the vertical component of the free-stream, i.e. $q_\infty \sin \alpha \sim q_\infty \alpha$, if camber and twist are neglected.

Note that this is already implying that the downwash will be uniform along the span, which is known to be a property of elliptic loading—see Equation (11.20). The present model is very highly simplified and in particular hasn't any resolution along the chord, so it doesn't have any way of representing taper or twist; recall from Section 11.6 that these were brought back in to the classical lifting line model via the external relation between the sectional lift coefficient and the effective incidence. In the discrete context, information on the planform is better incorporated by also discretizing in the chordwise direction, along the lines of the lumped vortex method of Chapter 7 in two dimensions. That is taken up in shortly in Section 13.3 and

Listing 13.2 Octave output from Listing 13.1; cf. the Toeplitz matrix in Equation (13.9).

```
ans =

    1    0    0    0
   -1    1    0    0
    0   -1    1    0
    0    0   -1    1
    0    0    0   -1
```

then seriously in Chapter 14, but for the moment let us complete the implementation of the finite-step lifting line with a single spanwise row of horseshoes.

Without twist or camber, the system of equations is

$$\frac{1}{4\pi} \sum_{j=1}^{n} a_{ij} \Gamma_j = -q_\infty b\alpha \qquad (i = 1, 2, \ldots, n)$$

where the matrix of influence coefficients is defined by Equation (13.8) (with the constant factor b moved here to the right-hand side)

$$\begin{bmatrix} \frac{1}{z_1^{(c)}-z_1^{(v)}} & \frac{1}{z_1^{(c)}-z_2^{(v)}} & \cdots & \frac{1}{z_1^{(c)}-z_{n+1}^{(v)}} \\ \frac{1}{z_2^{(c)}-z_1^{(v)}} & \frac{1}{z_2^{(c)}-z_2^{(v)}} & \cdots & \frac{1}{z_2^{(c)}-z_{n+1}^{(v)}} \\ & & \ddots & \\ \frac{1}{z_n^{(c)}-z_1^{(v)}} & \frac{1}{z_n^{(c)}-z_2^{(v)}} & \cdots & \frac{1}{z_n^{(c)}-z_{n+1}^{(v)}} \end{bmatrix} \begin{bmatrix} 1 & 0 & \cdots & \\ -1 & 1 & 0 & \cdots \\ 0 & -1 & 1 & 0 & \cdots \\ \vdots & \ddots & \ddots & \ddots \\ & & 0 & -1 & 1 \\ & & & 0 & -1 \end{bmatrix}$$

and computed by the product of a matrix with the above Toeplitz matrix. A complete stand-alone Octave function to solve this problem for a lifting line with span divided equally into n horseshoes is given in Listing 13.3; since the right-hand side is constant, the computed vortex strengths are those for $q_\infty\alpha = 1$ and the full values can be recovered by subsequent rescaling.

The spanwise variation predicted by the model can be inspected by plotting the computed normalized horseshoe vortex strengths at their collocation points, which produces Figure 13.4, which may be compared with the exact elliptic lift loading in Figure 11.5.

The normalized horseshoe vortex strengths calculated in Listing 13.3 can be related to the sectional lift coefficient by the Kutta–Joukowsky theorem in the usual way:

$$\frac{\Gamma}{q_\infty b\alpha} = \frac{\ell}{\rho q_\infty^2 b\alpha} = \frac{cC_\ell}{2b\alpha},$$

where c is the chord (which reappears here via the defintion of C_ℓ). Thus the lift coefficient can be recovered from

$$\frac{C_L}{A\!R\alpha} = \frac{2}{n} \sum_{j=1}^{n} \frac{\Gamma_j}{q_\infty b\alpha}.$$

Listing 13.3 `line_horseshoe.m`.

```
function [Gamma, zc] = line_horseshoe (n)
  zv = 0.5 - (0:n) / n;
  zc = zv(1:n)' - 0.5 / n;
  A = 0.25 ./ bsxfun (@minus, zc, zv) / pi ...
      * toeplitz ([1; -1; zeros(n-1, 1)], [1, zeros(1, n-1)]);
  Gamma = A \ -ones (n, 1);
```

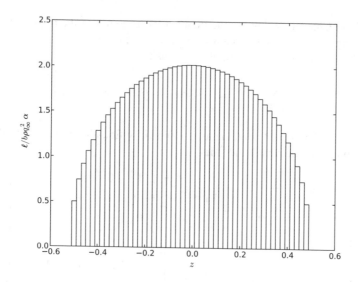

Figure 13.4 The spanwise loading predicted by a row of $n = 50$ horseshoe vortices of equal breadth, as implemented in Listing 13.3

Listing 13.4 Demonstrating the convergence of the lift coefficient predicted by **function** line_horseshoe from Listing 13.3 in comparison with Equation (11.20).

```
for n = 2 .^ (1:9)
  C = 2 * mean (line_horseshoe (n));
    fprintf ('%3d  %5.3f  %5.3f\n', n, C, C - pi)
end% for
```

According to Equation (11.20) of the continuous lifting-line theory, the quantity on the left-hand side here is π for elliptic loading. The present numerical result does approach π as n is increased, but only linearly (i.e. the product of n and the error is roughly constant as n is increased) and about a hundred horseshoe vortices are required to get within 1% of the exact answer. This is demonstrated in Listing 13.4, which was used to produce Table 13.1.

13.3 An Improved Discrete Horseshoe Model

A suggestion for improving the single horseshoe model developed in Section 13.1 may have occurred to the reader of Chapter 7: even though the lifting-line model was designed for wings of high aspect ratio, so that the chord is insignificant in comparison with the span, why not impose the condition of impermeability at the three-quarter chord point instead of right at the bound vortex? Correspondingly, if the chord is not to be neglected altogether, the bound vortex might be put along the quarter-chord line. This is illustrated in Figure 13.5.

Thus, let us shift the bound vortex from $x = 0$ to $x = \frac{c}{4}$, where c is the chord, the horseshoe vertices to $\left(\frac{c}{4}, 0, \pm\frac{b}{2}\right)$, and the collocation point to $\left(\frac{3c}{4}, 0, 0\right)$. We need to calculate the velocity

Table 13.1 Convergence of the numerical single row of equal-breadth horseshoe vortices

n	$\dfrac{C_L}{\mathcal{R}\alpha}$	$\dfrac{C_L}{\mathcal{R}\alpha} - \pi$
2	4.712	1.571
4	3.927	0.785
8	3.534	0.393
16	3.338	0.196
32	3.240	0.098
64	3.191	0.049
128	3.166	0.025
256	3.154	0.012
512	3.148	0.006

Source: Using Listing 13.4.

induced at the collocation point by the three legs of the horseshoe vortex. Equation (10.9) cannot be used for the two semi-infinite legs since the collocation point is no longer in their terminal normal plane; instead we need to go back to the general Equation (10.6).

For the incoming leg, we have $\hat{\ell} = -\mathbf{i}$, $r = \frac{3c}{4}\mathbf{i}$, and, say, $r' = r + \frac{b}{2}\mathbf{k}$ if for convenience we take the representative point r' on the vortex line as the one closest to the collocation point. Thus

$$r - r' = \frac{-b}{2}\mathbf{k}$$

$$\hat{\ell} \times (r - r') = \frac{-b}{2}\mathbf{j}$$

$$\frac{\hat{\ell} \times (r - r')}{|\hat{\ell} \times (r - r')|^2} = \frac{-2}{b}\mathbf{j}$$

$$\cos \beta_1 = 1$$

$$\cos \beta_2 = \frac{-1}{\sqrt{1 + \left(\frac{b}{c}\right)^2}}$$

$$q(r) = \frac{-\Gamma\mathbf{j}}{2\pi b}\left(1 + \frac{1}{\sqrt{1 + \mathcal{R}^2}}\right),$$

where we have introduced the aspect ratio $\mathcal{R} \equiv \frac{b}{c}$.

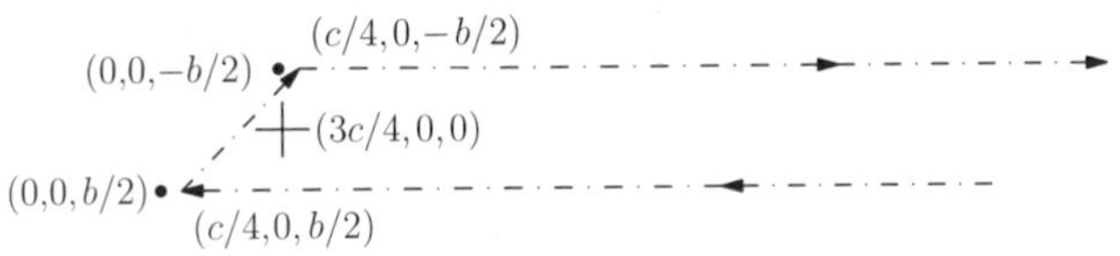

Figure 13.5 Modification of the single lumped horseshoe of Figure 13.1 by separating and shifting the bound vortex segment and the collocation point chordwise

The outgoing semi-infinite segment contributes the same, by symmetry.
For the bound vortex segment we have $\hat{\ell} = -\mathbf{k}$, and

$$r - r' = \frac{c}{2}\mathbf{i}$$

$$\hat{\ell} \times (r - r') = \frac{-c}{2}\mathbf{j}$$

$$\frac{\hat{\ell} \times (r - r')}{|\hat{\ell} \times (r - r')|^2} = \frac{-2\mathbf{j}}{c}$$

$$\cos \beta_1 = \frac{\mathcal{R}}{\sqrt{\mathcal{R}^2 + 1}}$$

$$\cos \beta_2 = -\cos \beta_1 \qquad \text{(by symmetry)}$$

$$q(r) = \frac{-\Gamma \mathbf{j}}{\pi c}\left(\frac{\mathcal{R}}{\sqrt{\mathcal{R}^2 + 1}}\right).$$

Adding the three contributions at the three-quarter chord point, which are all pure down-washes, gives:

$$q(r) = \frac{-\Gamma}{\pi b}\left(1 + \frac{1}{\sqrt{1 + \mathcal{R}^2}}\right)\mathbf{j} + \frac{-\Gamma}{\pi c}\left(\frac{\mathcal{R}}{\sqrt{\mathcal{R}^2 + 1}}\right)\mathbf{j}$$

$$= \frac{-\Gamma}{\pi c}\left\{\frac{1}{\mathcal{R}} + \sqrt{\frac{1}{\mathcal{R}^2} + 1}\right\}\mathbf{j}$$

The oncoming free-stream has velocity $q_\infty(\mathbf{i}\cos\alpha + \mathbf{j}\sin\alpha)$, so the circulation required to balance that is

$$\Gamma = \frac{\pi q_\infty c\alpha}{\mathcal{R}^{-1} + \sqrt{\mathcal{R}^{-2} + 1}}.$$

This tends to the two-dimensional flat-plate result for the circulation around a given by Equation (5.20), $\Gamma = \pi c q_\infty \alpha$, as the wingspan tends to infinity, as it should.

Notice that by introducing the chord into the model by shifting the collocation point off the bound vortex, we have not had to import the two-dimensional Kutta–Joukowsky theorem to determine the circulation.

Further multiplying this circulation by ρq_∞ to get the lift per unit span and then b to get the total lift force, and finally normalizing by $\rho q_\infty bc/2$ to get the lift coefficient, we get

$$\ell = \frac{\pi \rho q_\infty^2 c\alpha}{\mathcal{R}^{-1} + \sqrt{\mathcal{R}^{-2} + 1}}$$

$$L = \frac{\pi \rho q_\infty^2 bc\alpha}{\mathcal{R}^{-1} + \sqrt{\mathcal{R}^{-2} + 1}}$$

$$C_L = \frac{2\pi\alpha}{\mathcal{R}^{-1} + \sqrt{\mathcal{R}^{-2} + 1}}.$$

Expressed as the ratio of the lift–incidence slope at finite aspect ratio to that at infinity, this is

$$\frac{C_L(\mathcal{R})}{C_L(\infty)} = \frac{1}{\mathcal{R}^{-1} + \sqrt{\mathcal{R}^{-2} + 1}}$$

which is of the same form as Helmbold's equation (Anderson 2007),

$$\frac{\mathrm{d}C_L(\mathcal{R})/\mathrm{d}\alpha}{\mathrm{d}C_L(\infty)/\mathrm{d}\alpha} = \frac{1}{\frac{m}{\pi\mathcal{R}} + \sqrt{\left(\frac{m}{\pi\mathcal{R}}\right)^2 + 1}},$$

which was formerly much used as a quick approximation for the effect of aspect ratio on lift–incidence slope for small aspect ratios ($\mathcal{R} < 4$) instead of Equation (11.23) from lifting line theory.

13.4   Implementing Horseshoe Vortices in Octave

The calculation via the formulae of Section 10.5.2 of the velocity induced by a polygonal chain of finite or infinite rectilinear vortex segment is straightforward but tedious and therefore ideally suited to automatic computation.

Once the semi-infinite leading and trailing segments have been handled, as discussed in Section 10.5.6, either by truncation or with proper formulae, the only special cases to watch for are any evaluation points collinear with inducing segments. This may be handled by first computing the cross product $\hat{\ell} \times (r - r')$ of the unit vector along the segment and the vector from any convenient point on the segment to the point in question, this cross product being required in any case to compute the velocity, and then checking that the magnitude (or square magnitude) is sufficiently above zero; that it should be zero would mean that the cross product has vanished which means collinearity. With an eye to subsequent developments (Chapter 14), the subroutine should be able to accept multiple segments and evaluation points.

Say we have p points at which to evaluate the velocity and s segments (each of which has two terminal points) of vortex lines inducing it, and so $3(p + 2s)$ coordinates. Let's store these in one $3 \times p$ array (r) and two $3 \times s$ arrays (r1 and r2). The columns of these arrays then are 3×1 column vectors, representing (x, y, z). The difference r2 - r1 is a $3 \times s$ array of the s column vectors representing the segments.

One operation we'll need to be able to do a few times is normalize an array of column vectors. This is done with normalize, as defined in Listing 13.5.

With this in hand, the velocity induced at any number of points by any number of segments of unit strength can be implemented in Octave fairly directly from Equations (10.6) and (10.7). This is Listing 13.6. The reason for omitting the vortex strength is because in many cases

Listing 13.5   normalize: divide a matrix of columns by their norms.

```
function u = normalize (v)
  u = bsxfun (@rdivide, v, sqrt (dot (v, v)));
```

Listing 13.6 `vortex_segment`: compute the velocity induced at the points in the columns of `r` by the unit-strength rectilinear vortex segments running from the points in the columns of `r1` to the points in the corresponding columns of `r2`.

```
function q = vortex_segment (r, r1, r2)
  lhat = normalize (r2 - r1);
  h = cross (lhat, r - r1);
  normh2 = dot (h, h);
  q = repmat (dot (lhat / 4 / pi, ...
                   normalize(r - r1) - normalize(r - r2)) ...
               ./ normh2, [3, 1, 1] ) .* h;
  q(repmat (normh2 < eps, [3, 1, 1])) = 0; % r collinear
```

Listing 13.7 `vortex_semiinf`: compute the velocity induced at the points in the columns of `r` by the unit-strength semi-infinite rectilinear vortex segments beginning from the points in the columns of `v` and extending indefinitely in the directions in the columns of `l`.

```
function q = vortex_semiinf (r, v, l)
  q = vortex_segment (r, v, v + 1e3 * l);
```

the strengths are to be determined as part of the problems, as in Listing 13.3, and so it is the influence coefficients (analogous to those of Listings 7.3 and 8.5) that are required.

A variant for the semi-infinite case, for which the second point of segment is defined instead by a direction from the first, is given in Listing 13.7. It uses the truncation approximation, but its internals could be modified to use Equation 10.9 while preserving its interface and therefore not necessitating any revision of any of the other functions, such as `vortex_segment` of Listing 13.6 or `vortex_horseshoe` of Listing 13.8. (See Exercise 5 at the end of the chapter.)

Now, if we have p points and h horseshoes, this means $p \times h$ influence coefficients, except that in three dimensions each influence coefficient is a vector of three components, so a $3 \times p \times h$ array is required. These $p \times h$ columns are most conveniently dealt with by

Listing 13.8 `vortex_horseshoe`: compute the velocity induced at the points in the columns of `r` induced by the unit-strength horseshoe vortices with corners in the columns of `r1` and `r2` and trailing direction `lhat`; calls `vortex_segment` and `vortex_semiinf` from Listings 13.6 and 13.7.

```
function q = vortex_horseshoe (r, r1, r2, lhat)
  vsize = [1, 1, size(r, 2)];
  R = repmat (r, [1, 1, size(r1, 2)]);
  R1 = permute (repmat (r1, vsize), [1, 3, 2]);
  R2 = permute (repmat (r2, vsize), [1, 3, 2]);
  LHAT = permute (repmat (lhat, vsize), [1, 3, 2]);
  q = vortex_segment (R, R1, R2) ...
      - vortex_semiinf (R, R1, LHAT) ...
      + vortex_semiinf (R, R2, LHAT);
```

Table 13.2 4π times the downwash influence coefficients of three 'yawed' horseshoe vortices

$-x$	$\psi = 30°$	$45°$	$60°$
−60.00	−0.2664	−0.2664	−0.2664
−10.00	−.2577	−.2577	−.2576
−5.00	−.2395	−.2395	−.2391
−3.00	−.2172	−.2176	−.2175
−2.00	−.1980	−.1991	−.1999
−1.00	−.1715	−.1737	−.1765
−.20	−.1454	−.1486	−.1535
.20	−.1314	−.1351	−.1409
1.00	−.1038	−.1077	−.1147
2.00	−.0742	−.0773	−.0838
3.00	−.0526	−.0546	−.0592

Source: after Campbell (1951, table I, at p. 17).

replicating the array of points h times and the arrays of the quantities defining the horseshoes p times, using Octave's **repmat** and **permute** built-in functions. This is done in Listing 13.8.

13.4.1 Example: Yawed Horseshoe Vortex Coefficients

The real motivation for building such generality into Listing 13.8 will only become apparent in Chapter 14, but it is still quite usable for simpler problems, as the reader may verify by trying the exercises at the end of this chapter.

As a test of the correctness of the functions, and to demonstrate their usage, they may be used to reproduce the downwash coefficients tabulated by Campbell (1951, table I) for 'yawed' horseshoe vortices, having corners at $\left(\pm \frac{1}{2} \tan \psi, 0, \pm \frac{1}{2} \right)$ and trailing vortices extending in the x-direction, for angles $\psi = 30°, 45°$ and $60°$, for various points in the zx-plane.

For example, the subtable for $z = -4$ at the bottom of p. 17 is extracted here in Table 13.2 and reproduced in Octave using Listing 13.9 as Listing 13.10.

Listing 13.9 `campbell51TableI.m`: Octave script to reproduce Table 13.2.

```
x = [60, 10, 5, 4, 2, 1, 0.2, -[0.2, 1:3]];
r = [x; zeros(size (x)); -4*ones(size (x))];
psi = pi ./ [6, 4, 3];
r2 = [tan(psi); repmat([0; 1], [1, length(psi)])];
q = vortex_horseshoe (r, -r2, r2, ...
                      repmat ([1; 0; 0], [1, size(r2, 2)]));
reshape (4 * pi * q(2,:)', size (r, 2), length (psi))
```

Listing 13.10   Octave output for Listing 13.9; cf. the rightmost three columns of Table 13.2.

```
ans =

   -0.266391   -0.266391   -0.266391
   -0.257747   -0.257724   -0.257628
   -0.239490   -0.239495   -0.239143
   -0.230401   -0.230517   -0.230193
   -0.198000   -0.199069   -0.199941
   -0.171478   -0.173704   -0.176455
   -0.145362   -0.148647   -0.153472
   -0.131375   -0.135058   -0.140883
   -0.103775   -0.107657   -0.114745
   -0.074207   -0.077285   -0.083850
   -0.052589   -0.054560   -0.059206
```

Campbell (1951) concluded sixty years ago:

Solutions of this type are somewhat cumbersome if performed with the usual type of manually operated computing equipment but are readily adaptable to relay-type digital computing machines.

and today we can only agree.

13.5   Exercises

1. (a) Show that the two arguments passed to the **toeplitz** function in Listing 13.1 do generate the leading column and row, respectively of the matrix in Equation (13.9) and Listing 13.2.
   (b) Write an Octave function (of integer n) to generate the same bidiagonal toeplitz matrix without using **toeplitz**; e.g. by adding two calls to **diag**.

2. Identify the steps of a general discrete singularity method from Section 7.4 in Listing 13.3.

3. Modify Listing 13.3 to:
   (a) exploit the symmetry of the spanwise loading;
   (b) account for twist.

4. Use the Octave code of Listing 13.6 to verify numerically the results obtained analytically for Exercise 10.6.2. Check at at least one point collinear with the segment and two not.

5. Write a version of Listing 13.7 that has the same interface but using Equation (10.9) instead of truncating. Compare the answers for Exercise 10.6.3 for some representative points to assess the difference in implementations.

6. Using the improved discrete horseshoe model of Section 13.3, with a single discrete horseshoe per wing, consider the mutual influence of:
   (a) two wings flying abreast, with collinear bound vortices;

  (b) two wings flying in tandem, with collinear trailing vortices;
  (c) two wings in a biplane arrangement, as in Section 7.6 but now taking into account the
       three-dimensional effects arising from finite span
  (d) three wings in a coplanar vee-formation, with neighbouring trailing vortices collinear.

7. Obtain a copy of NACA Research Memorandum No. L50L13 (Campbell 1951) and repro-
    duce table I in its entirety.

13.6   Further Reading

Several deductions about the case of uniform spanwise loading, which is equivalent to the
assumption of a single lumped horseshoe vortex, are discussed by Milne-Thomson (1973,
Chapter 11).

The discrete representation of a lifting line by a row of horseshoe vortices was originally
know as a 'finite-step' method (Campbell 1951; Blackwell 1969). Today it is more important
as a predecessor of the vortex lattice method, and it is in this context that its implementation
was discussed by Katz and Plotkin (2001, section 12.1) and Bertin and Smith (1979).

References

Anderson, J.D. (2007) *Fundamentals of Aerodynamics*, 4th edn. New York: McGraw-Hill.

Bertin, J.J. and Smith, M.L. (1979) *Aerodynamics for Engineers*, 1st edn. Englewood Cliffs, New Jersey: Prentice-Hall.

Blackwell, Jr., J.A. (1969) A finite-step method for calculation of theoretical load distributions for arbitrary lifting-
    surface arrangements at subsonic speeds. Technical Note TN D-5335, NASA.

Campbell, G.S. (1951) A finite-step method for the calculation of span loadings of unusual plan forms. Research
    Memorandum RM L50L13, NACA.

Katz, J. and Plotkin, A. (2001) *Low-Speed Aerodynamics*, 2nd edn. Cambridge: Cambridge University Press.

Milne-Thomson, L.M. (1973) *Theoretical Aerodynamics*, 4th edn. New York: Dover.

14

The Vortex Lattice Method

14.1 Meshing the Mean Lifting Surface of a Wing

The `vortex_horseshoe` program given in Listing 13.8 for computing influence coefficients is very general, being able to handle an arbitrary number of independently aligned horseshoe vortices and an arbitrary number of evaluation points, but to put it to proper use, the grid of horseshoe vortices and collocation points has to be generated; since the layout of the horseshoe vortices has to follow that of the wing they are used to model, this can get arbitrarily complicated. Here a short demonstration program is developed that allows taper, sweep, dihedral, twist, and camber, as follows:

- constant linear spanwise taper from root to tip, with positive tip chord (so avoiding the coalescence of leading and trailing edges at the zero tip chord of a triangular wing)
- constant linear sweep, Λ, measured along the leading edge
- dihedral, as constant linear spanwise variation of the elevation of the leading edge, and again measured along the leading edge
- geometric twist, as constant linear variation of the angle between the sectional chord and the root chord, which is taken as the x-axis
- a camber line for each spanwise section as described by an arbitrary function (provided by the caller) of position along the chord.

The construction begins from the origin $(0, 0, 0)$ at the leading edge of the root chord.

The first step is laying the root chord along the positive x-axis, out to $(c_r, 0, 0)$.

The second step is to draw the leading edge rightwards over $0 > z > \frac{-b}{2}$ and leftwards over $0 < z < \frac{b}{2}$, using the absolute value of z for left–right symmetry, sweeping the leading edge back along the direction of the root chord with $\frac{x}{|z|} = \tan \Lambda$, and similarly $\frac{y}{|z|}$ given by the tangent of the constant leading edge dihedral angle. Note that a dihedral angle does not mean so much a rotation of the half-wing about the root chord as a shear, since the semispan is not measured along the wing itself but rather on its projection in the zx-plane, and the sections are still taken as parallel to the xy-plane of symmetry. In Octave, if z holds a row-vector of spanwise stations, the x- and y-coordinates of the points along the leading edge can be computed from the z-coordinate with:

Theory of Lift: Introductory Computational Aerodynamics in MATLAB®/Octave, First Edition. G. D. McBain.
© 2012 John Wiley & Sons, Ltd. Published 2012 by John Wiley & Sons, Ltd.

```
lex = tan (sweep) * abs (z);
ley = tan (dihedral) * abs (z);
```

The third step is to draw chords back from the leading edge, roughly in the positive *x*-direction parallel to the root chord, but twisted by an angle linearly varying along the semispans. Also the length of the chords (which are measured along them rather than their projections) are allowed to vary similarly. The chord lengths are computed with (the verification of this formula is left as Exercise 14.4.2):

```
c = 2 * (1 - 2 * abs (z) * (1 - taper)) / (1 + taper) / AR;
```

but the laying out of the chord lines themselves is deferred since they will be bent into the camber lines.

The fourth step is to trace the camber line for each of the chords. Let xi be a coordinate (or rather a column vector of values of this coordinate) running from zero to one along the universal chord, and assume the user has provided a function camber to compute the (normalized) camber line ordinate as a function of this. Then the normalized universal camber line is computed with:

```
meanline = [xi, camber(xi)];
```

and this is twisted and tapered with

```
c = 2 * (1 - 2 * abs (z) * (1 - taper)) / (1 + taper) / AR;
twists = 2 * twist * abs (z);
dx = meanline * ([c; c] .* [cos(twists); -sin(twists)]);
```

Finally, these sectional camber lines are added to the leading edge (suitably replicated using repmat), in the last three lines of the complete Listing 14.1.

This system only permits *geometric* rather than *aerodynamic* twist, since the camber line is the same function of chord in each section, while the local chord does have its own inclination; generalization to let the camber function also depend on *z* would not be difficult, but is not required to deal with the preliminary examples presented here.

14.1.1   Plotting the Mesh of a Mean Lifting Surface

A quick check of the correctness of the implementation of Listing 14.1 and of the input in any particular application can be made by plotting a set of three third-angle orthographic

Listing 14.1   `meshwing`: construct a rectangular grid for a lifting surface.

```
function [x, y, z] = meshwing (AR, nchord, nspan, sweep, ...
                              dihedral, taper, twist, ...
                              camber)
  z = linspace (0.5, -0.5, nspan + 1);
  lex = tan (sweep) * abs (z);
  ley = tan (dihedral) * abs (z);
  xi = linspace (0, 1, nchord + 1)';
  meanline = [xi, camber(xi)];
  c = 2 * (1 - 2 * abs (z) * (1 - taper)) / (1 + taper) / AR;
  twists = 2 * twist * abs (z);
  dx = meanline * ([c; c] .* [cos(twists); -sin(twists)]);
  dy = meanline * ([c; c] .* [sin(twists); +cos(twists)]);
  x = repmat (lex, size (xi)) + dx;
  y = repmat (ley, size (xi)) + dy;
  z = repmat (z, size (xi));
```

projections of the spanwise and chordwise lines, as shown in Listing 14.2, the former being the columns of the arrays returned by `meshwing` and the latter being the rows, and so the columns of the transposes. Alternatively, one could use **plot3** to obtain axonometric projections. Flat wings (without dihedral, twist, or camber) don't have useful *yz* elevations or *xy* end elevations, so only their plan is drawn. Note that `meshwing_plot` deliberately does not draw the axes to scale since the default unequal stretching tends to give a clearer picture. As an example, Figure 14.1 was produced by Listing 14.3.

Listing 14.2   `meshwing_plot`: draw three third-angle orthographic projections of a lifting surface mesh produced by `meshwing` of Listing 14.1.

```
function meshwing_plot (x, y, z)
  Y = [min(y(:)), max(y(:))];
  if (Y(2) - Y(1) < eps)              % flat
    plot (x, z, 'ko-', x', z', 'kx-') % planform
    xlabel ('x')
    ylabel ('z')
  else                                % third-angle ortho.
    subplot (2, 2, 1), plot (z, x, 'ko-', z', x', 'kx-')
    ylabel ('x')

    subplot (2, 2, 3), plot (z, y, 'ko-', z', y', 'kx-')
    xlabel ('z'), ylabel ('y')

    subplot (2, 2, 4), plot (x, y, 'ko-', x', y', 'kx-')
    xlabel ('x')
  end%if
```

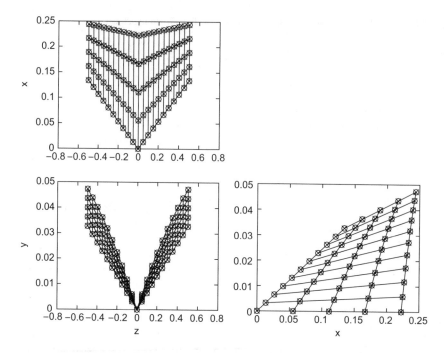

Figure 14.1 Third-angle orthographic projections of the mesh of the mean surface of a swept tapered twisted uncambered wing with dihedral, produced by Listing 14.3

Listing 14.3 Script used to produce Figure 14.1, demonstrating Listings 14.1 and 14.2.

```
[x, y, z] = meshwing (6.0, 4, 20, pi/12, ...
                      pi/48, 0.5, pi/24, ...
                      @ (x) 0*x);
meshwing_plot (x, y, z);
```

14.2   A Vortex Lattice Method

A vortex lattice method consists of representing the wing with a finite rectangular array of horseshoe vortices with strengths determined by enforcing impermeability by collocation at an equal number of points. There are several variants of the method, as described in the references in the Further Reading section at the end of the chapter, but here we describe a simple effective method, well suited to implementation in Octave.

The layout follows directly from the meshing of the wing implemented in Section 14.1. The mean surface of the wing is now divided into a lattice of quadrilaterals. One horseshoe is placed on each. Both trailing vortices are taken in the **i** direction, which is assumed to be that of the flow off the trailing edge—far downstream it must be parallel to the oncoming stream but it is the nearer parts of the wake which most influence the downwash on the lifting surface and at small angles of incidence it makes little difference anyway.

Listing 14.4 `meshwing_vlm`: set up the collocation points and bound vortex segment vertices for the vortex lattice method, using lattice points from `meshwing` in Listing 14.1.

```
function [r, r1, r2] = meshwing_vlm (x, y, z)
  xyz = permute (cat (3, x, y, z), [3, 1, 2]);
  r1 = (3*xyz(:,1:(end-1),1:(end-1)))  ...
       + xyz(:,1:(end-1),2:end) ) / 4;
  r1 = (3*xyz(:,1:(end-1),1:(end-1)))  ...
       + xyz(:,2:end,1:(end-1)) ) / 4;
  r2 = (3*xyz(:,1:(end-1),2:end) + xyz(:,2:end,2:end)) / 4;
  r = ((r1 + r2)/2 + xyz(:,2:end,1:(end-1))  ...
                   + xyz(:,2:end,2:end) ) / 3;
```

As in the two-dimensional lumped vortex method of Chapter 7 and the improved discrete horseshoe model of Section 13.3, the bound vortex segment is laid along the quarter-line. This is found by linearly interpolating one quarter of the way from each of the two forward points towards the two aft points. That is, in the three coordinate arrays returned by `meshwing` of Listing 14.1, the ijth element refers to the ith chordwise position (counted from leading to trailing, with x increasing) and the jth spanwise position (counted from left edge to right edge, with z decreasing). Thus the ijth panel has $r_{i,j}$ and $r_{i+1,j}$ as its leading and trailing left points and $r_{i,j+1}$ and $r_{i+1,j+1}$ as its leading and trailing right points, and its bound vortex segment runs from

$$r_{ij}^{(1)} \equiv \frac{3r_{i,j} + r_{i+1,j}}{4}$$

to

$$r_{ij}^{(2)} \equiv \frac{3r_{i,j+1} + r_{i+1,j+1}}{4}.$$

Again as in Chapter 7 and Section 13.3, the collocation point is placed at the three-quarter point. This is computed by going two-thirds of the way from the midpoint of the bound vortex (just computed) to the midpoint of the trailing edge of the panel; i.e.

$$r_{ij}^{(c)} \equiv \frac{\frac{r_{ij}^{(1)} + r_{ij}^{(2)}}{2} + r_{i+1,j} + r_{i+1,j+1}}{3}$$

If the arrays x, y, and z are as computed by `meshwing` in Listing 14.1, then the collocation points and bound segment vertices can be computed using these formulae, as shown in Listing 14.4.

14.2.1 The Vortex Lattice Equations

The vortex lattice equations are equal in number to the horseshoes and to the collocation points; they are that the lifting surface be impermeable, i.e. that the normal component of the sum of the velocities induced by all the horseshoes balances the normal component of the free-stream. Now it will be convenient in solving the problem to store the vortex strengths in a column vector rather than the two-dimensional array reflecting their geometrical layout, so the horseshoes are renumbered with a single index rather than two, e.g. let the index J of the Jth horseshoe, being previously described as the ijth, be given by

$$J = i + (j - 1)n_{\text{chord}},$$

but hereafter it will be referred to in lowercase again as j. Then, if

- $\boldsymbol{q}_\infty = q_\infty(\mathbf{i}\cos\alpha + \mathbf{j}\sin\alpha)$ is the free-stream,
- $\hat{\boldsymbol{n}}_i$ is the unit normal to the lifting surface at the ith collocation point, and
- $\boldsymbol{q}_{ij}\Gamma_j$ the velocity induced at the ith collocation point by the jth horseshoe vortex, which has strength Γ_j,

then the velocity at the ith collocation point is

$$q_\infty(\mathbf{i}\cos\alpha + \mathbf{j}\sin\alpha) + \sum_{j=1}^{n} \boldsymbol{q}_{ij}\Gamma_j,$$

and the vanishing of the normal component of this is

$$\sum_{j=1}^{n} (\hat{\boldsymbol{n}}_i \cdot \boldsymbol{q}_{ij})\Gamma_j = -q_\infty(\hat{\boldsymbol{n}}_i \cdot \mathbf{i}\cos\alpha + \hat{\boldsymbol{n}}_i \cdot \mathbf{j}\sin\alpha),$$

i.e.

$$\sum_{j=1}^{n} a_{ij}\Gamma_j = b_i,$$

where the matrix coefficients are

$$a_{ij} = \hat{\boldsymbol{n}}_i \cdot \boldsymbol{q}_{ij}$$

and the elements of the right-hand side vector are

$$b_i = -q_\infty(\hat{\boldsymbol{n}}_i \cdot \mathbf{i}\cos\alpha + \hat{\boldsymbol{n}}_i \cdot \mathbf{j}\sin\alpha).$$

If the wing is flat in the zx-plane then $\hat{\boldsymbol{n}}_i = \mathbf{j}$ for all i, and if the incidence is small these coefficients are approximately

$$a_{ij} = v_{ij}$$

and the elements of the right-hand side vector are

$$b_i = -q_\infty\alpha.$$

This is the form that we will work with here. Note its similarity to the matrix–vector Equation 7.4 of the two-dimensional lumped vortex method.

14.2.2   Unit Normals to the Vortex-lattice

In the general case next we should construct a normal vector to the panel at the collocation point so as to enforce the impermeability condition there, but in the examples to follow we shall, for simplicity, only consider flat lifting surfaces; i.e. as generated by Listing 14.1 with zero dihedral, twist, and camber. As a hint for further work in this direction, a vector normal to a surface can be constructed from the cross-product of two nonparallel vectors tangent to it; e.g. chordwise and spanwise vectors, or the diagonals of the quadrilateral.

14.2.3   Spanwise Symmetry

Although both left and right halves of the wing have been constructed symmetrically in Listing 14.1, and it is simple to pass the whole mesh of points to Listing 14.4 to compute the locations of the bound vortices and collocation points, the usual spanwise symmetry should be exploited before solving for the vortex strengths.

A simple way to do that using the above routines is to assemble one matrix which gives the influence of the (say) right horseshoes on the right collocation points and then another for the influence of the left horseshoes on the same right collocation points. Then knowing that the strengths of the left horseshoes will just be the same as their right mirror images, this second influence matrix can be rearranged and added to the first.

14.2.4   Postprocessing Vortex Lattice Methods

The first result required from a vortex lattice method is its estimate of the lift. This is most easily and directly calculated by applying a slight three-dimensional generalization of the Kutta–Joukowsky theorem to each of the bound vortex segments. That is, each bound vortex segment of strength Γ_i and spanwise extent b_i contributes an amount $\rho q_\infty \Gamma_i b_i$ to the total lift force and so

$$L = \rho q_\infty \sum_i \Gamma_i b_i$$

$$C_L \equiv \frac{L}{\frac{1}{2}\rho q_\infty^2 b\bar{c}} = 2 \mathbb{R} \sum_i \left(\frac{\Gamma_i}{q_\infty b}\right)\left(\frac{b_i}{b}\right). \tag{14.1}$$

Note that it is only the spanwise extent of the bound vortex segment that contributes, rather than its actual length. The chordwise component of its length, due to sweep, is too parallel to the free-stream to induce any force in the linear approximation. The y-component, due to dihedral, induces a inboard side-force, but this will be cancelled by the equal and opposite contribution from the other wing in a spanwise symmetric configuration and so not pursued further here.

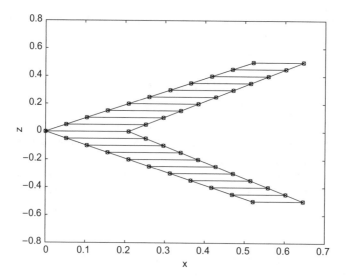

Figure 14.2   Planform of the flat wing with $\mathcal{R} = 5$, taper ratio 0.6, and $\Lambda_{c/4} = 45°$, as considered by Campbell (1951), here generated from Listings 14.5 and 14.2

14.3   Examples of Vortex Lattice Calculations

14.3.1   Campbell's Flat Swept Tapered Wing

In an early report on the use of the vortex lattice method, Campbell (1951, tables II, III) presented detailed results of a vortex lattice calculation for a single row of twenty yawed horseshoe vortices a wing with taper ratio 0.6, $\mathcal{R} = 6$, and sweep measured at the quarter-chord of $\Lambda_{c/4} = 45°$, as shown in Figure 14.2 which was produced by the call `meshwing_plot (x,y,z)` in Listing 14.5.

With the machinery developed in Chapters 13 and 14, this example can now easily be replicated, e.g. as shown in Listing 14.5. Note that in the implementation of the lift coefficient Equation (14.1):

- the leading-edge sweepback angle Λ is computed in terms of that at the quarter-chord, $\Lambda_{c/4}$, using a simple trigonometic formula (see Exercise 14.4.1)
- the lift coefficient has a leading factor 4 rather than 2 to account for the equal contribution of the wing omitted by symmetry; and
- the dimensionless spanwise extent of each bound vortex is $\frac{b_i}{b} = \frac{1}{n}$.

The results are shown in Listing 14.6, with the vortex strengths multiplied by two to agree with the slightly different nondimensionalization used by Campbell (1951). The original results are reproduced in Table 14.1. The slight differences, occurring typically in the third decimal place, are most likely due to the higher round-off error the computers of Campbell (1951) would have incurred by retaining less precision in the intermediate steps of the solution of the linear system of equations; nevertheless, the final lift coefficient–incidence slopes, 3.548 and 3.563, agree to within 0.5%.

Listing 14.5 `vlm_demo_campbell.m`: Octave code to demonstrate Listing 14.4 and calculate the vortex strengths and lift–incidence slope for a flat wing with $\!R = 6$, taper ratio 0.6, and $\Lambda_{c/4} = 45°$ quarter-chord sweep using 1×20 horseshoes, chordwise by spanwise; see Figure 14.2 and Listing 14.6 for output.

```
AR = 6.0;
taper = 0.6;
sweep = atan ((1-taper)/(1+taper)/AR + tan (pi/4));
n = 2 * 10;
[x, y, z] = meshwing (AR, 1, n, sweep, 0, taper, ...
                      0, @ (z) 0*z);
meshwing_plot (x, y, z);
[r, r1, r2] = meshwing_vlm (x, y, z);
ihat = repmat ([1; 0; 0], 1, n);
qrr = vortex_horseshoe (r(:,n/2+1:n), ...
                        r1(:,n/2+1:n), ...
                        r2(:,n/2+1:n), ...
                        ihat(:,1:n/2) );
vrr = squeeze (qrr(2,:,:));
qrl = vortex_horseshoe (r(:,n/2+1:n),  ...
                        fliplr (r1(:,1:n/2)), ...
                        fliplr (r2(:,1:n/2)), ...
                        ihat(:,1:n/2) );
vrl = squeeze (qrl(2,:,:));
Gamma = (vrr + vrl) \ -ones (n/2, 1);
disp (reshape (2 * flipud (Gamma), n/4, 2))
CLa = 4 * AR * sum (Gamma) / n
```

Listing 14.6   Octave output for Listing 14.5.

```
   0.38735    0.64522
   0.50738    0.65563
   0.56634    0.65848
   0.60276    0.65224
   0.62786    0.63562
CLa =   3.5633
```

Table 14.1   Vortex strengths reported for Figure 14.2

$\frac{2\Gamma}{bq_\infty\alpha}$	$\frac{2\Gamma}{bq_\infty\alpha}$
0.3860	0.6429
0.5049	0.6542
0.5634	0.6578
0.5936	0.6516
0.6241	0.6368

Source: after Campbell (1951, table III).

14.3.2   Bertin's Flat Swept Untapered Wing

In his presentation of the vortex lattice method, Bertin (2002) carried out in even more detail a very similar example for a flat untapered wing with $R = 5$ and $\Lambda = 45°$. One times eight horseshoes were taken along the chord and span, with symmetry again assumed so only the strengths of the four right horseshoes were calculated explicitly. This example can be repeated here with Listing 14.7, using exactly the same code from Listing 14.5 used to reproduce the example of Campbell (1951) with just the variation of the first four lines defining the input.

For further demonstration of the use of the `vortex_horseshoe` function of Listing 13.8 in vortex lattice methods, the same calculation without exploiting spanwise symmetry is performed with the code shown in Listing 14.8. Of course the answers for vortex strengths and lift coefficients, as given in Listing 14.9, are indistinguishable. Bertin (2002) quotes the four vortex strengths (normalized by $bq_\infty\alpha$, as here) as 4π times 0.0273, 0.0287, 0.0286, and 0.0250; these agree to their three significant figures with the output of the present program.

Listing 14.7   First four lines of `vlm_demo_bertin`, for reproducing Bertin's vortex lattice example; the rest is as in Listing 14.5.

```
AR = 5.0;
taper = 1.0;
sweep = pi/4;
n = 2 * 4;
```

Listing 14.8   Redoing Bertin's example from Listing 14.7 (which is invoked here as `vlm_demo_bertin`) without exploiting spanwise symmetry.

```
vlm_demo_bertin
q = vortex_horseshoe (r(:,:), ...
                      r1(:,:), ...
                      r2(:,:), ...
                      ihat );
v = squeeze (q(2,:,:));
Gamma = v \ -ones (n, 1);
disp (Gamma(n/2+1:n) / 4 / pi)
CLa = 2 * AR * sum (Gamma) / n
```

Listing 14.9   Output of Listing 14.8, for Bertin's vortex lattice method example.

```
    0.027302
    0.028733
    0.028636
    0.024962
CLa =   3.4442
```

14.3.3   Spanwise and Chordwise Refinement

The vortex lattice is easily refined spanwise by increasing the third argument of `meshwing` from 20 in Listing 14.5. Chordwise refinement isn't much more difficult and is demonstrated here in Listing 14.10 for two by eight lattice panels, following the pattern of Listing 14.8 and not exploiting symmetry. The exploitation of symmetry requires a small amount of bookkeeping with multidimensional array indices and is left as Exercise 14.4.5. The increasing refinement does not much affect the final answer for the slope of the lift coefficient with respect to incidence, going from 3.4442 to 3.4389 to 3.4369 as the number of vortices chordwise increases from one to three, but it does begin to illustrate the chordwise distribution of lift loading, as shown in Figure 14.3, generated as in Listing 14.11. Notice that the chordwise lift loading begins to qualitatively resemble the exact two-dimensional distribution obtained for a flat plate from conformal mapping in Equation (5.13) and plotted in Figure 5.4.

Listing 14.10   `vlm_demo_bertin2`: extending Listing 14.8 by doubling the lattice chordwise.

```
AR = 5.0;
taper = 1.0;
sweep = pi/4;
nchord = 2;
nspan = 8;
[x, y, z] = meshwing (AR, nchord, nspan, sweep, ...
                      0, taper, 0, @ (z) 0*z);
meshwing_plot (x, y, z);
[r, r1, r2] = meshwing_vlm (x, y, z);
ihat = repmat ([1; 0; 0], 1, nchord * nspan);
q = vortex_horseshoe (r(:,:), r1(:,:), r2(:,:), ihat);
v = squeeze (q(2,:,:));
Gamma = v \ -ones (nchord * nspan, 1);
CLa = 2 * AR * sum (Gamma) / nspan
```

Listing 14.11   Octave code to produce the upper drawing in Figure 14.3; the `vlm_demo_bertin2` referred to is Listing 14.10.

```
vlm_demo_bertin2

plot3 (z, x, y, 'k-', z', x', y', 'k-')
hold ('on')
rv = (r1 + r2) / 2;
quiver3 (rv(3,:), rv(1,:), rv(2,:), ...
         0*r(3,:), 0*r(1,:), Gamma, 'ko-')
xlabel ('z'), ylabel ('x'), zlabel ('y')
```

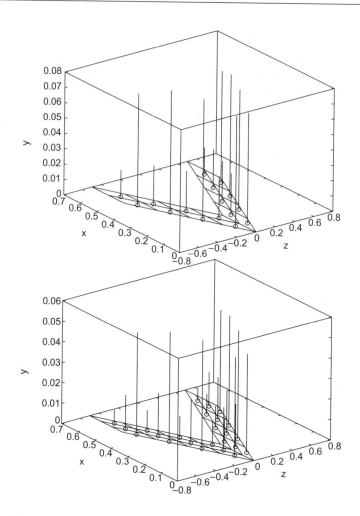

Figure 14.3 The aerodynamic force vectors acting on the midpoints of the bound vortex segments of the lattice for two (above, as plotted by Listing 14.11) and three (below) panels chordwise. The true inclination rotates downstream with incidence but is drawn here in the vertical limit approached for $\alpha \sim 0$

14.4 Exercises

1. The sweep angle Λ is defined above in terms of the leading edge; however some authorities (Campbell 1951; Bertin 2002, figure 7.23) use the quarter-chord line, though not all (Belotserkovskii 1967; Bertin 2002, figures 7.24 and 7.31).

   Verify that the leading edge sweep Λ is given in terms of the quarter-chord sweep $\Lambda_{c/4}$ by

   $$\Lambda = \arctan \left\{ \frac{1 - \lambda}{(1 + \lambda)\mathcal{R}} + \tan \Lambda_{c/4} \right\},$$

   for taper ratio λ and aspect ratio $\mathcal{R}$.

2. Verify that the chord c in the section z of a linearly tapering wing is given in terms of the span b, aspect ratio $\mathcal{R}$, and taper ratio λ by

$$\frac{c}{b} = \frac{2\left\{1 - \frac{2|z|}{b}(1 - \lambda)\right\}}{(1 + \lambda)\mathcal{R}}.$$

3. Modify Listing 14.1 to mesh the upper and lower surfaces of a wing rather than its mean surface. Work by analogy with the two-dimensional case in Section 8.1.2; i.e. assume the caller will provide a function giving the thickness as a function of position along the chord. (This is not required for a vortex-lattice method but might generate input for a three-dimensional panel method.)

4. Write an alterative lifting-surface mesh generator to that of Exercise 3 which takes the universal wing section in the form of an array of points, as in Table 1.1 or in the UIUC (n.d.) Airfoil Database, and extrudes them left and right along the span, accounting for sweep, dihedral, taper, and twist.

5. Rewrite Listing 14.10 to exploit spanwise symmetry in the same way that Listing 14.5 does.

6. Having in Exercise 5 fixed Listing 14.10 to exploit spanwise symmetry, investigate the convergence of the lift coefficient with the refinement of the vortex lattice along the lines of Figure 12.2, except that now there are two independent variables for the discretization: the number of vortices chordwise and spanwise.

   Does chordwise or spanwise refinement have more effect on the predicted lift–incidence slope? For a given number of unknowns, what is the optimal choice of chordwise and spanwise refinement? Say for definiteness that 128 horseshoes shall be used; compare $1 \times 128, 2 \times 64, 4 \times 32, \ldots$, chordwise–spanwise lattices. (Note that if this optimum is found with respect to lift–incidence slope, it might not be the optimum for other measures, such as those related to the pitching moment, and often vortex lattice methods are used to predict those too.)

14.5 Further Reading

A brief early history of the vortex-lattice method was given by DeYoung (1976a) along with commentary, analysis, and extensions by DeJarnette (1976), Hough (1976), and DeYoung (1976b). Anderson (2007) gives an overview of the method, but no details or quantitative examples; he defers to Bertin and Smith (1979) for 'a worked example that clearly illustrates the salient points of the technique'. The vortex lattice method described in the present text is essentially the one of Campbell (1951) (using the 'yawed' bound vortex segments), Belotserkovskii (1967) (who called them 'oblique'), and Bertin (2002). All three of these last references give thoroughly detailed elementary descriptions of the method, as do Gray and Schenk (1953), who used rectangular (rather than yawed/oblique) horseshoe vortices.

   Educational FORTRAN programs for vortex-lattice methods are listed by Moran (2003) for a flat rectangular wing and by Katz and Plotkin (2001) for a more general case. A program intended for serious use is listed by Margason and Lamar (1971). More recent implementations of the vortex lattice method, much more elaborate than that presented here, include LinAir Durstan (1993), TORNADO (Melin 2000), and AVL (Drela and Youngren n.d.).

14.5.1 Three-dimensional Panel Methods

The natural combination of the ideas of the vortex lattice methods developed in Chapters 13–14 with the two-dimensional panel methods of Chapter 8 is a three-dimensional panel method. There are numerous possibilities for this, but a simple and effective idea is to replace the infinite line vortices of Chapter 8 with horseshoes of finite breadth and then stack a number of these panel models along the span, extruding the segments of the boundary in two dimensions into quadrilaterals in three; this plan is described in detail by Hess (1972). For references to this and several variants, see Paraschivoiu (1998) and Katz and Plotkin (2001), who also provide a FORTRAN implementation.

References

Anderson, J.D. (2007) *Fundamentals of Aerodynamics*, 4th edn. New York: McGraw-Hill.

Belotserkovskii, S.M. (1967) *The Theory of Thin Wings in Subsonic Flow*. New York: Plenum.

Bertin, J.J. (2002) *Aerodynamics for Engineers*, 4th edn. Upper Saddle River, New Jersey: Prentice Hall.

Bertin, J.J. and Smith, M.L. (1979) *Aerodynamics for Engineers*, 1st edn. Englewood Cliffs, New Jersey: Prentice-Hall.

Campbell, G.S. (1951) A finite-step method for the calculation of span loadings of unusual plan forms. Research Memorandum RM L50L13, NACA.

DeJarnette, F. (1976) Arrangement of vortex lattices on subsonic wings. *Vortex-Lattice Utilization*, SP-405 NASA.

DeYoung, J. (1976a) Historical evolution of vortex-lattice methods. *Vortex-Lattice Utilization*, SP-405 NASA.

DeYoung, J. (1976b) Optimum lattice arrangement developed from a rigorous analytical basis. *Vortex-Lattice Utilization*, SP-405 NASA.

Drela, M. and Youngren, H. (n.d.) AVL. http://web.mit.edu/drela/Public/web/avl. Accessed 3 Dec. 2011.

Durstan, D.A. (1993) LinAir: A multi-element discrete vortex Weissinger aerodynamic prediction method. Technical Memorandum 108786, NASA.

Gray, W.L. and Schenk, K.M. (1953) A method for calculating the subsonic steady-state loading on an airplane with a wing of arbitrary plan form and stiffness. Technical Note 3030, NACA.

Hess, J.L. (1972) Calculation of potential flow about arbitrary three-dimensional lifting bodies. Final Technical Report MDC J5679-01, Douglas Aircraft Company, Long Beach, California.

Hough, G.R. (1976) Lattice arrangements for rapid convergence. *Vortex-Lattice Utilization*, SP-405 NASA.

Katz, J. and Plotkin, A. (2001) *Low-Speed Aerodynamics* 2nd edn. Cambridge: Cambridge University Press.

Margason, R.J. and Lamar, J.E. (1971) Vortex-lattice FORTRAN program for estimating subsonic aerodynamic characteristics of complex planforms. Technical Note TN D-6142, NASA.

Melin, T. (2000) *A vortex lattice MATLAB implementation for linear aerodynamic wing applications*. Master's thesis, Department of Aeronautics, KTH.

Moran, J. (2003) *An Introduction to Theoretical and Computational Aerodynamics*. New York: Dover.

Paraschivoiu, I. (1998) *Aérodynamique Subsonique*. Éditions de l'École de Montréal.

UIUC (n.d.) UIUC Airfoil Coordinates Database. http://www.ae.illinois.edu/m-selig/ads/coord_database.html. Accessed 3 Dec. 2011.

Part Three

Nonideal Flow in Aerodynamics

15

Viscous Flow

The models based on ideal flow discussed thus far have done surprisingly well in explaining and quantitatively predicting the generation of lift by two-dimensional aerofoils and three-dimensional wings; however, that discussion has had to remain within certain bounds which it was not able to explain. No drag was predicted in two dimensions, which is at odds with experience. An important source of two-dimensional drag is skin friction which arises from viscosity. Further, the ideal theories have been restricted to small angles of incidence because of the phenomenon of stall. Stall arises when the flow fails to follow the contour of the aerofoil into regions of higher pressure because near the wall the air is retarded by viscous skin friction. Both these important aerodynamical effects can be explained by the large effects that even a small amount of viscosity can have on a flow in the vicinity of a solid wall. This extension to ideal aerodynamics is called boundary layer theory and is introduced in Chapters 15–17.

15.1   Cauchy's First Law of Continuum Mechanics

In the derivation in Chapter 2 of Euler's Equation (2.6), which governs the flow of a perfect fluid, we began from the principle that the rate of change of momentum of a particle is equal to the resultant of the forces acting on it. These forces can be external, such as gravity, or internal.

We assume that the fluid is such that internal forces are only transmitted by contact, rather than by action at a distance. Therefore, they act on a fluid particle only through its surface; i.e. as stresses. This was the case for the pressure forces in a perfect fluid, pressure being an isotropic stress, but here we consider more general fluids; in particular, we relax the assumption that the fluid cannot sustain shearing stresses.

Stress is a second-order tensor. Just as velocity, a vector (i.e. a first-order tensor), is a quantity with magnitude and direction, stress has a magnitude and two directions. If we specify (i) the orientation of a surface and (ii) a particular direction, the stress gives us the component of force in direction (ii) from the internal fluid forces acting through a surface with orientation (i). Stresses are described in a coordinate system with a double set of components; thus σ_{xy} gives the component of force (per unit area) acting on a surface perpendicular to the x-axis and acting in the y-direction. Another way of thinking about this is that the stress is a vector-valued function of direction: given one direction (the orientation of a surface element), it gives a vector, being the resultant force from the internal fluid interactions across the surface.

Theory of Lift: Introductory Computational Aerodynamics in MATLAB®/Octave, First Edition. G. D. McBain.
© 2012 John Wiley & Sons, Ltd. Published 2012 by John Wiley & Sons, Ltd.

Stresses are conventionally reckoned positive in tension (although this is not universal).

If we consider again the fluid initially occupying an infinitesimal rectangle in two dimensions from Figure 2.3, there we had that the only contributions to the force in the x-direction were from the pressure on the left face $p\delta x \delta y$ and from the right face $-\left(p + \frac{\partial p}{\partial x}\delta x\right)\delta y$, giving a resultant $-\frac{\partial p}{\partial x}$ per unit area. For a perfect fluid, the Cartesian components of the stress in two dimensions are simply given by

$$\sigma_{xx} = -p \qquad\qquad\qquad \sigma_{xy} = 0$$
$$\sigma_{yx} = 0 \qquad\qquad\qquad \sigma_{yy} = -p,$$

or in matrix form

$$\begin{bmatrix} \sigma_{xx} & \sigma_{xy} \\ \sigma_{yx} & \sigma_{yy} \end{bmatrix} = -p \begin{bmatrix} 1 & 0 \\ 0 & 1 \end{bmatrix}. \tag{15.1}$$

The appearance of the identity matrix here is equivalent to the statement above that pressure stresses are isotropic.

For a more general fluid, with a general anisotropic stress state, the contributions to the x-component of force on an infinitesimal two-dimensional particle are $-\sigma_{xx}\delta y$ from the left face, $\left(\sigma_{xx} + \frac{\partial \sigma_{xx}}{\partial x}\delta x\right)\delta y$ from the right face, $-\sigma_{yx}\delta x$ from the bottom, and $\left(\sigma_{yx} + \frac{\partial \sigma_{yx}}{\partial y}\right)\delta x$ from the top. The component of the resultant in the x-direction is therefore

$$\left(\frac{\partial \sigma_{xx}}{\partial x} + \frac{\partial \sigma_{yx}}{\partial y}\right)\delta x \delta y.$$

Similarly, stresses acting over the four faces produce a component of force

$$\left(\frac{\partial \sigma_{xy}}{\partial x} + \frac{\partial \sigma_{yy}}{\partial y}\right)\delta x \delta y$$

in the y-direction. Notice the pattern: the first index of the stress component matches the direction of differentiation while the second matches the direction of action of the resulting force.

Replacing the pressure terms in the Euler Equations (2.6) with these more general stress terms gives

$$\rho \frac{Dq}{Dt} = -\nabla P + \rho f \tag{15.2}$$

$$\text{or} \quad \rho\left(\frac{\partial u}{\partial t} + u\frac{\partial u}{\partial x} + v\frac{\partial u}{\partial y}\right) = \frac{\partial \sigma_{xx}}{\partial x} + \frac{\partial \sigma_{yx}}{\partial y} + \rho f_x$$

$$\rho\left(\frac{\partial v}{\partial t} + u\frac{\partial v}{\partial x} + v\frac{\partial v}{\partial y}\right) = \frac{\partial \sigma_{xy}}{\partial x} + \frac{\partial \sigma_{yy}}{\partial y} + \rho f_y.$$

These are (the two-dimensional form of) *Cauchy's First Law of continuum mechanics*.

15.2 Rheological Constitutive Equations

To apply Cauchy's First Law of continuum mechanics to actual problems, we need a relation between the stress and the motion of the material; such a relation is called a *rheological equation of state* or a *rheological constitutive equation*.

15.2.1 Perfect Fluid

For a perfect fluid, the stress is given by Equation (15.1), and substituting Equation (15.1) into Equation (15.2) leads us back to the Euler Equation (2.6).

15.2.2 Linearly Viscous Fluid

Aerodynamics, at least at low and moderate speeds, is concerned with fluids that are almost perfect, but imperfect in that they have a slight viscosity. It is convenient then to write the stress as a perfect fluid part and an extra part, called the *deviatoric stress*:

$$
\begin{bmatrix} \sigma_{xx} & \sigma_{xy} \\ \sigma_{yx} & \sigma_{yy} \end{bmatrix} = -p \begin{bmatrix} 1 & 0 \\ 0 & 1 \end{bmatrix} + \begin{bmatrix} \tau_{xx} & \tau_{xy} \\ \tau_{yx} & \tau_{yy} \end{bmatrix}.
$$

This split is rendered unique by requiring $\tau_{xx} + \tau_{yy} = 0$ (i.e. that the *trace* of the deviatoric stress is zero).

To accord with the elementary definition of viscosity, the shear component τ_{yx} of the deviatoric stress should reduce to

$$
\tau_{yx} = \mu \frac{\partial u}{\partial y}
$$

in a one-dimensional shear flow $q = u(y)\mathbf{i}$. Here, as in Section 1.4.2, μ is the *coefficient of viscosity*.

The proper generalization of this to two dimensions is

$$
\begin{bmatrix} \tau_{xx} & \tau_{xy} \\ \tau_{yx} & \tau_{yy} \end{bmatrix} = \mu \begin{bmatrix} 2\frac{\partial u}{\partial x} & \left(\frac{\partial u}{\partial y} + \frac{\partial v}{\partial x}\right) \\ \left(\frac{\partial v}{\partial x} + \frac{\partial u}{\partial y}\right) & 2\frac{\partial v}{\partial y} \end{bmatrix}, \tag{15.3}
$$

which is called the *linear viscous stress* tensor. The derivation of Equation (15.3) is beyond the scope of this course; see the suggestions for further reading at the end of the chapter.

Notice that the stress tensor is symmetric: $\tau_{xy} = \tau_{yx}$. That the stress tensor is generally symmetric is a consequence of Cauchy's Second Law of continuum mechanics. Notice too that $\tau_{xx} + \tau_{yy}$ is proportional to the divergence and does vanish, as required.

15.3   The Navier–Stokes Equations

Once the linear viscous stress tensor as given by Equation (15.3) is introduced into Cauchy's Equation (15.2), we obtain the (two-dimensional) Navier–Stokes equations. The x-component is

$$\frac{\partial u}{\partial t} + u\frac{\partial u}{\partial x} + v\frac{\partial u}{\partial y} - f_x = \frac{1}{\rho}\left(\frac{\partial \sigma_{xx}}{\partial x} + \frac{\partial \sigma_{yx}}{\partial y}\right)$$

$$= \frac{1}{\rho}\left(\frac{\partial}{\partial x}\left\{-p + 2\mu\frac{\partial u}{\partial x}\right\} + \mu\frac{\partial}{\partial y}\left\{\frac{\partial v}{\partial x} + \frac{\partial u}{\partial y}\right\}\right)$$

$$= -\frac{1}{\rho}\frac{\partial p}{\partial x} + \frac{\mu}{\rho}\left(\frac{\partial}{\partial x}\left\{2\frac{\partial u}{\partial x}\right\} + \frac{\partial}{\partial y}\left\{\frac{\partial v}{\partial x} + \frac{\partial u}{\partial y}\right\}\right)$$

$$= -\frac{1}{\rho}\frac{\partial p}{\partial x} + v\left(\frac{\partial^2 u}{\partial x^2} + \frac{\partial^2 u}{\partial x^2} + \frac{\partial^2 v}{\partial x\partial y} + \frac{\partial^2 u}{\partial y^2}\right)$$

$$= -\frac{1}{\rho}\frac{\partial p}{\partial x} + v\left(\frac{\partial^2 u}{\partial x^2} + \frac{\partial}{\partial x}\left\{\frac{\partial u}{\partial x} + \frac{\partial v}{\partial y}\right\} + \frac{\partial^2 u}{\partial y^2}\right),$$

where we have introduced the coefficient of kinematic viscosity $v \equiv \frac{\mu}{\rho}$, from Equation (1.6).

Now, the expression in braces here is just the two-dimensional divergence, and so vanishes for any flow of a fluid of uniform density (any *incompressible* fluid), in which case

$$\frac{\partial u}{\partial t} + u\frac{\partial u}{\partial x} + v\frac{\partial u}{\partial y} - f_x = -\frac{1}{\rho}\frac{\partial p}{\partial x} + v\left(\frac{\partial^2 u}{\partial x^2} + \frac{\partial^2 u}{\partial y^2}\right). \tag{15.4a}$$

Similarly the y-component of the Navier–Stokes equation for an incompressible fluid is

$$\frac{\partial v}{\partial t} + u\frac{\partial v}{\partial x} + v\frac{\partial v}{\partial y} - f_y = -\frac{1}{\rho}\frac{\partial p}{\partial y} + v\left(\frac{\partial^2 v}{\partial x^2} + \frac{\partial^2 v}{\partial y^2}\right). \tag{15.4b}$$

These must be supplemented with the conservation of mass Equation (2.3)

$$\frac{\partial u}{\partial x} + \frac{\partial v}{\partial y} = 0,$$

just as the Euler equation is.

15.4   The No-Slip Condition and the Viscous Boundary Layer

If the fluid is only slightly viscous, so that v is small, Equation (15.4) is close to the Euler Equation (2.6) in the sense that they only differ by a term that must be small as it contains the small coefficient v as a factor.

The big difference though is that with the inclusion of the viscous terms, the equations are second-order in the velocity rather than first, and therefore require two boundary conditions on the velocity instead of one. For the inviscid fluid, the usual condition at a solid is impermeability; this still has to hold for a viscous fluid. The new condition is zero slip: if the fluid has any viscosity at all, and so can sustain some shearing stress, then the tangential component of velocity must be continuous across a solid–fluid interface.

This means that although the Euler and Navier–Stokes equations are close for slightly viscous fluids, the flow of an inviscid and a slightly viscous fluid in the same geometry might be quite different, as the slightly viscous fluid has to adhere to all solid boundaries whereas the inviscid fluid can slip over them freely.

Since the viscous stress is proportional to the velocity gradient, the action of viscosity is to oppose or reduce velocity gradients. If the fluid is very viscous, as molasses and oils are, velocity gradients are quickly eliminated, or are only sustained by the application of significant external tractions. On the other hand, if the fluid is only slightly viscous, as air and water are, steep velocity gradients are possible. Thus, given a geometry, the flow of a slightly viscous fluid might be almost the same as that of an inviscid fluid almost everywhere with the exception of a thin layer along the no-slip boundaries, within which the component of velocity tangential to the solid surface steeply changes from zero (at the wall) to a value similar to that prevailing near the wall in the inviscid fluid. Such a layer is called a *viscous boundary layer*.

The difference between a fluid with zero viscosity and one with some finite viscosity is illustrated by a special example in Section 15.6, below.

15.5   Unidirectional Flows

The Navier–Stokes Equation (15.4) is very difficult to solve, in general, even numerically; however, some progress can be made for unidirectional flows, i.e. flows of the form

$$u = u(y, t)$$
$$v = 0.$$

The fact that u is independent of x is a necessary consequence of the incompressibility Equation (2.3). With these restrictions on the velocity components, the y-component of the Navier–Stokes equation reduces to

$$-f_y = -\frac{1}{\rho}\frac{\partial p}{\partial y},$$

so that in the absence of external forces, the pressure is independent of y. Further, since no value of x is different to any other, there is no reason to suppose that the pressure shouldn't be anything other than a constant, or perhaps that the pressure gradient in the x-direction should be constant (in which case it can be included in f_x). The x-component reduces to

$$\frac{\partial u}{\partial t} - f_x = \nu \frac{\partial^2 u}{\partial y^2}. \tag{15.5}$$

Notice that the nonlinear acceleration terms have disappeared; this makes the equation much easier to solve, as demonstrated in Sections 15.5.1 and 15.6.

15.5.1   *Plane Couette and Poiseuille Flows*

For example,

$$u = \frac{Uy}{h} \tag{15.6}$$

describes steady flow between a stationary plate at $y = 0$ and a plate at $y = h$ sliding in the direction of increasing x with speed U. This is called *plane Couette flow*, and is important in the theory of lubrication.

As a second example,

$$u = U(h^2 - y^2) \tag{15.7}$$

satisfies the unidirectional Navier–Stokes equation if the external force is $f = 2\nu U\mathbf{i}$. This is called *plane Poiseuille flow*, and describes the gravity-driven flow through a vertical fissure.

15.6 A Suddenly Sliding Plate

Consider a fluid of kinematic viscosity ν occupying the half-space $y > 0$ above a solid plane $y = 0$ and free from any external forces. Assume that the fluid is initially stagnant but that at $t = 0$, the solid plane begins to move parallel to itself in the direction of increasing x with constant speed U.

If we assume that $v = 0$ in the fluid for all time, the remaining component u has to satisfy the initial–boundary value problem

$$\frac{\partial u}{\partial t} = \nu \frac{\partial^2 u}{\partial y^2} \tag{15.8a}$$

$$
\begin{aligned}
u(0, t) &= U & (t > 0) \\
u(\infty, t) &\sim 0 & (t > 0) \\
u(y, 0) &= 0 & (y > 0).
\end{aligned}
\tag{15.8b}
$$

Of course, if the fluid were perfect, we could drop the no-slip condition implied by Equation (15.8b) and then $u \equiv 0$ would be a solution; the plate could freely slip under the fluid without disturbing it.

15.6.1 Solution by Similarity Variable

Although this problem can be solved by separating variables, since there is no explicit length scale, it seems possible that the two independent variables might be combined into one. This indeed works. The technique is called *similarity* transformation. It is something like a powerful version of dimensional analysis.

To begin with, we have u depending on y, t, U, and ν. The obvious choice for nondimensionalizing u is

$$u^* = \frac{u}{U},$$

in terms of which the problem becomes

$$\frac{\partial u^*}{\partial t} = \nu \frac{\partial^2 u^*}{\partial y^2} \tag{15.9}$$

$$
\begin{aligned}
u^*(0, t) &= 1 & (t > 0) \\
u^*(\infty, t) &\sim 0 & (t > 0) \\
u^*(y, 0) &= 0 & (y > 0).
\end{aligned}
$$

Now we have u^* depending on y, t, and ν, which is three quantities involving two dimensions: length and time. (Mass is absent.) Thus, Buckingham's Π-Theorem states that we should be able to eliminate two of these and have u^* as a function of a single variable. This is very appealing, as it would mean the partial differential Equation (15.9) would have to be reduced to an ordinary differential equation.

Let's begin using dimensional analysis, as in Section 1.5.1, with $y[=]$m. We have $t[=]$s and $\nu[=]$m^2/s, which means $\nu t[=]$m^2 and $\sqrt{\nu t}[=]$m. Therefore,

$$\eta \equiv \frac{y}{\delta(t)} \tag{15.10}$$

is a dimensionless coordinate with the time-dependent length scale

$$\delta(t) \equiv 2\sqrt{\nu t}.$$

The numerical factor 2 here isn't necessary, but is introduced for later convenience.

Alternatively, if we began instead with Equation (15.10), then the differential operators become

$$\frac{\partial}{\partial t} = \frac{\partial \eta}{\partial t}\frac{d}{d\eta} = \frac{-y}{\{\delta(t)\}^2}\frac{d\delta(t)}{dt}\frac{d}{d\eta}$$

$$\frac{\partial^2}{\partial y^2} = \left(\frac{\partial \eta}{\partial y}\frac{d}{d\eta}\right)\left(\frac{\partial \eta}{\partial y}\frac{d}{d\eta}\right) = \left(\frac{1}{\delta(t)}\frac{d}{d\eta}\right)\left(\frac{1}{\delta(t)}\frac{d}{d\eta}\right) = \frac{1}{\{\delta(t)\}^2}\frac{d^2}{d\eta^2}$$

and the differential Equation (15.9) becomes

$$\frac{-y}{\{\delta(t)\}^2}\frac{d\delta(t)}{dt}\frac{du^*}{d\eta} = \frac{\nu}{\{\delta(t)\}^2}\frac{d^2u^*}{d\eta^2}$$

$$\left(\frac{-y}{\nu}\frac{d\delta(t)}{dt}\right)\frac{du^*}{d\eta} = \frac{d^2u^*}{d\eta^2}.$$

This works provided y and t don't appear explicitly (only via the similarity variable η); this depends on the factor in parentheses:

$$\frac{-y}{\nu}\frac{d\delta(t)}{dt} = \frac{-\delta(t)\eta}{\nu}\frac{d\delta(t)}{dt}$$

$$= \frac{-\eta}{\nu}\frac{d\left[\frac{1}{2}\{\delta(t)\}^2\right]}{dt};$$

and will occur if

$$\frac{d\left[\frac{1}{2}\{\delta(t)\}^2\right]}{dt} = \nu \times \text{const.}$$

$$\frac{1}{2}\{\delta(t)\}^2 = \nu t \times \text{const.}$$

$$\delta(t) \propto \sqrt{\nu t},$$

and so again the form of Equation (15.10) is indicated. With Equation (15.10), the governing partial differential Equation (15.9) is indeed reduced to an ordinary differential equation:

$$-2\eta \frac{du^*}{d\eta} = \frac{d^2 u^*}{d\eta^2}. \qquad (15.11)$$

Note that the initial condition at $t = 0$ and the far-field condition at $y \to \infty$ both correspond to $\eta \to \infty$, so it's necessary that $u^* \sim 0$ in both cases, which it does; otherwise the method wouldn't be applicable. Thus the two boundary conditions for Equation (15.11) are

$$u^*(0) = 1$$
$$u^*(\infty) \sim 0.$$

Although the ordinary differential Equation (15.11) is second-order in u^*, it is only first-order in $\frac{du^*}{d\eta}$. It can therefore be integrated easily enough to get:

$$\frac{d^2 u^*/d\eta^2}{du^*/d\eta} = -2\eta$$

$$\frac{d}{d\eta} \ln \left(\frac{du^*}{d\eta} \right) = -2\eta$$

$$\ln \left(\frac{du^*}{d\eta} \right) = -\eta^2 + \text{const.}$$

$$\frac{du^*}{d\eta} = \text{const.} \times e^{-\eta^2}$$

$$u^* = \text{const.} \times \int e^{-\eta^2} \, d\eta.$$

To match the boundary condition at infinity, we choose the limits of integration as

$$u^*(\eta) = \text{const.} \times \int_\eta^\infty e^{-t^2} \, dt,$$

and then since

$$\int_0^\infty e^{-t^2} \, dt = \frac{\sqrt{\pi}}{2},$$

we must have the constant as $2/\sqrt{\pi}$; thus

$$u^*(\eta) = \frac{2}{\sqrt{\pi}} \int_\eta^\infty e^{-t^2} \, dt.$$

This function arises so often, in applications ranging from thermal conduction to statistics, that it has a special name: the *complementary error function*:

$$\text{erfc } \eta \equiv \frac{2}{\sqrt{\pi}} \int_\eta^\infty e^{-t^2} \, dt.$$

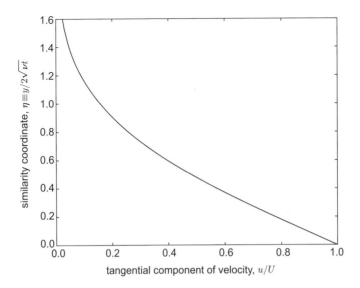

Figure 15.1 Tangential velocity profile on a suddenly started sliding infinite plate in a viscous fluid, as given by Equation (15.12), in terms of the similarity variable defined by Equation (15.10)

It is built in to most modern scientific programming languages, including Octave. Using erfc, the solution is

$$u^*(\eta) = \text{erfc}\,\eta \qquad\qquad (15.12)$$

or in dimensional terms

$$u(y, t) = U\ \text{erfc}\,\frac{y}{2\sqrt{vt}}.$$

The velocity profile is plotted in Figure 15.1.

15.6.2 The Diffusion of Vorticity

The important thing to observe in the similarity solution is not so much the shape of the tangential velocity profile but the fact that if we consider the distance from the plate at which fluid is moving at, say, half the velocity of the plate, this distance increases with the square-root of the time.

Since for $\eta > 1$ (or 2), erfc $\eta < 0.16$ (or 0.005), we can more or less say that the fluid outside the layer $0 < \eta < 1$ (or 2) is unaffected by the slipping of the plate and hence is the same whether or not the fluid is viscous. In terms of y this condition is

$$y < \delta(t)$$

(or 2δ). Thus we have a layer of thickness one or two $\delta(t)$ along the boundary which is affected by viscosity. As the viscosity tends to zero, the layer vanishes, and outside the layer the fluids acts as though it were inviscid. It is a simple explicit example of a viscous boundary layer.

15.7 Exercises

1. Show that the perfect fluid's constitutive Equation (15.1) indeed converts the Cauchy Equation (15.2) into the Euler Equation (2.6).

2. *Stream function for incompressible plane viscous flow.* Show that if a two-dimensional velocity field is derived from a stream function ψ then it is divergence-free. Express the vorticity in terms of the stream function and deduce that the flow is not necessarily irrotational. What is the condition on the stream function that the flow be irrotational?

3. Construct stream functions for:
   (a) plane Couette flow and
   (b) plane Poiseuille flow.

4. *A plate oscillating in its plane.* Show that if the partial differential Equation (15.8a) has a solution of the form

$$u(y, t) \propto e^{-2\pi y/\delta} \cos\{2\pi(ft - y/\delta)\}$$

   where f is a constant frequency, and δ a constant length then there must be a relation between δ and f:

$$f\delta^2 = 4v\pi$$

   What is the velocity at the plate? What is the velocity far from the plate? How far from the plate is it that the fluid never moves with 1% of the maximum velocity of the plate?

5. *Pressure-driven Poiseuille flow.* Show that the plane Poiseuille parabolic velocity profile given by Equation (15.7) can also apply in the absence of external body forces (or if these are conservative and treated as described in Section 2.4.4) if there is a constant uniform pressure gradient. What is the relation between the rate of flow (per unit span) and the pressure drop per unit length?

6. *Combined plane Couette and Poiseuille flow.* Show that a linear combination of Equations (15.6) and (15.7) still satisfies the partial differential Equation (15.5) and therefore describes the flow in a fissure with a longitudinal pressure gradient and relative sliding motion between the two walls.

7. *Skin friction.* What is the shear stress on a unit area of each of the two walls in plane Couette and plane Poiseuille flow? Carry out a force balance on the fluid contained in a unit span and length of channel in each case.

15.8 Further Reading

Properties of the stream function in incompressible plane viscous flow are derived by Abbott and von Doenhoff (1959).

For other introductions to the linear model of viscous rheology, see Kuethe and Chow (1998); Bertin (2002); or Anderson (2007). More detailed derivations are given by Stokes (1851), Lamb (1932), Schlichting (1960), Truesdell and Rajagopal (2000), and Moran (2003).

For classical treatments of plane Couette and Poiseuille flows, with references to the original experiments, see Lamb (1932).

The most cited texts on boundary layer theory are the ones by Goldstein (1938) and Schlichting (1960).

The problem of the suddenly sliding plate (Section 15.6) and its solution are discussed by many writers on Navier–Stokes flow, including Lamb (1932), Dryden, Murnaghan and Bateman (1956), Whitham (1963), Lagerstrom (1964), and Truesdell and Rajagopal (2000). It is sometimes called *Stokes's first problem*, after G. G. Stokes, who first considered it in 1850, and sometimes *Rayleigh's problem*, after Lord Rayleigh, who realized its usefulness as an illustrative example in the present context.

For the thermal and statistical applications of the complementary error function, see, e.g. Carslaw and Jaeger (1959) and Weatherburn (1949).

References

Abbott, I.H. and von Doenhoff, A.E. (1959) *Theory of Wing Sections*. New York: Dover.

Anderson, J.D. (2007) *Fundamentals of Aerodynamics*, 4th edn. New York: McGraw-Hill.

Bertin, J.J. (2002) *Aerodynamics for Engineers*, 4th edn. Upper Saddle River, New Jersey: Prentice Hall.

Carslaw, H.S. and Jaeger, J.C. (1959) *Conduction of Heat in Solids*, 2nd edn. Oxford: Oxford University Press.

Dryden, H.L. Murnaghan, F.D. and Bateman, H. (1956) *Hydrodynamics*. New York: Dover.

Goldstein, S. (1938) *Modern Developments in Fluid Dynamics*. Oxford: Clarendon Press.

Kuethe, A.M. and Chow, C.Y. (1998) *Foundations of Aerodynamics*, 5th edn. New York: John Wiley & Sons, Ltd..

Lagerstrom, P.A. (1964) *Laminar Flow Theory*. New Jersey: Princeton University Press.

Lamb, H. (1932) *Hydrodynamics*, 6th edn. Cambridge: Cambridge University Press.

Moran, J. (2003) *An Introduction to Theoretical and Computational Aerodynamics*. New York: Dover.

Schlichting, H. (1960) *Boundary-Layer Theory*, 4th edn. New York: McGraw-Hill.

Stokes, G.G. (1851) On the effect of the internal friction of fluids on the motion of pendulums. *Transactions of the Cambridge Philosophical Society* **9**(10).

Truesdell, C. and Rajagopal, K.R. (2000) *An Introduction to the Mechanics of Fluids*. Basel: Birkhäuser.

Weatherburn, C.E. (1949) *A First Course in Mathematical Statistics*, 2nd edn. Cambridge: Cambridge University Press.

Whitham, G.B. (1963) The Navier–Stokes equations of motion. In *Laminar Boundary Layers* (ed. L. Rosenhead) Fluid Motion Memoirs. Oxford: Oxford University Press, chapter 3.

16

Boundary Layer Equations

16.1 The Boundary Layer over a Flat Plate

Consider steady plane flow of a linearly viscous incompressible fluid at speed u_∞ parallel to a solid surface lying in the plane $y = 0$. The flow will be governed by the conservation of mass Equation (2.3) and the Navier–Stokes Equation (15.4) without the terms involving partial derivatives with respect to time and without the body-force terms:

$$u\frac{\partial u}{\partial x} + v\frac{\partial u}{\partial y} = -\frac{1}{\rho}\frac{\partial p}{\partial x} + \nu\left(\frac{\partial^2 u}{\partial x^2} + \frac{\partial^2 u}{\partial y^2}\right) \tag{16.1a}$$

$$u\frac{\partial v}{\partial x} + v\frac{\partial v}{\partial y} = -\frac{1}{\rho}\frac{\partial p}{\partial y} + \nu\left(\frac{\partial^2 v}{\partial x^2} + \frac{\partial^2 v}{\partial y^2}\right) \tag{16.1b}$$

Following the solution for the suddenly started sliding plate (Section 15.6), it's plausible that the effects of viscosity might be largely limited to a thin layer along the plate, $0 < y < \delta$, where the thickness δ may depend on x. Outside this layer ($\delta \ll y$), we expect the flow to be essentially the same as that of a perfect fluid.

Assume that any changes in the flow in the tangential (x) direction occur over a length scale comparable to the distance from the leading edge; i.e. x.

16.1.1 Scales in the Conservation of Mass

In Equation (2.3) for the conservation of mass, if the free-stream speed u_∞ is a representative scale for the tangential speed, then $\partial u/\partial x$ should be of order u_∞/x (by application of the mean value theorem, for example). Similarly if V is a typical magnitude of the normal component of velocity, then (expressing the same idea in mathematical notation)

$$\frac{\partial v}{\partial y} = O\left(\frac{V}{\delta}\right)$$

Theory of Lift: Introductory Computational Aerodynamics in MATLAB®/Octave, First Edition. G. D. McBain.
© 2012 John Wiley & Sons, Ltd. Published 2012 by John Wiley & Sons, Ltd.

where δ is again the boundary layer thickness scale. If the two terms of the continuity equation are to balance each other, they must have comparable magnitudes and therefore

$$O\left(\frac{\partial u}{\partial x}\right) + O\left(\frac{\partial v}{\partial y}\right) = 0$$

$$O\left(\frac{u_\infty}{x}\right) + O\left(\frac{V}{\delta}\right) = 0.$$

It follows that the velocity scales are related to the length scales by

$$\frac{V}{u_\infty} = O\left(\frac{\delta}{x}\right). \tag{16.2}$$

There is an obvious geometric interpretation of this relation in terms of the slope of the streamlines (V/u_∞) and the shape of the boundary layer (δ thick and x long).

If the boundary layer is thin, so that $\delta \ll x$, then Equation (16.2) implies $V \ll u_\infty$, which means that the flow is basically parallel to the solid surface, as expected.

16.1.2   Scales in the Streamwise Momentum Equation

Now apply the same technique, *scale analysis*, to the streamwise (x) component of the steady Navier–Stokes momentum Equation (16.1a). Let Δp be the pressure scale, then the terms of Equation (16.1a) have scales:

$$u\frac{\partial u}{\partial x} = O\left(\frac{u_\infty^2}{x}\right)$$

$$v\frac{\partial u}{\partial y} = O\left(\frac{u_\infty V}{\delta}\right) = O\left(\frac{u_\infty^2}{x}\right)$$

$$\frac{1}{\rho}\frac{\partial p}{\partial x} = O\left(\frac{\Delta p}{\rho x}\right)$$

$$v\frac{\partial^2 u}{\partial x^2} = O\left(\frac{v u_\infty}{x^2}\right) \tag{16.3}$$

$$v\frac{\partial^2 u}{\partial y^2} = O\left(\frac{v u_\infty}{\delta^2}\right). \tag{16.4}$$

If the boundary layer is thin ($\delta \ll x$) then the streamwise viscous term as given in Equation (16.3) must be smaller than the transverse viscous term in Equation (16.4); dropping it gives the *boundary layer momentum equation*:

$$u\frac{\partial u}{\partial x} + v\frac{\partial u}{\partial y} = -\frac{1}{\rho}\frac{\partial p}{\partial x} + v\frac{\partial^2 u}{\partial y^2}. \tag{16.5}$$

16.1.3   The Reynolds Number

Now, if we are to have a balance between the viscous term (of magnitude $\nu u_\infty/\delta^2$) and the inertial terms (of magnitude u_∞^2/x) in the tangential momentum balance Equation (16.5), then

$$O\left(\frac{\nu u_\infty}{\delta^2}\right) = O\left(\frac{u_\infty^2}{x}\right)$$

so that the boundary layer thickness scale must satisfy

$$\frac{\delta}{x} = O\left\{\left(\frac{u_\infty x}{\nu}\right)^{-1/2}\right\}. \tag{16.6}$$

The dimensionless quantity in parentheses on the right-hand side is the Reynolds number based on tangential length scale x:

$$\mathrm{Re}_x \equiv \frac{u_\infty x}{\nu}.$$

This Reynolds number is typically large in aerodynamical applications—consider flight of a wing section of chord $c = 1$ m at $u_\infty = 100$ m/s in air of kinematic viscosity $\nu = 1.5 \times 10^{-5}$ m^2/s, for which it is $\mathrm{Re}_x \approx 7 \times 10^6 \times \frac{x}{c}$ therefore the assumption that the boundary layer is thin is supported by Equation (16.6): this value of Re_x implies that the boundary layer thickness at the trailing edge is of order 4×10^{-4} times the tangential length scale c.

In summary, the boundary layer shape and the streamline slopes scale with the inverse square-root of the Reynolds number:

$$\frac{\delta(x)}{x} = O\left(\frac{V}{u_\infty}\right) = \mathrm{Re}_x^{-1/2}.$$

16.1.4   Pressure in the Boundary Layer

If the pressure term in the tangential momentum balance Equation (16.5) is to be comparable with the inertial terms, the pressure must be of magnitude

$$\Delta p = O(\rho u_\infty^2), \tag{16.7}$$

as may have been expected from Bernoulli's Equation (2.14). Numerical factors like the half in Bernoulli's equation are usually omitted in order relations like Equation (16.7).

16.1.5   The Transverse Momentum Balance

The pressure term in the transverse (y) component of the steady Navier–Stokes Equation (16.1b) is then of order:

$$\frac{1}{\rho}\frac{\partial p}{\partial y} = O\left(\frac{\Delta p}{\rho \delta}\right) = O\left(\frac{u_\infty^2 \mathrm{Re}_x^{1/2}}{x}\right)$$

whereas the inertial terms are of order

$$u\frac{\partial v}{\partial x} = O\left(\frac{u_\infty V}{x}\right) = O\left(\frac{u_\infty^2}{x\mathrm{Re}_x^{1/2}}\right)$$

$$v\frac{\partial v}{\partial y} = O\left(\frac{V^2}{\delta}\right) = O\left(\frac{u_\infty^2}{x\mathrm{Re}_x^{1/2}}\right)$$

and so negligible for large Reynolds numbers. Further, the viscous terms have scales:

$$\nu\frac{\partial^2 v}{\partial x^2} = O\left(\frac{\nu V}{x^2}\right) = O\left(\frac{u_\infty^2}{x\mathrm{Re}_x^{3/2}}\right)$$

$$\nu\frac{\partial^2 v}{\partial y^2} = O\left(\frac{u_\infty^2}{x\mathrm{Re}_x^{1/2}}\right).$$

Thus for large Reynolds numbers, all terms are small except the pressure term which implies

$$0 = \frac{-1}{\rho}\frac{\partial p}{\partial y};$$

i.e. that the pressure is uniform across the boundary layer.

This is an extremely useful result. It means that p is only a function of x; i.e. that the pressure at any point in the boundary layer is the same as that at the same longitudinal position x outside the boundary layer, and is therefore the same as that which would prevail near the surface if the fluid were perfect. Assuming we can solve the flow problem for the perfect fluid in the given geometry, the pressure term in the tangential momentum equation is then known.

16.1.6   The Boundary Layer Momentum Equation

Since the boundary layer momentum Equation (16.5) is second-order in y, two boundary conditions are required in the normal direction; these are usually that the fluid adhere to the solid surface and that the boundary layer flow smoothly merge with the external inviscid flow outside the boundary layer

$$u = 0, \qquad\qquad\qquad\qquad (y = 0)$$

$$u \sim u_\infty, \qquad\qquad\qquad\qquad (y \gg \delta)$$

where u_∞ is the tangential component of velocity at x that would be obtained by a perfect fluid in the same geometry.

Note that the tangential velocity and pressure obtained as outer conditions on the boundary layer analysis ($y \gg \delta$) are those at $y = 0$ in the perfect fluid analysis. This is because the boundary layer is very thin, on any meaningful scale in the perfect fluid flow. One way to make this clearer is to use y in the perfect fluid analysis and introduce a new stretched normal coordinate $\eta = y/\delta$ in the boundary layer analysis, as was done in Section 15.6.

Since the equation is first-order in x, only one boundary condition is required in the tangential direction. Usually information is prescribed about the upstream state and then the equations integrated downstream to study the longitudinal development of the boundary layer. This situation is quite different to the full Navier–Stokes equations which are second-order in all directions and require boundary conditions both upstream and downstream. This simplification is an essential feature of the boundary layer approximation.

16.1.7   Pressure and External Tangential Velocity

Since the flow outside the viscous boundary layer is presumed to be the same as that of a perfect fluid, the pressure and velocity there are related by Bernoulli's Equation (2.14), so, since $q = u$ for $y = 0$ in the perfect fluid (as the surface is impermeable),

$$\frac{-1}{\rho}\frac{dp}{dx} = u_\infty \frac{du_\infty}{dx}. \tag{16.8}$$

This is often more convenient; with it, the boundary layer momentum Equation (16.5) becomes

$$u\frac{\partial u}{\partial x} + v\frac{\partial u}{\partial y} = u_\infty \frac{du_\infty}{dx} + v\frac{\partial^2 u}{\partial y^2}. \tag{16.9}$$

16.1.8   Application to Curved Surfaces

Although the analysis thus far has been for a plane surface, provided the radius of curvature of the surface is much less than the boundary layer thickness, it applies well to curved surfaces. In this case, x is to be interpreted as the tangential coordinate and y the normal. This means that acceptable results can be obtained for typical wing sections by using the present model.

16.2   Momentum Integral Equation

Although the boundary layer momentum Equation (16.9) can be solved analytically in special configurations using various mathematical techniques, or numerically for almost any geometry, much of the most useful information about the boundary layer (such as the skin friction) can be approximated from the *momentum integral equation*. We consider this for steady boundary layers.

To obtain the momentum integral equation, integrate both sides of Equation (16.9) normal to the surface from 0 to ∞ (by which we mean to a distance far outside the boundary layer on the scale of the boundary layer thickness, but still very near the surface on the scale of the perfect fluid flow):

$$\int_0^\infty u\frac{\partial u}{\partial x}\,dy + \int_0^\infty v\frac{\partial u}{\partial y}\,dy = \int_0^\infty u_\infty \frac{du_\infty}{dx}\,dy + v\int_0^\infty \frac{\partial^2 u}{\partial y^2}\,dy.$$

We want to eliminate v from this, which we can do, using the conservation of mass: integrate Equation (2.3) from 0 to y:

$$\int_0^y \left(\frac{\partial u}{\partial x} + \frac{\partial v}{\partial y}\right) dy' = 0$$

$$\int_0^y \frac{\partial u}{\partial x}(x, y')\,dy' + v = 0,$$

(noting that $v = 0$ at $y = 0$ for impermeability) so that

$$\int_0^\infty v \frac{\partial u}{\partial y} \, dy = -\int_0^\infty \left\{ \int_0^y \frac{\partial u}{\partial x}(x, y') \, dy' \right\} \frac{\partial u}{\partial y} \, dy$$

which can be integrated by parts to give

$$-\int_0^\infty \left\{ \int_0^y \frac{\partial u}{\partial x}(x, y') \, dy' \right\} \frac{\partial u}{\partial y} \, dy$$

$$= -\left[\left\{ \int_0^y \frac{\partial u}{\partial x}(x, y') \, dy' \right\} u \right]_0^\infty + \int_0^\infty u \frac{\partial u}{\partial x} \, dy$$

$$= -u_\infty \int_0^\infty \frac{\partial u}{\partial x} \, dy + \int_0^\infty u \frac{\partial u}{\partial x} \, dy.$$

For the viscous term,

$$\nu \int_0^\infty \frac{\partial^2 u}{\partial y^2} \, dy = \nu \left[\frac{\partial u}{\partial y} \right]_0^\infty = -\nu \left. \frac{\partial u}{\partial y} \right|_{y=0}$$

since $\partial u / \partial y \sim 0$ for $y \gg \delta$. The right-hand side can be written as $-\tau_w / \rho$, where τ_w stands for the shear stress component τ_{xy} at the wall (w for wall). In general $\tau_{xy} = \mu \left(\frac{\partial v}{\partial x} + \frac{\partial u}{\partial y} \right)$, but v vanishes along the impermeable solid surface so that on $y = 0$ we have $\tau_{xy} = \mu \frac{\partial u}{\partial y}$, as asserted.

Thus, the integrated momentum equation is

$$\int_0^\infty \left(2u \frac{\partial u}{\partial x} - u_\infty \frac{\partial u}{\partial x} - u_\infty \frac{du_\infty}{dx} \right) dy = -\frac{\tau_w}{\rho}.$$

Now,

$$\frac{\partial}{\partial x} \{ u(u - u_\infty) \} = 2u \frac{\partial u}{\partial x} - u_\infty \frac{\partial u}{\partial x} - u \frac{du_\infty}{dx}$$

so the momentum integral equation can be rewritten

$$\int_0^\infty \left\{ \frac{\partial}{\partial x} [u(u - u_\infty)] + (u - u_\infty) \frac{du_\infty}{dx} \right\} dy = -\frac{\tau_w}{\rho}$$

$$\frac{d}{dx} \int_0^\infty u(u - u_\infty) \, dy + \frac{du_\infty}{dx} \int_0^\infty (u - u_\infty) \, dy = -\frac{\tau_w}{\rho}$$

$$\frac{d}{dx} \left\{ u_\infty^2 \int_0^\infty \frac{u}{u_\infty} \left(1 - \frac{u}{u_\infty} \right) dy \right\} + u_\infty \frac{du_\infty}{dx} \int_0^\infty \left(1 - \frac{u}{u_\infty} \right) dy = \frac{\tau_w}{\rho}.$$

16.3   Local Boundary Layer Parameters

16.3.1   The Displacement and Momentum Thicknesses

Let's define

$$\delta^* \equiv \int_0^\infty \left(1 - \frac{u}{u_\infty}\right) dy \tag{16.10}$$

$$\theta \equiv \int_0^\infty \frac{u}{u_\infty} \left(1 - \frac{u}{u_\infty}\right) dy. \tag{16.11}$$

We call δ^* the *displacement thickness* and θ the *momentum thickness*. In terms of them, the momentum integral equation can be written as

$$\frac{d}{dx}(u_\infty^2 \theta) + \delta^* u_\infty \frac{du_\infty}{dx} = \frac{\tau_w}{\rho}.$$

Both δ^* and θ depend on x, and both parameterize the boundary layer tangential velocity profile $u(y)$, at that value of x.

The physical significance of the displacement thickness can be understood as follows. If the fluid were perfect, the tangential velocity throughout the boundary layer right down to the solid surface would be u_∞, and the volume flow (per unit span) along the boundary layer would be $\int_0^\delta u_\infty \, dy$; however, in the boundary layer the actual value is $\int_0^\delta u \, dy$ so that the difference is

$$\int_0^\infty (u_\infty - u) \, dy = u_\infty \delta^*$$

(using the fact that the integrand is negligible for $y > \delta$). Thus, the retardation of the flow in the boundary layer causes a defect in the volumetric flow-rate the same as the subtraction of a layer of thickness δ^*.

For almost all boundary layer profiles $u(y)$, the tangential velocity remains in $0 < u < u_\infty$; therefore, the momentum thickness will be less than the displacement thickness, since the integrand of θ contains the additional weighting factor $\frac{u}{u_\infty}$. The ratio of the two thicknesses

$$H \equiv \frac{\delta^*}{\theta},$$

called the *shape factor*, is therefore greater than one. This is illustrated in the examples of Section 16.3.3.

16.3.2   The Skin Friction Coefficient

Define a dimensionless version of the shear stress at the wall as

$$c_f \equiv \frac{\tau_w}{\frac{1}{2}\rho u_\infty^2} = \frac{2\nu \left.\frac{\partial u}{\partial y}\right|_{y=0}}{u_\infty^2};$$

c_f is called the *(local) skin-friction coefficient*.

Expressing the wall shear stress in terms of c_f, the momentum integral equation is

$$\frac{d\theta}{dx} + \frac{\theta}{u_\infty}(2+H)\frac{du_\infty}{dx} = \frac{c_f}{2}. \tag{16.12}$$

This is the most common and useful form of the *Kármán integral relation*. It provides one equation linking the three unknown local boundary layer parameters, H, θ, and c_f, to the known external speed variation $u_\infty(x)$.

To proceed further, another two relations between the boundary layer thicknesses and skin friction are required. Two popular approaches for deriving these are:

- the assumption of a velocity profile;
- the use of (possibly empirical) correlations.

Local Skin Friction in a Turbulent Boundary Layer

Although the above derivation of the Kármán integral relation was in the context of laminar flow, the relation holds as well for turbulent boundary layers. In the laminar case, the local skin friction can be obtained from an assumed velocity profile simply by evaluating its derivative at the wall; in the turbulent case it's more difficult, as turbulent boundary layers contain a number of different sublayers and a profile function accurately describing the variation of the tangential velocity over most of the boundary layer thickness might be of no use at all in determining the skin friction coefficient. An example is the much-used *one-seventh profile*, discussed below in Section 16.3.3. Basically, if one takes a velocity profile approach to turbulent boundary layer momentum integral analysis, a separate correlation is required for the local skin friction.

16.3.3 Example: Three Boundary Layer Profiles

Example: Linear Profile

Consider the linear profile:

$$\frac{u}{u_\infty} = \begin{cases} \frac{y}{\delta}, & 0 < y < \delta \\ 1, & \delta < y. \end{cases}$$

The displacement and momentum thicknesses are

$$\delta^* = \int_0^\infty \left(1 - \frac{u}{u_\infty}\right) dy = \int_0^1 (1-\eta) \, d\eta \times \delta = \frac{\delta}{2}$$

$$\theta = \int_0^\infty \frac{u}{u_\infty}\left(1 - \frac{u}{u_\infty}\right) dy = \int_0^1 \eta(1-\eta) \, d\eta \times \delta = \frac{\delta}{6}$$

and the shape factor is therefore $H = 3$. The integrands are plotted in Figure 16.1.

The local skin friction coefficient of the linear profile is

$$c_f = \frac{2\nu \left.\frac{\partial u}{\partial y}\right|_{y=0}}{u_\infty^2} = \frac{2\nu \frac{u_\infty}{\delta}}{u_\infty^2} \equiv \frac{2}{\text{Re}_\delta},$$

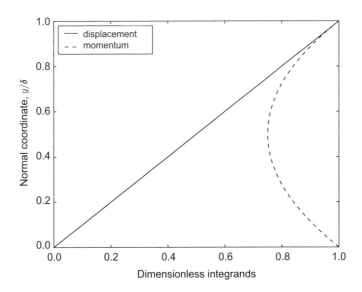

Figure 16.1 Linear boundary layer velocity profile: the displacement thickness is the area to the right of the solid curve (the velocity profile), the momentum thickness to the right of the dashed curve

where

$$\mathrm{Re}_\delta \equiv \frac{u_\infty \delta}{\nu}$$

is a Reynolds number based on the boundary layer thickness.

Example: Error Function Profile

Consider the error function profile, which arose in the suddenly started sliding plate problem:

$$\frac{u}{u_\infty} = \mathrm{erf}\,\frac{y}{\delta}.$$

(Note that we use erf η instead of erfc η as here we take the plate to be fixed and the far fluid to be moving at u_∞.)

The displacement thickness can be carried out analytically (Gautschi 1965):

$$\frac{\delta^*}{\delta} = \int_0^\infty (1 - \mathrm{erf}\,\eta)\,\mathrm{d}\eta = \int_0^\infty \mathrm{erfc}\,\eta\,\mathrm{d}\eta = \frac{1}{\sqrt{\pi}} \doteq 0.56419.$$

The (dimensionless) momentum thickness (θ/δ) can be obtained by integrating numerically, as shown in Listing 16.1, which prints

```
deltastar =   0.56419
theta =   0.23369
H =   2.4142
```

Listing 16.1   Octave code to calculate the boundary layer parameters for the error function profile.

```
deltastar = quadgk (@ (y) 1 - erf (y), 0, Inf)
theta = quadgk (@ (y) erf (y) .* (1 - erf (y)), 0, Inf)
H = deltastar / theta
```

The thickness integrands are illustrated in Figure 16.2.
   To calculate the skin friction, note:

$$\frac{d\,\mathrm{erf}\,\eta}{d\eta} = \frac{2e^{-\eta^2}}{\sqrt{\pi}}$$

$$\left.\frac{\partial}{\partial y}\left(u_\infty\,\mathrm{erf}\,\frac{y}{\delta}\right)\right|_{y=0} = \frac{2u_\infty}{\sqrt{\pi}\delta};$$

thus

$$c_f = \frac{4/\sqrt{\pi}}{\mathrm{Re}_\delta}.$$

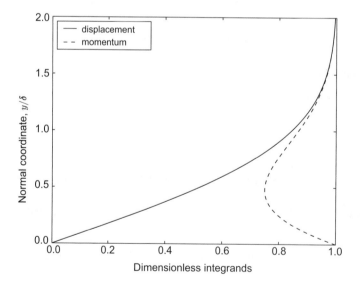

Figure 16.2   Boundary layer velocity profile $u = u_\infty\,\mathrm{erf}\,y/\delta$: the displacement thickness is the area to the right of the solid curve (the velocity profile), momentum thickness to the right of dashed curve

Just as for the linear profile, the local skin friction coefficient is inversely proportional to Re_δ. In each case it would also have been proportional to the Reynolds number based on the local boundary layer momentum thickness

$$\mathrm{Re}_\theta \equiv \frac{u_\infty \theta}{\nu}$$

and this latter result applies to many laminar boundary layers; it is exploited in Thwaites's method for solving the momentum integral equation (ibid.; see also Section 17.3, below).

Example: One-seventh Profile

A popular simple profile used to describe turbulent boundary layers is

$$\frac{u}{u_\infty} = \begin{cases} \left(\frac{y}{\delta}\right)^{1/7}, & y < \delta \\ 1, & y > \delta. \end{cases}$$

The displacement and momentum thicknesses are

$$\frac{\delta^*}{\delta} = \int_0^1 \left(1 - \eta^{1/7}\right) \mathrm{d}\eta = \frac{1}{8}$$

$$\frac{\theta}{\delta} = \int_0^1 \eta^{1/7} \left(1 - \eta^{1/7}\right) \mathrm{d}\eta = \frac{7}{72}$$

which means the shape function is $H = \frac{9}{7}$. The profile, illustrated in Figure 16.3, is much 'fuller' than the two previous, which are more typical of laminar profiles. This is apparent in

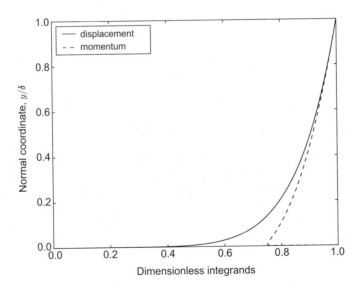

Figure 16.3 Boundary layer velocity profile $u = u_\infty(y/\delta)^{1/7}$: the displacement thickness is the area to the right of the solid curve (the velocity profile), momentum thickness to the right of dashed curve

the much smaller shape factor. This fullness is caused by the effectiveness of the turbulent eddies in transferring momentum across the boundary layer. A consequence, and another apparent feature, is the steeper relative gradient at the wall; however, while this is qualitatively realistic, the one-seventh profile is useless for estimating the local skin friction, since it implies an infinite velocity gradient at the wall.

16.4 Exercises

1. *Blasius's boundary layer.* Consider a semi-infinite flat plate lying on the positive x-axis, parallel to a free-stream of speed u_∞ of an incompressible fluid of kinematic viscosity v. The pressure gradient is zero and there are no external forces.

   Assume the flow field has a steady two-dimensional boundary layer structure and so, as in the suddenly slid plate problem (Section 15.6) try nondimensionalizing the normal coordinate y by forming

   $$\eta \equiv \frac{y}{\delta},$$

   but allow the boundary layer thickness δ to vary with x rather than t.

   Show by dimensional analysis (Section 1.5) that $\delta(x)$ must have the form x times some function of

   $$\text{Re}_x \equiv \frac{u_\infty x}{v}.$$

   Assuming it has the form $\delta(x) = x\text{Re}_x^b$ for some constant b, show that the equation governing the dimensionless stream function $f \equiv \psi/u_\infty\delta(x)$ (the boundary layer momentum equation, with the velocity components expressed in terms of the stream function) is an ordinary differential equation in the *similarity coordinate* η, if b is chosen appropriately. Show that the boundary conditions $u = v = 0$ on $y = 0$ and $u \sim u_\infty$ as $y \gg \delta(x)$ can also be written purely in terms of f and η.

   Why is $u_\infty\delta(x)$ a logical choice for the scale of the stream function?

   Show that the zero pressure gradient condition is consistent with the pressure variation far from the plate (outside the viscous boundary layer).

2. *The momentum thickness integral.* Each of the curves for the momentum integrands in Figures 16.1, 16.2, and 16.3 displays a turning point, i.e. a minimum with respect to the normal coordinate, though it is a little harder to see with the last because it is so close to the wall. Show that this turning point occurs where the tangential velocity passes through one-half its freestream value, and so that it is a general feature for any tangential velocity profile increasing continuously (though not even necessarily monotonically) from zero to the freestream value.

   Calculate its location for each of these three profiles, exactly for the linear and one-seventh laws and numerically for the error function. [*Ans.:* For the linear and one-seventh profiles, it occurs at $y/\delta = 1/2$ and $1/128$, respectively; and for the error function near 0.47694.]

3. (a) Write an Octave function which computes the displacement thickness, momentum thickness, and shape factor, given a function generating the profile; that is, the call

```
[deltastar, theta, H] = blparams (@erf)
```

should produce output equivalent to that of Listing 16.1.

   (b) Demonstrate the function on the linear and one-seventh profiles, checking the answers against the exact ones given in Section 16.3.3.

4. *Piecewise linear profile.* Calculate the displacement and momentum intergrals for a tangential velocity profile composed of two linear parts, i.e.

$$
\frac{u}{u_\infty} = \begin{cases} \frac{u_{1/2}}{u_\infty} \frac{y}{\delta}, & 0 < \frac{y}{\delta} < \frac{1}{2} \\ \frac{u_{1/2}}{u_\infty} + \left(\frac{y}{\delta} - \frac{1}{2} \right) \left(1 - \frac{u_{1/2}}{u_\infty} \right), & \frac{1}{2} < \frac{y}{\delta} < 1 \\ 1, & y > \delta, \end{cases}
$$

where $u_{1/2}$ is a parameter, being the tangential velocity halfway through the boundary layer.

Prepare a plot of the shape factor H as a function of the fraction $u_{1/2}/u_\infty$; for what halfway velocity values does this piecewise profile match the shape factor of the error function and one-seventh profiles?

16.5   Further Reading

The 'asymptotic matching principle' which provides the outer boundary condition in Section 16.1.6 is a key concept in boundary layer theory; see Ashley and Landahl (1985).

The Blasius boundary layer is treated in many more general books on the mechanics of viscous fluids as well as such aerodynamical textbooks as Abbott and von Doenhoff (1959), Kuethe and Chow (1998), Anderson (2007), and Bertin (2002).

The boundary layer momentum equation, and associated concepts such as

- displacement and momentum thicknesses
- shape factor
- (local) skin-friction coefficient
- Kármán integral relation

are treated by Abbott and von Doenhoff (1959) and Moran (2003). This integral approach is very powerful and in particular can be extended to turbulent boundary layers. That lies beyond the scope of this text, but laminar boundary layers are studied along these lines in Chapter 17.

References

Abbott, I.H. and von Doenhoff, A.E. (1959) *Theory of Wing Sections*. New York: Dover.
Anderson, J.D. (2007) *Fundamentals of Aerodynamics*, 4th edn. New York: McGraw-Hill.

Ashley, H. and Landahl, M. (1985) *Aerodynamics of Wings and Bodies*. New York: Dover.

Bertin, J.J. (2002) *Aerodynamics for Engineers*, 4th edn. Upper Saddle River, New Jersey: Prentice Hall.

Gautschi, W. (1965) Error function and Fresnel integrals. In *Handbook of Mathematical Functions* (ed. M. Abramowitz and I. Stegun) New York: Dover, chapter 7.

Kuethe, A.M. and Chow, C.Y. (1998) *Foundations of Aerodynamics*, 5th edn. New York: John Wiley & Sons, Ltd.

Moran, J. (2003) *An Introduction to Theoretical and Computational Aerodynamics*. New York: Dover.

17

Laminar Boundary Layers

17.1 Boundary Layer Profile Curvature

If we evaluate the boundary layer momentum Equation (16.9)

$$u\frac{\partial u}{\partial x} + v\frac{\partial u}{\partial y} = u_\infty\frac{\mathrm{d}u_\infty}{\mathrm{d}x} + \nu\frac{\partial^2 u}{\partial y^2},$$

right at the wall ($y = 0$) so that $u = v = 0$ (for no-slip and impermeability), it reduces to

$$0 = u_\infty\frac{\mathrm{d}u_\infty}{\mathrm{d}x} + \nu\frac{\partial^2 u}{\partial y^2}, \qquad (17.1)$$

or, replacing the free-stream acceleration term with the free-stream pressure gradient via Bernoulli's Equation (16.8),

$$0 = \frac{-1}{\rho}\frac{\mathrm{d}p}{\mathrm{d}x} + \nu\frac{\partial^2 u}{\partial y^2},$$

so that the profile curvature $\frac{\partial^2 u}{\partial y^2}$ at the wall has the same sign as the pressure gradient given by Bernoulli's Equation (16.8)

$$\frac{\mathrm{d}p}{\mathrm{d}x} = \mu\frac{\partial^2 u}{\partial y^2}.$$

Towards the outer edge of the boundary layer, u will generally be increasing asymptotically up to its free-stream value u_∞, so (assuming $u_\infty > 0$), $u > 0$ and $\frac{\partial u}{\partial y} > 0$ but $\frac{\partial^2 u}{\partial y^2} < 0$ there. This means that if the pressure gradient is positive, there must be an inflexion point in the velocity profile, somewhere between the wall and the outside of the boundary layer, where $\frac{\partial^2 u}{\partial y^2}$ changes sign.

Theory of Lift: Introductory Computational Aerodynamics in MATLAB®/Octave, First Edition. G. D. McBain.
© 2012 John Wiley & Sons, Ltd. Published 2012 by John Wiley & Sons, Ltd.

17.1.1 Pressure Gradient and Boundary Layer Thickness

In Kármán's integral Equation (16.12)

$$\frac{d\theta}{dx} + \frac{\theta}{u_\infty}(2 + H)\frac{du_\infty}{dx} = \frac{c_f}{2},$$

neglecting the possibility of boundary layer separation for the moment, the skin friction coefficient will be positive. Basically this tends to make the boundary layer thickness increase with x, corresponding to the first term on the left-hand side, but in general this quantity is to be balanced by both terms on the left-hand side. If the free-stream is decelerating, which, via Bernoulli's Equation (16.8), implies a positive or *adverse* pressure gradient, the second term is negative and the boundary layer thickness increases more rapidly with x; if the free-stream is accelerating (*favourable* pressure gradient), the second term is positive and the boundary layer thickness will increase more slowly or may even decrease if the favourable pressure gradient is large enough.

17.2 Pohlhausen's Quartic Profiles

From Equation (17.1), there is a simple direct link between the curvature of the boundary layer profile at the wall and the local pressure gradient. To explore the effect of this using the assumed-velocity method in the Kármán integral relation, we need a family of profiles with curvature at the wall. Such a family was developed by Pohlhausen (1921) based on quartic polynomials. The other four coefficients of the quartic were chosen to satisfy

- the no-slip condition at the wall $u(0) = 0$,
- the matching condition at the edge of the boundary layer $u(\delta) = u_\infty$,
- a smoothness condition there $\frac{\partial u}{\partial x} = 0$, and
- another smoothness condition there $\frac{\partial^2 u}{\partial x^2} = 0$.

With $\eta \equiv y/\delta$, the requisite dimensionless quartic is

$$\frac{u}{u_\infty} = 2\eta - 2\eta^3 + \eta^4 + \frac{\Lambda}{6}\eta(1 - \eta)^3,$$

which satisfies

$$u = 0 \qquad (\eta = 0)$$

$$\frac{\partial^2 u}{\partial y^2} = -\frac{u_\infty \Lambda}{\delta^2} \qquad (\eta = 0)$$

$$u = u_\infty \qquad (\eta = 1)$$

$$\frac{\partial u}{\partial y} = \frac{\partial^2 u}{\partial^2 y} = 0 \qquad (\eta = 1).$$

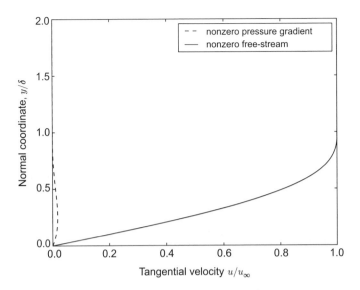

Figure 17.1 The two ingredients of Pohlhausen's quartic profile: $2\eta - 2\eta^3 + \eta^4$ which is 1 at $\eta = 1$ and $\eta(1-\eta)^3/6$ which has -1 curvature at $\eta = 0$

The pressure gradient parameter defined here is

$$\Lambda \equiv -\frac{\delta^2}{u_\infty \mu}\frac{\mathrm{d}p}{\mathrm{d}x} = \frac{\delta^2}{\nu}\frac{\mathrm{d}u_\infty}{\mathrm{d}x}. \tag{17.2}$$

Pohlhausen's profile consists of two parts, $2\eta - 2\eta^3 + \eta^4$ which accounts for the nonzero free-stream velocity and the part multiplied by Λ which accounts for the curvature at the wall. The two parts are plotted in Figure 17.1 and various combinations are shown in Figure 17.2.

The various boundary layer parameters are

$$\frac{\delta^*}{\delta} = \frac{36 - \Lambda}{120}$$

$$\frac{\theta}{\delta} = \frac{5328 - 48\Lambda - 5\Lambda^2}{45360}$$

so that the shape factor is

$$H = \frac{378(\Lambda - 36)}{5\Lambda^2 + 48\Lambda - 5328}.$$

The local skin friction coefficient is

$$c_f \equiv \frac{2\frac{\partial}{\partial\eta}\frac{u}{u_\infty}}{\mathrm{Re}_\delta} = \frac{12 + \Lambda}{3\mathrm{Re}_\delta}.$$

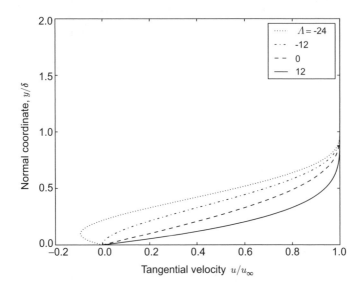

Figure 17.2   Some of Pohlhausen's quartic profiles, including that for zero pressure gradient $\Lambda = 0$ and the separation profile $\Lambda = -12$

Thus at $\Lambda = -12$, the skin friction is zero, and if $\Lambda < -12$, it's negative which corresponds to reversed flow and *boundary layer separation*. This is illustrated in Figure 17.2 for $\Lambda = -24$.

Whenever $\Lambda < 0$, the pressure gradient is adverse and the free-stream is decelerating. Whenever $\Lambda > 0$, the pressure gradient is favourable and the free-stream is accelerating.

17.3   Thwaites's Method for Laminar Boundary Layers

In Pohlhausen's family of quartic profiles, δ^*, θ, H, and c_f are all functions of the local pressure gradient, via Pohlhausen's pressure gradient parameter Λ. It turns out that if the shape factor H is plotted against another pressure gradient parameter

$$\lambda \equiv \frac{\theta^2}{\nu} \frac{\mathrm{d}u_\infty}{\mathrm{d}x} = \frac{-\theta^2}{u_\infty \mu} \frac{\mathrm{d}p}{\mathrm{d}x} \tag{17.3}$$

for several profiles obtained from exact laminar solutions of the boundary layer equations, then $H(\lambda)$ seems to be a universal function. Notice that Thwaites's pressure gradient parameter defined by Equation (17.3) differs from Pohlhausen's, as given by Equation (17.2), only in using the momentum thickness θ instead of the plain boundary layer thickness δ (which is always somewhat arbitrary):

$$\lambda \equiv \left(\frac{\theta}{\delta}\right)^2 \Lambda.$$

Further, the dimensionless skin friction parameter

$$l \equiv \frac{\mathrm{Re}_\theta \, c_f}{2} \equiv \frac{\theta}{u_\infty} \left.\frac{\partial u}{\partial y}\right|_{y=0} \tag{17.4}$$

also seems to be a universal function of λ for laminar boundary layers.

To apply Thwaites's ideas to Kármán's integral Equation (16.12), first substitute l for c_f from Equation (17.4) and multiply through by Re_θ

$$\frac{u_\infty \theta}{\nu} \frac{\mathrm{d}\theta}{\mathrm{d}x} + \frac{\theta^2}{\nu}(2+H)\frac{\mathrm{d}u_\infty}{\mathrm{d}x} = l,$$

and then replace the free-stream velocity gradient using Equation (17.3)

$$\frac{u_\infty \theta}{\nu} \frac{\mathrm{d}\theta}{\mathrm{d}x} + (2+H)\lambda = l,$$

and rearrange

$$\frac{u_\infty}{\nu} \frac{\mathrm{d}\left(\theta^2\right)}{\mathrm{d}x} = 2\{l - (2+H)\lambda\}. \tag{17.5}$$

So far this is equivalent to Kármán's integral relation; the novel part of Thwaites's method is assuming that the right-hand side is a universal function of λ (which is known, at least approximately).

17.3.1 $F(\lambda) \approx 0.45 - 6\lambda$

A simple choice that performs well is

$$2\{l - (2+H)\lambda\} \approx 0.45 - 6\lambda.$$

Substituting this into Equation (17.5) gives

$$u_\infty \frac{\mathrm{d}\left(\theta^2\right)}{\mathrm{d}x} = 0.45\nu - 6\theta^2 \frac{\mathrm{d}u_\infty}{\mathrm{d}x}$$

$$u_\infty \frac{\mathrm{d}\left(\theta^2\right)}{\mathrm{d}x} + 6\theta^2 \frac{\mathrm{d}u_\infty}{\mathrm{d}x} = 0.45\nu$$

$$u_\infty^6 \frac{\mathrm{d}\left(\theta^2\right)}{\mathrm{d}x} + 6\theta^2 u_\infty^5 \frac{\mathrm{d}u_\infty}{\mathrm{d}x} = 0.45\nu u_\infty^5$$

$$\frac{\mathrm{d}\left(\theta^2 u_\infty^6\right)}{\mathrm{d}x} = 0.45\nu u_\infty^5. \tag{17.6}$$

For known $u_\infty(x)$, this can be integrated downstream (analytically or numerically) to get $\theta(x)$, then $\lambda(x)$ can be calculated from Equation (17.3). Finally, the other parameters of interest, such as the shape factor H and in particular the skin friction parameter l are obtained from approximations to their universal functions of λ.

17.3.2 Correlations for Shape Factor and Skin Friction

The functions $H(\lambda)$ and $l(\lambda)$ have been tabulated, but more convenient are the formulae:

$$H(\lambda) \approx \begin{cases} 2.088 + \frac{0.0731}{0.14+\lambda}, & -0.1 < \lambda < 0 \\ 2.61 - 3.75\lambda + 5.24\lambda^2, & 0 < \lambda < 0.1 \end{cases}$$

$$l(\lambda) \approx \begin{cases} 0.22 + 1.402\lambda + \frac{0.018\lambda}{0.107+\lambda}, & -0.1 < \lambda < 0 \\ 0.22 + 1.57\lambda - 1.8\lambda^2, & 0 < \lambda < 0.1. \end{cases}$$

Notice that the skin friction parameter $l(\lambda)$ vanishes near $\lambda = -0.090$, and recall that $l = 0$ corresponds to boundary layer separation.

17.3.3 Example: Zero Pressure Gradient

As a simple example, let's apply Thwaites's method to a flat plate with zero pressure gradient; assume the boundary layer begins with zero thickness at the leading edge

$$\theta(0) = 0.$$

The zero pressure gradient corresponds to the boundary layer on a thin flat plate at zero incidence.

In the ordinary differential equation

$$\frac{d(\theta^2 u_\infty^6)}{dx} = 0.45\nu u_\infty^5,$$

u_∞ is constant so

$$\frac{d(\theta^2)}{dx} = \frac{0.45\nu}{u_\infty}$$

which can be integrated to give

$$\theta = \sqrt{\frac{0.45\nu x}{u_\infty}},$$

the constant of integration being zero if the boundary layer thickness begins at zero at the leading edge (this is the case for a sharp leading edge on a flat plate).

This can be written nondimensionally as

$$\frac{\theta}{x} = \frac{0.671}{\sqrt{Re_x}}$$

or

$$Re_\theta = 0.671 Re_x^{1/2}.$$

From the correlations of Section 17.3.2, noting that $\lambda = 0$ throughout for zero pressure gradient, we have $H \approx 2.61$ and $l \approx 0.22$; the latter means the skin friction coefficient is

$$c_f \approx \frac{0.44}{\mathrm{Re}_\theta} = \frac{0.656}{\sqrt{\mathrm{Re}_x}}.$$

This turns out to be very close to the exact answer, $0.664\mathrm{Re}_x^{-1/2}$, which can be obtained by a similarity transformation (see the exercise in Section 16.4).

17.3.4 Example: Laminar Separation from a Circular Cylinder

The full aerodynamics of viscous flow over a circular cylinder is very complicated, involving large-scale unsteadiness in the form of vortex shedding; however, Thwaites's method can be used to show that the laminar boundary layer will fail to reach the rear stagnation point.

The complex velocity for the flow without circulation with free-stream speed q_∞ over a circular cylinder of radius a is (cf. Equation 3.6)

$$w = q_\infty \left(1 - \frac{a^2}{z^2} \right), \tag{17.7}$$

so that the azimuthal velocity on the surface $|z| = a$ is

$$v_\theta = -\Im \left\{ e^{i\theta} w(ae^{i\theta}) \right\} = -2q_\infty \sin\theta$$

(here, for the moment, θ is the polar coordinate).

Take the longitudinal boundary layer coordinate x to begin at the forward stagnation point ($z = -a$) and to follow the upper surface; i.e. $x = a(\pi - \theta)$. It is convenient to nondimensionalize this as $\xi \equiv x/a$. Then, relative to this, the external tangential speed for the boundary layer is

$$u_\infty(\xi) = 2q_\infty \sin(\pi - \xi) \equiv 2q_\infty \sin\xi.$$

Thus over the forward part of the upper surface ($0 < \xi < \frac{\pi}{2}$) the inviscid surface speed is accelerating,

$$\frac{du_\infty}{d\xi} = 2q_\infty \cos\xi$$

and so the pressure gradient

$$\frac{dp}{d\xi} = -\rho u_\infty \frac{du_\infty}{d\xi} = -2\rho q_\infty \sin 2\xi$$

is favourable (negative) whereas it becomes unfavourable (positive) thereafter ($\frac{\pi}{2} < \xi < \pi$). Notice that x and ξ follow the curve of the boundary, and similarly the profiles and thicknesses are measured normal to the boundary; as mentioned in Section 16.1.8, this curvature does not introduce any new terms into the equations provided the radius of curvature is much greater than the boundary layer thickness.

In terms of ξ, Thwaites's form of the momentum integral Equation (17.6) is (hereafter θ is the momentum thickness again)

$$\frac{d(\theta^2 u_\infty^6)}{d\xi} = 0.45 v a u_\infty^5$$

$$\frac{d(\theta^2 \sin^6 \xi)}{d\xi} = \frac{0.45 v a}{2 q_\infty} \sin^5 \xi$$

which can be integrated to give

$$\theta^2 \sin^6 \xi = \frac{-0.45 v a}{2 q_\infty} \left(\cos \xi - \frac{2}{3} \cos^3 \xi + \frac{1}{5} \cos^5 \xi + \text{const.} \right)$$

$$\theta^2 = \frac{-0.45 v a}{2 q_\infty} \frac{\cos \xi - \frac{2}{3} \cos^3 \xi + \frac{1}{5} \cos^5 \xi + \text{const.}}{\sin^6 \xi}.$$

The integration constant must be $\frac{-8}{15}$ if θ is to be finite at the forward stagnation point $\xi = 0$, so

$$\theta^2 = \frac{-0.45 v a}{2 q_\infty} \frac{\cos \xi - \frac{2}{3} \cos^3 \xi + \frac{1}{5} \cos^5 \xi + \frac{8}{15}}{\sin^6 \xi}$$

$$\frac{\theta^2}{a^2} = \frac{0.015}{\text{Re}_a} \left(\frac{8 - 15 \cos \xi + 10 \cos^3 \xi - 3 \cos^5 \xi}{\sin^6 \xi} \right), \tag{17.8}$$

where

$$\text{Re}_a \equiv \frac{q_\infty a}{\nu}$$

is the Reynolds number based on the cylinder's radius. The predicted momentum thickness is displayed around the cylinder in Figure 17.3.

Given this solution for $\theta(\xi)$, Thwaites's pressure gradient parameter as defined by Equation (17.3) can be computed as

$$\lambda(\xi) = 0.03 \left\{ \frac{8 - 15 \cos \xi + 10 \cos^3 \xi - 3 \cos^5 \xi}{\sin^6 \xi} \right\} \cos \xi,$$

and then the variation along the boundary layer of the shape factor H and skin friction factor l computed; the results are plotted in Figure 17.4. It can be seen that the skin friction factor l vanishes near $103°$. At this point the boundary layer would separate.

That the boundary layer separates means that the outer supposedly inviscid region is changed. This invalidates the procedure used, since our assumption of a perfect fluid solution holding everywhere except in a thin layer near the boundary is untrue. This means that we do not obtain a realistic prediction for the laminar boundary layer which does occur on the forward part of a circular cylinder, neither for the location of the separation point. What we do obtain is a very good argument that the flow picture cannot consist of the basic perfect fluid flow plus thin laminar boundary layers. In fact, the displacement of the exterior irrotational flow by the separated region pushes the points of minimum pressure forwards from $90°$, and the actual

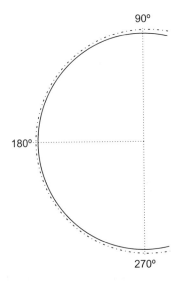

Figure 17.3 Growth of the laminar boundary layer (momentum) thickness circular cylinder in a uniform stream (moving from left-to-right), as predicted by Equation (17.8) for $Re_a = 100$. The model assumes, wrongly, that the external flow is given by the complex velocity of Equation (17.7), $w = q_\infty \left\{ 1 - (a/z)^2 \right\}$

laminar separation points forwards too. The failure to obtain a quantitative result should not be attributed to Thwaites's method, but rather to the assumption about the surrounding inviscid flow; any boundary layer method requires an accurate description of the prevailing longitudinal pressure gradient as input.

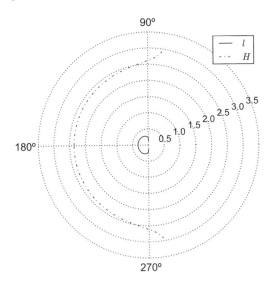

Figure 17.4 Variation of shape factor and skin friction factor along laminar boundary layer on a circular cylinder (assuming attached flow)

Another question to be addressed in this and similar flow configurations is whether and where the laminar boundary layers undergo transition to turbulence. This is outside the scope of this course; see the suggestions for further reading in Section 17.5.

In a sense only a negative result has been obtained here, but it does highlight the importance of adverse pressure gradients in aerodynamics. Notice that in the circular cylinder problem, where the minimum pressure was supposed to occur at 90°, the laminar boundary layer did not proceed far past this into the adverse region without separating.

17.4   Exercises

1. Plot the following velocity profiles:

    (a) $\dfrac{u}{u_\infty} = \mathrm{erf}\, \eta$

    (b) $\dfrac{u}{u_\infty} = 1 - e^{-\eta}$

    and use them to solve Kármán's integral relation for a flat plate at zero incidence.

    (c) Which of Pohlhausen's quartics is appropriate for this problem? Use it.

2. For constant η_0 which of the conditions satisfied by Pohlhausen's quartic boundary layer velocity profiles are satisfied by

$$\frac{u}{u_\infty} = \begin{cases} \dfrac{\sin \frac{\pi(\eta-\eta_0)}{2(1-\eta_0)} + \sin \frac{\pi\eta_0}{2(1-\eta_0)}}{1 + \sin \frac{\pi\eta_0}{2(1-\eta_0)}}, & 0 < \eta < 1; \\[4mm] 1 & 1 < \eta < \infty? \end{cases}$$

    (a) Give a relation between the constant η_0 and the pressure gradient.

    (b) What value of η_0 should be used for the flat plate at zero incidence? Use it to solve Kármán's integral relation.

3. Consider the steady plane ideal flow

$$u - iv = \kappa(x + iy),$$

    with a positive real κ constant.

    (a) What are the dimensions of κ?

    (b) Show that $y = 0$ is a streamline.

    (c) How does the pressure vary along $y = 0$?

4. Assume $y = 0$ is a solid surface. Show that if we assume a Pohlhausen quartic profile with constant thickness δ then the pressure gradient parameter is constant too. Show that Kármán's integral relation reduces to a polynomial in the pressure gradient parameter. Show that this polynomial has a root in the range $0 < \Lambda < 10$, and try and locate it more closely.

5. The plane ideal flow outside a solid two-dimensional wedge occupying the sector $(2 - \beta)\pi < \theta < 2\pi$ with $0 < \beta \le 1$, should satisfy the boundary condition that there be no flow component normal to the surface of the wedge.

If r and θ are the polar coordinates in the plane (the latter measured anticlockwise from the positive x-axis so that $0 < \theta < 2\pi$) and a is a constant, consider the two-dimensional velocity field defined for $0 < \theta < (2 - \beta)\pi$ with speed

$$\sqrt{u^2 + v^2} = ar^{\beta/(2-\beta)}$$

and directional slope

$$\frac{v}{u} = \tan\frac{-\beta\theta}{2 - \beta}.$$

(a) Sketch the flow.
(b) What are the dimensions of a?
(c) Show that it is irrotational, divergence-free, and satisfies the boundary conditions.
(d) Describe the variation of velocity over the surface.
(e) Using Thwaites's method, show that the viscous boundary layer along the surface $y = 0$ grows away from the vertex with momentum thickness $\theta(x)$ such that the pressure gradient parameter

$$\lambda \equiv \frac{\theta^2}{\nu}\frac{du_\infty}{dx}$$

is constant. Assume for the purpose of calculating $\theta(x)$ that $\theta(0)$ is finite.
Does Pohlhausen's pressure gradient parameter vary along the boundary layer?
(f) Assuming that the skin friction coefficient c_f is related to the pressure gradient parameter λ in such a way that c_f is positive whenever $\lambda > -0.09$, show that laminar separation is not to be expected for any wedge, regardless of the angle it includes.
(g) How do the skin friction coefficient and surface shear stress vary with distance from the vertex?
Show that $\sqrt{Re_x}c_f$ is uniform along the surface.

6. (a) Try developing a family of cubic profiles, dropping the second smoothness condition at the outside of the boundary layer.
(b) What about quadratic?
(c) The only reasonable linear profile is that of Section 16.3.3; analyse it in this light.

17.5   Further Reading

See also Goldstein (1938) for the application of Pohlhausen's quartic profiles to laminar boundary layers.

For Thwaites's method, see Thwaites (1987) and Moran (2003).

The transition from laminar to turbulent in aerodynamical boundary layers is discussed by Abbott and von Doenhoff (1959), Thwaites (1987), Kuethe and Chow (1998), Bertin (2002), Moran (2003), and Anderson (2007).

For descriptions of the varieties of viscous flow over circular cylinders possible at different Reynolds numbers, including vortex shedding, see Goldstein (1938), Kuethe and Chow (1998), or Anderson (2007).

References

Abbott, I.H. and von Doenhoff, A.E. (1959) *Theory of Wing Sections*. New York: Dover.

Anderson, J.D. (2007) *Fundamentals of Aerodynamics*, 4th edn. New York: McGraw-Hill.

Bertin, J.J. (2002) *Aerodynamics for Engineers*, 4th edn. Upper Saddle River, New Jersey: Prentice Hall.

Goldstein, S. (1938) *Modern Developments in Fluid Dynamics*. Oxford: Clarendon Press.

Kuethe, A.M. and Chow, C.Y. (1998) *Foundations of Aerodynamics*, 5th edn. New York: John Wiley & Sons, Ltd.

Moran, J. (2003) *An Introduction to Theoretical and Computational Aerodynamics*. New York: Dover.

Pohlhausen, K. (1921) Zur näherungsweisen Integration der Differentialgleichung der laminaren Grenzschicht. *Z. Angew. Math. Mech.* **1**(4).

Thwaites, B. (ed.) 1987 *Incompressible Aerodynamics*. New York: Dover.

18

Compressibility

While restricting speeds to a small fraction of that of sound might render the flow incompressible and the aerodynamics more susceptible to analysis, there are practical imperatives for increasing the speed of flight—passengers and payload want to get where they're going and proprietors want to maximize the utilization of their aircraft—and therefore the theory must be extended. The primary weakness of the model developed thus far as the speed increases is its assumption of uniform density. It is apparent from the differential Bernoulli Equation (2.13) that increasing speed leads to decreasing pressure, and, except for the fictitious 'perfect fluid' of Section 1.4.1 or equally the fictitious linearly viscous incompressible fluid of Section 15.2.2, a reduction in the pressure on a body allows it to expand. That is, unless the density is artificially constrained in the model to be constant, decreasing pressure should lead to decreasing density. In general, pressure is linked to density by the *equation of state* and this limitation of the incompressible theory is expected. This chapter introduces the additional physics required for extensions to the theory to permit the raising of its speed limit.

18.1 Steady-State Conservation of Mass

At steady-state, the net mass of fluid inside a given fixed region $\mathcal{R}$ will be constant; therefore the net flux of mass through its bounding surface $\mathcal{S}$ must be zero; i.e.

$$\int_{\mathcal{S}} \hat{\boldsymbol{n}} \cdot (\rho \boldsymbol{q}) \, \mathrm{d}S = 0,$$

or by the divergence theorem (Equation 9.7)

$$\int_{\mathcal{R}} \nabla \cdot (\rho \boldsymbol{q}) \, \mathrm{d}V = 0.$$

Since the region $\mathcal{R}$ is arbitrary, this can only hold if $\rho \boldsymbol{q}$ is divergence-free:

$$\nabla \cdot (\rho \boldsymbol{q}) = 0. \tag{18.1}$$

Theory of Lift: Introductory Computational Aerodynamics in MATLAB®/Octave, First Edition. G. D. McBain.
© 2012 John Wiley & Sons, Ltd. Published 2012 by John Wiley & Sons, Ltd.

Expand the steady conservation of mass Equation (18.1) using the product rule for the divergence (Equation 9.4):

$$\rho \nabla \cdot \boldsymbol{q} + \boldsymbol{q} \cdot \nabla \rho = 0.$$

Then take the steady Euler Equation (9.11) and assume that the pressure is some function of density (i.e. the fluid is *barotropic*) and use the vector-calculus chain-rule Equation (9.2) to expand the gradient of that function:

$$\rho \boldsymbol{q} \cdot \nabla \boldsymbol{q} = -\nabla p = -\frac{\mathrm{d}p}{\mathrm{d}\rho} \nabla \rho.$$

Now eliminate the density gradient between the last two equations,

$$\nabla \cdot \boldsymbol{q} = \left(\frac{\mathrm{d}p}{\mathrm{d}\rho} \right)^{-1} \boldsymbol{q} \cdot (\nabla \boldsymbol{q}) \cdot \boldsymbol{q}.$$

The left-hand side is the divergence of the velocity field, the vanishing of which was the Equation (2.3) of the conservation of mass in the incompressible case. The right-hand side is divided by the derivative of pressure with respect to density. This factor can also be expressed in terms of the *compressibility* κ, which is defined as the fractional rate of change of density with respect to pressure:

$$\kappa \equiv \frac{1}{\rho} \frac{\mathrm{d}\rho}{\mathrm{d}p}; \tag{18.2}$$

thus:

$$\nabla \cdot \boldsymbol{q} = \rho \kappa \, \boldsymbol{q} \cdot (\nabla \boldsymbol{q}) \cdot \boldsymbol{q}.$$

If the compressibility is low, i.e. the density varies little with pressure, the right-hand side will be small.

Further, the dimensions of the rate of change of pressure with respect to density are the square of those of velocity; thus it can be called the square of a speed. This speed is called the *speed of sound* and is denoted by a.

$$\nabla \cdot \boldsymbol{q} \equiv \frac{1}{a^2} \, \boldsymbol{q} \cdot (\nabla \boldsymbol{q}) \cdot \boldsymbol{q}. \tag{18.3}$$

$$\text{where } a^2 \equiv \frac{\mathrm{d}p}{\mathrm{d}\rho}. \tag{18.4}$$

Acoustics is not of central interest here, but this speed is. It is not necessary to digress into the analysis of the propagation of sound waves, which are oscillations of pressure and density, to appreciate the importance of this speed in aerodynamics. For the present purposes, the name is a conventional convenience. The significance applies equally to the steady-state flows of present interest, since, pressure signals travel at the speed of sound, so if the flow from one point to another further down its streamline is supersonic, there will be neither convective nor compressive information from the second point back to the first, and so no influence whatever from the second point to the first. This is in strong contradistinction to the modelling of incompressible flows in Chapters 3–12 by means of sources and vortices with fields of influence extending to infinity in all directions (as shown in Figures 3.3 and 3.6).

Observe that the conservation of mass reduces to the divergence-free Equation (2.3) when the speed of sound tends to infinity, which is another way of interpreting incompressibility. This reinforces the idea that the divergence-free condition is a good approximation when the relevant speeds are much less than the speed of sound. In an incompressible fluid, the pressure signals are propagated instantaneously throughout the domain and can propagate upstream against any flow no matter how fast.

It is important to note that in Equation (18.3), the speed of sound a is not a constant but depends on the pressure or density, both of which are variables in compressible flow (though linked together by some definite function if the fluid can be treated as barotropic).

18.2   Longitudinal Variation of Stream Tube Section

In order to conserve mass, the mass flow-rate is constant along any stream tube, so if its cross-sectional area is A and the average velocity over that section is q,

$$d(\rho A q) = 0$$

$$\rho A dq + \rho q dA + q A d\rho = 0.$$

If the fluid is barotropic, the speed of sound may be introduced, and then Bernoulli's differential Equation (2.13), to reduce this to a relation between the change in speed and the change in area:

$$\rho A dq + \rho q dA + \frac{q A}{a^2} dp = 0$$

$$A dq + q dA - \frac{q^2 A}{a^2} dq = 0$$

$$A \left(1 - \text{Ma}^2\right) dq + q dA = 0,$$

where, as in Equation 1.12,

$$\text{Ma} \equiv \frac{q}{a}$$

is the *Mach number*: the ratio of the speed to the speed of sound.

For *subsonic* flow $\text{Ma} < 1$ and a decrease in area corresponds to an increase in velocity. This is called the *Venturi effect* and applies as well in incompressible flow ($\text{Ma} = 0$) in which $dp = 0$ and so the velocity is inversely proportional to the stream tube sectional area of the stream tube, as in Equation (10.2). A particularly relevant instance is the air passing over a thick cambered wing section: the obstruction to the stream tube posed by the aerofoil causes acceleration and a corresponding decrease in pressure.

As the speed and Mach number increase from zero, the increase in velocity into the contraction is no longer inversely proportional the area, as the coefficient $1 - \text{Ma}^2$ decreases. In fact, the same contraction causes more acceleration at higher (though still subsonic) Mach number. This is because if the fluid is compressible, the decrease in pressure with acceleration leads to a decrease in density and therefore even more volume of fluid must be passed through the section to maintain the constant mass flow-rate.

For *supersonic* flow, Ma > 1 and the sign of the differential coefficient changes; i.e. when flowing faster than the speed of sound, a stream tube expands if the fluid is accelerating. An immediate consequence of this simple analysis is that thickness and camber are of no use in generating suction above the wing in supersonic flight, and indeed, wings designed for high speeds are generally thinner and flatter.

18.2.1   The Design of Supersonic Nozzles

Another practical application of these considerations is in the design of high-speed nozzles, e.g. for turbines or rockets. It follows that the shape of a nozzle through which fluid is to issue from a compressed reservoir must first converge and then diverge, if the fluid is to finally exit at a speed above that of sound. In such a converging–diverging nozzle shape, the speed of sound can only be attained at the *throat*, or section of minimum area, and then only if the inlet pressure is sufficient; moreover, the speed at the throat cannot exceed that of sound, which phenomenon is called *choking* and can be exploited in the analysis of supersonic nozzles.

18.3   Perfect Gas Thermodynamics

The thermodynamics of incompressible fluids are rather trivial, which is another way of saying that the incompressible model is only capable of reproducing a limited range of behaviours, but the behaviour of flows in which the density varies dynamically is strongly dependent on the thermal and caloric equations of state.

18.3.1   Thermal and Caloric Equations of State

As noted, in barotropic fluids, the pressure depends only density, but the pressure and density of a substance are more generally related by a *thermal equation of state*, which also involves the temperature.

A *thermally* perfect gas obeys as equation of state the ideal gas equation

$$p = \rho RT, \tag{18.5}$$

where R is a constant for each gas, being the ratio of the *universal gas constant* (≈ 8314.472 J/K kmol) to the molar mass. For air the latter is about 28.97 kg/kmol so $R \approx 287.0$ J/kg K.

A thermally perfect gas also has an isochoric specific heat c_v which is independent of pressure. If the gas is also *calorically* perfect, this coefficient is independent of temperature too, so changes in the specific internal energy u are simply proportional to changes in temperature:

$$du = c_v dT.$$

The internal energy is a quantity defined by the equivalence between heat and mechanical work; i.e. the First Law of Thermodynamics.

18.3.2   The First Law of Thermodynamics

The First Law of Thermodynamics states that the increase in internal energy is the sum of the heat δQ and work added to a system. The work done in ideal flow is the product of the pressure and the decrease in volume, i.e. for a unit mass, for which the volume is the reciprocal of the density, the work done is

$$-p\mathrm{d}\left(\frac{1}{\rho}\right) = +\left(\frac{p}{\rho^2}\right)\mathrm{d}\rho,$$

so for a system consisting of a unit mass of calorically perfect gas,

$$c_\mathrm{v}\mathrm{d}T = \delta Q + \frac{p\mathrm{d}\rho}{\rho^2}. \tag{18.6}$$

18.3.3   The Isochoric and Isobaric Specific Heat Coefficients

If the unit mass system of gas is held in a rigid container, no work is done because the walls cannot move. Equivalently, its volume cannot change and the density is constant ($\mathrm{d}\rho = 0$); either way, $\delta Q = c_\mathrm{v}\mathrm{d}T$, which is why c_v is called the *isochoric* specific heat: it is the amount of energy in the form of heat required to raise the temperature of a unit mass of the substance by one unit under conditions of constant volume.

So how does the amount of heat which is required to raise the temperature by a fixed amount change if the gas is subject to constant pressure rather than held in a container of fixed volume? In general the density changes, in fact in accordance with the thermal Equation (18.5) of state,

$$\mathrm{d}p = R\rho\mathrm{d}T + RT\mathrm{d}\rho, \tag{18.7}$$

but in an *isobaric* process $\mathrm{d}p = 0$ so this reduces to

$$T\mathrm{d}\rho = -\rho\mathrm{d}T,$$

and so the First Law (Equation 18.6) becomes

$$c_\mathrm{v}\mathrm{d}T = \delta Q - \frac{p\mathrm{d}T}{\rho T}$$

$$\left(c_\mathrm{v} + \frac{p}{\rho T}\right)\mathrm{d}T = \delta Q$$

$$(c_\mathrm{v} + R)\,\mathrm{d}T = \delta Q.$$

Thus the *isobaric specific heat* c_p of a perfect gas is

$$c_\mathrm{p} = c_\mathrm{v} + R \tag{18.8}$$

18.3.4   Isothermal and Adiabatic Processes

In certain circumstances, the variation of temperature can be related to the variation of pressure and density and so eliminated from the equation of state, leaving it barotropic.

The simplest example would be *isothermal* processes, for which the temperature is constant and therefore the pressure would simply be proportional to the density: $p(\rho) = RT\rho$.

However, for a process to be isothermal, any heat generated must be instantaneously conducted away to the environment, i.e. on a timescale which is short compared to those of interest—isothermal processes are slow processes, or processes are asymptotically isothermal in the limit that they are carried out ever more slowly. Since air is a fairly poor conductor, this tends not to be a good model for aerodynamics; instead, a much better description is that the parcels of air flow by a wing (increasing in pressure as it approaches a stagnation point, or decreasing in pressure as it passes over the wing) before any appreciable thermal conduction has had time to occur. This leads to a consideration of the opposite limit: the *adiabatic* process.

18.3.5 Adiabatic Expansion

Without the addition of heat, $\delta Q = 0$ and the First Law for a perfect gas (Equation 18.6) reduces to

$$c_{\mathrm{v}} \mathrm{d}T = \frac{p\,\mathrm{d}\rho}{\rho^2}.$$

The temperature differential can be expressed in terms of those of pressure and density using the differential thermal equation of state (Equation 18.7), so

$$c_{\mathrm{v}} \left(\frac{1}{R\rho}\,\mathrm{d}p - \frac{T}{\rho}\,\mathrm{d}\rho \right) = \frac{p\,\mathrm{d}\rho}{\rho^2}$$

$$\frac{c_{\mathrm{v}}}{R\rho}\,\mathrm{d}p = \frac{T}{\rho} \left(c_{\mathrm{v}} + \frac{p}{\rho T} \right)\,\mathrm{d}\rho$$

$$= \frac{T c_{\mathrm{p}}}{\rho}\,\mathrm{d}\rho$$

$$c_{\mathrm{v}}\,\mathrm{d}p = RT c_{\mathrm{p}}\,\mathrm{d}\rho.$$

Then using the thermal Equation (18.5) of state and defining the ratio of specific heats

$$\gamma \equiv \frac{c_{\mathrm{p}}}{c_{\mathrm{v}}}, \tag{18.9}$$

the pressure and density are related adiabatically by

$$\rho\,\mathrm{d}p = \gamma p\,\mathrm{d}\rho. \tag{18.10}$$

The ratio of specific heats for air is very near $\frac{7}{5}$ at conditions relevant for lower atmospheric flight. It decreases slightly with heating, but remains within a few percent of that value up to temperatures of several hundred degrees (Celsius or Kelvin). Insofar as γ can be taken to be independent of temperature, Equation (18.10) is a differential equation between pressure and density and so a fluid obeying it is barotropic, as required above in the derivation of the nonlinear continuity Equation (18.3).

18.3.6   The Speed of Sound and Temperature

From the definition of the speed of sound in Equation (18.4)

$$a \equiv \sqrt{\frac{\mathrm{d}p}{\mathrm{d}\rho}},$$

the adiabatic differential Equation (18.10)

$$\frac{\mathrm{d}p}{\mathrm{d}\rho} = \frac{\gamma p}{\rho},$$

and the ideal gas Equation (18.5)

$$\frac{p}{\rho} = RT,$$

it may be seen that the speed of sound in a given gas depends solely on temperature:

$$a = \sqrt{\gamma RT}. \tag{18.11}$$

18.3.7   The Speed of Sound and the Speed

Where possible, however, the introduction of temperature as an another dependent variable is avoided, so it is more convenient to try and express the speed of sound in terms of purely mechanical variables.

Differentiating the definition of the speed of sound (Equation 18.4), using the quotient rule, and finally using the adiabatic differential Equation (18.10) gives

$$\mathrm{d}a^2 = \mathrm{d}\left(\frac{\mathrm{d}p}{\mathrm{d}\rho}\right) = \mathrm{d}\left(\frac{\gamma p}{\rho}\right) = \frac{\gamma}{\rho^2}(\rho\,\mathrm{d}p - p\,\mathrm{d}\rho) = (\gamma - 1)\frac{\mathrm{d}p}{\rho}.$$

This can be combined with the differential Bernoulli Equation (2.13) to express the adiabatic variation of the speed of sound in terms of the local velocity.

$$\frac{\mathrm{d}q^2}{2} = \frac{-\mathrm{d}p}{\rho}$$

$$= \frac{-\mathrm{d}a^2}{\gamma - 1}.$$

If the streamlines originate from a uniform upstream where $q = q_\infty$ and $a = a_\infty$, this can be integrated from there to any point downstream in the flow to give

$$\frac{q^2 - q_\infty^2}{2} = \frac{a_\infty^2 - a^2}{\gamma - 1}. \tag{18.12}$$

18.3.8   Thermodynamic Characteristics of Air

At very high (*hypersonic*) speeds, the extreme temperatures generated dynamically lead to chemical changes in the nature of the air and so its gas constant R is no longer constant; also these additional modes of energy conversion lead to dramatic changes in the specific heats. Thus over high temperature ranges, air is neither thermally nor calorically perfect; however, in the troposphere and stratosphere, below hypersonic speeds, air behaves much as a perfect gas with a constant molar mass and a constant ratio of specific heats.

18.3.9   Example: Stagnation Temperature

If a stream of air at 900 km/hr, 9 km up in the International Standard Atmosphere, is brought to rest adiabatically, what temperature does it attain?

Solution: Recall that from Equation (18.11) the speed of sound is just a function of temperature for a given gas. This function can be inverted:

$$T = \frac{a^2}{\gamma R}.$$

The speed of sound at rest, $a = a_0$, can be computed from the free-stream speed q_∞ and speed of sound a_∞ using Equation (18.12), with $q = q_0 = 0$. The free-stream speed is $q_\infty = 900$ km/hr $= 250$ m/s, and the free-stream speed of sound at the temperature 229.65 K prevailing at an altitude of 9 km according to Equations (1.8) and (1.4) is $a_\infty = 303.77$ m/s, so:

$$a_0 = \sqrt{a_\infty^2 + \frac{\gamma - 1}{2} q_\infty^2} = 323.69 \text{ m/s}.$$

Thus $T_0 = 260.76$ K; i.e. the air is heated by a little over 30 K.

18.4   Exercises

1. Show that the dimensions of the rate of change of pressure with respect to density are the square of those of velocity.

2. From its definition in Equation (18.2), what would be the SI units of compressibility?

3. The compressibility of condensed phases (i.e. liquids and solids as opposed to gases) is often tabulated in terms of its reciprocal, the *bulk modulus*.
   (a) Look up the compressibility or bulk modulus and density of liquid water at normal conditions (e.g. 1 atmosphere and 15°C). What would be the speed of sound in water?
   (b) Find typical values of the compressibility of some common solids, e.g. steel, and estimate their speed of sound.

4. Derive from Equation (1.4) and Equation (18.12) a formula for the temperature along an adiabatic streamline in terms of the free-stream velocity and temperature, the local speed, the ratio of specific heats, and the gas constant. [*Ans.:* $T = T_\infty + (\gamma - 1)(q_\infty^2 - q^2)/2\gamma R$.]

5. How hot might the forward stagnation point of a wing get flying steadily at 920 km/hr, 8800 m up? (Neglect viscous dissipation, which adds to the heating, on the one hand and thermal conduction, which reduces it, on the other.)

6. While the isochoric and isobaric specific heat coefficients of air (and many other gases) are well tabulated, show that for any perfect gas they can be simply computed from the gas constant and ratio of specific heats by

$$c_\mathrm{v} = \frac{\gamma R}{\gamma - 1}$$

and then Equation (18.9). [*Hint:* Use Equation (18.8).]

7. Say you would like to raise the temperature of your 3 m × 4 m × 5 m study from 15 to 25°C. Assuming the walls are sealed, rigid, and well insulated, and neglecting the heat capacity of the furnishings and occupants, as well as any heat generated by the latter, how long would you have to run a 2 kW heater?

8. If the air at an adiabatic stagnation point not far above sea-level in the International Standard Atmosphere has reached 25°C, what must the speed and Mach number be in the free-stream? [*Ans.:* 141 m/s, Ma = 0.42.]

18.5   Further Reading

Other cursory outlines of theories of the effects of compressibility in aerodynamics are given by Abbott and von Doenhoff (1959), Milne-Thomson (1973) and Moran (2003). More detail may be found in Jones and Cohen (1960), Kuethe and Chow (1998), and Anderson (2007).

Deeper investigations of compressbility necessarily more closely intertwine aerodynamics and thermodynamics (Liepmann and Roshko 1957; Bers 1958). An excellent general introduction to thermodynamics is given by Callen (1960); see also Van Wylen and Sonntag (1985).

References

Abbott, I.H. and von Doenhoff, A.E. (1959) *Theory of Wing Sections*. New York: Dover.
Anderson, J.D. (2007) *Fundamentals of Aerodynamics*, 4th edn. New York: McGraw-Hill.
Bers, L. (1958) *Mathematical Aspects of Subsonic and Transonic Gas Dynamics*. New York: John Wiley & Sons, Ltd.
Callen, H.B. (1960) *Thermodynamics*. New York: John Wiley & Sons, Ltd.
Jones, R.T. and Cohen, D. (1960) *High Speed Wing Theory*. Princeton Aeronautical Paperbacks, vol. 6. New Jersey: Princeton University Press.
Kuethe, A.M. and Chow, C.Y. (1998) *Foundations of Aerodynamics*, 5th edn. New York: John Wiley & Sons, Ltd.
Liepmann, H.W. and Roshko, A. (1957) *Elements of Gasdynamics*. New York: John Wiley & Sons, Ltd.
Milne-Thomson, L.M. (1973) *Theoretical Aerodynamics*, 4th edn. New York: Dover.
Moran, J. (2003) *An Introduction to Theoretical and Computational Aerodynamics*. New York: Dover.
Van Wylen, G.J. and Sonntag, R.E. (1985) *Fundamentals of Classical Thermodynamics*, 3rd edn. New York: John Wiley & Sons, Ltd.

19

Linearized Compressible Flow

Although somewhat simplistic in its thermodynamics, Chapter 18 has provided a sufficient basis to allow the extension of our theoretical aerodynamical framework to very much higher speeds, certainly well past the speed of sound. The key pieces to be taken from it are the nonlinear continuity Equation (18.3) for the scalar potential and Equation (18.12) to express the local speed of sound in terms of the local velocity.

While we now have enough thermodynamics for this raising of the speed limit, we content ourselves in the final chapter of this book, with showing that in fact the effort put into the fictitious incompressible perfect fluid in Chapters 2–14 was not merely a mathematical training exercise for real aerodynamics since a simple transformation converts a streamlined subsonic flow into a related incompressible flow.

The two pillars of aerodynamics are the conservation of mass and momentum. In the ideal theory these are represented by the velocity's being divergence-free and irrotational, and in plane ideal flow by the existence of an analytic complex velocity, or of a stream function and a potential. The potential can also be defined for three-dimensional flow, and requires only that the circulation of loops moving with the flow is preserved. In fact, Kelvin's theorem on the persistence of circulation (Section 9.5.2) follows from the Euler momentum Equation (9.11) and does not rely on the uniformity of density. That is, so long as the fluid is inviscid, it still holds that circulation persists, and so aerodynamical flows originating from irrotational uniform free-streams will still be irrotational and therefore possess a potential.

For an ideal flow, the potential is governed by the conservation of mass which leads to the Laplace Equation (9.10). Its derivation took the conservation of mass in terms of density and velocity and expressed the latter in terms of the potential. Its generalization for compressibility through Equation (18.1) to Equation (18.3) followed this same path but without assuming uniform density. In Chapter 19, we set about solving this generalized continuity equation.

19.1 The Nonlinearity of the Equation for the Potential

The great difference which enters when compressibility is admitted is not that the full continuity Equation (18.3) has a few more terms than its incompressible counterpart, Equation (9.10); it

Theory of Lift: Introductory Computational Aerodynamics in MATLAB®/Octave, First Edition. G. D. McBain.
© 2012 John Wiley & Sons, Ltd. Published 2012 by John Wiley & Sons, Ltd.

is that it is nonlinear. Even without the immense trove of analytic complex velocity solutions immediately opened up by the link with the Cauchy–Riemann Equations (2.22), it might still be possible to find a few simple solutions, more or less comparable to those of the incompressible case in Chapter 3, as for example the two-dimensional line-vortex was extended and deformed into the third dimension in Chapter 10, but these will be much less useful because nonlinearity means that it is no longer possible to combine multiple fundamental solutions to generate new solutions.

As discussed in Section 3.3, a linear operator $\mathcal{L}$ has the property of *additivity*: if ϕ_1 and ϕ_2 are operands (e.g. velocity potentials) and k_1 and k_2 are constants then $\mathcal{L}(k_1\phi_1 + k_2\phi_2) = k_1\mathcal{L}\phi_1 + k_2\mathcal{L}\phi_2$. Thus if ϕ_1 and ϕ_2 are solutions of the linear equation $\mathcal{L}\phi = 0$ then so is their linear combination:

$$\mathcal{L}(k_1\phi_1 + k_2\phi_2) = k_1\mathcal{L}\phi_1 + k_2\mathcal{L}\phi_2$$
$$= k_1 \times 0 + k_2 \times 0$$
$$= 0.$$

Partial derivatives are linear, which is why the Cauchy–Riemann Equations (2.22) and the three-dimensional incompressible continuity Equation (9.10) are too. Here, for the nonlinear compressible potential Equation (18.3), attempting to substitute a linear combination for the potential falls foul of the triple product in the second term, not to mention the speed-dependent Equation (18.12) for the speed of sound a.

This loss of linearity is very debilitating to the analysis, but it turns out that linearity can be recovered for the important case of the thin wing approximation as already used in the incompressible limit in two dimensions with the thin aerofoil theory of Chapter 6 and in three dimensions with the lifting line of Chapters 11 and 12 and the horseshoe vortex lattices of Chapters 13 and 14.

19.2   Small Disturbances to the Free-Stream

If a wing disturbs the free-stream only slightly, the properties of the air might be nearly constant at their free-stream values throughout the flow-field. Although a very severe restriction on the range of all possible compressible flows, this tiny subset is of particular practical interest in aerodynamics; as Jones (1947) put it:

> The theory of potential flows with small disturbances is particularly suited for application to aeronautical problems because the assumptions of small disturbances and isentropic flows on which this theory is based agree with the requirements for efficient flight. Theories of large disturbances, which deal with the formation of shock waves, are of lesser practical importance since such theories describe the losses of energy and the large drags associated with unsuitable forms.

Today this must be regarded as an overstatement since the imperative to faster flight has pushed regular commercial intercontinental flight into the transonic regime, but nevertheless the framework for linearized compressible flows that Jones referred to does serve well for subsonic and supersonic flight-speeds and so is developed in detail in this section.

19.3   The Uniform Free-Stream

If the speed of sound depends only the speed—e.g. via Equation (18.12)—then the speed of sound will be constant in a uniform stream. Therefore any uniform stream $q = q_\infty$ is a solution of the full continuity Equation (18.3), with its associated speed of sound a_∞, since the velocity gradient ∇q_∞ vanishes, by definition.

Of course, that the general uniform stream of Equation (3.5) and Figure 3.5 is still a valid solution is a necessary consequence of the invariance of the equations of motion with a constant velocity of the frame of reference (Section 1.1.1). Yet another way to show that the uniform stream is a valid solution is that it satisfies Equation (18.1) because the density will be uniform and so can be moved outside the divergence.

19.4   The Disturbance Potential

Now consider a thin wing placed in such a flow at a small incidence such as to only disturb it slightly. Thus the total flow might be of the form

$$q_\infty + \nabla\phi, \tag{19.1}$$

if the disturbance has a potential (i.e. is irrotational) where $|\nabla\phi| \ll q_\infty$. The conditions under which the total velocity will have a potential are the same as in the incompressible case: the circulation is preserved, and so, being irrotational in the uniform free-stream upstream, remains irrotational past the wing.

Substitute the above expression for the total velocity into the full continuity Equation (18.3) and drop terms quadratic or of higher order in the small disturbance potential.

$$\nabla^2\phi = \frac{1}{a_\infty^2}q_\infty \cdot \nabla(\nabla\phi) \cdot q_\infty.$$

If, as usual, the coordinate system is chosen so that $q_\infty = q_\infty(\mathbf{i}\cos\alpha + \mathbf{j}\sin\alpha)$ then

$$\nabla^2\phi = \mathrm{Ma}_\infty^2 \frac{\partial^2\phi}{\partial x^2}$$

where the small angle approximations have been applied to the incidence α, as in incompressible thin aerofoil theory (Section 6.1.3), and

$$\mathrm{Ma}_\infty \equiv \frac{q_\infty}{a_\infty}$$

is the Mach number of the free-stream.

Expressing the Laplacian in terms of the Cartesian coordinates, the linearized compressible continuity equation becomes

$$\left(1 - \mathrm{Ma}_\infty^2\right)\frac{\partial^2\phi}{\partial x^2} + \frac{\partial^2\phi}{\partial y^2} + \frac{\partial^2\phi}{\partial z^2} = 0. \tag{19.2}$$

Again we see that for small Mach numbers, i.e. free-stream velocity much less than the speed of sound, this reduces to the Laplace Equation (9.10). We can also see that the coefficient of the longitudinal second derivative changes sign relative to the other derivative terms if the Mach number exceeds unity. This is a mathematical indication (within the linearized framework) of the qualitatively different regime called *supersonic*.

19.5 Prandtl–Glauert Transformation

The linearized compressible potential Equation (19.2) differs from the Laplace Equation (9.10) only by the constant multiplier on the x-derivative term. Therefore compressible potentials can be obtained from incompressible solutions by dilating the dependence on x; i.e. consider, for a constant dimensionless dilation factor μ (not to be confused with the coefficient of viscosity)

$$\phi(x, y, z) = \Phi(\mu x, y, z) \equiv \Phi(X, Y, Z) \tag{19.3}$$

The transverse partial derivatives are identical:

$$\frac{\partial \phi}{\partial y} = \frac{\partial \Phi}{\partial y}, \qquad\qquad \frac{\partial^2 \phi}{\partial y^2} = \frac{\partial^2 \Phi}{\partial y^2}$$

$$\frac{\partial \phi}{\partial z} = \frac{\partial \Phi}{\partial z}, \qquad\qquad \frac{\partial^2 \phi}{\partial z^2} = \frac{\partial^2 \Phi}{\partial z^2}$$

but the derivatives with respect to x, via the chain rule, generate a constant factor:

$$\frac{\partial \phi}{\partial x} = \frac{dX}{dx}\frac{\partial \Phi}{\partial X} = \mu\frac{\partial \Phi}{\partial X},$$

$$\frac{\partial^2 \phi}{\partial x^2} = \left(\mu\frac{\partial}{\partial X}\right)^2 \Phi = \mu^2\frac{\partial^2 \Phi}{\partial X^2},$$

so the continuity Equation (19.2) becomes

$$\left(1 - Ma_\infty^2\right)\mu^2\frac{\partial^2 \Phi}{\partial X^2} + \frac{\partial^2 \Phi}{\partial Y^2} + \frac{\partial^2 \Phi}{\partial Z^2} = 0,$$

and this can be reduced to the Laplace equation

$$\frac{\partial^2 \Phi}{\partial X^2} + \frac{\partial^2 \Phi}{\partial Y^2} + \frac{\partial^2 \Phi}{\partial Z^2} = 0 \tag{19.4}$$

for the distorted potential Φ with the choice

$$\mu \equiv \frac{1}{\sqrt{1 - Ma_\infty^2}}.$$

This means that any known potential Φ which is a solution of the incompressible continuity equation generates a family of solutions at different free-stream Mach numbers

$$\phi(x, y, z) = \Phi\left(\frac{x}{\sqrt{1 - Ma_\infty^2}}, y, z\right);$$

Φ is the incompressible $Ma_\infty = 0$ member of the family.

19.5.1 Fundamental Linearized Compressible Flows

This dilation can be applied to any of the many two- or three-dimensional irrotational solutions obtained in Chapters 2–14.

To get an idea of what this dilation looks like, let's consider, as a first example, the zero Mach number two-dimensional point source of Section 3.1.6. Its complex velocity was $u - iv = 1/(x + iy)$ and its complex potential $\phi + i\psi = \ln(x + iy)$, so its scalar potential is $\phi = \ln |x + iy| = \ln \sqrt{x^2 + y^2} = \frac{1}{2} \ln |x^2 + y^2|$. Therefore the dilated scalar potential

$$\phi = \frac{1}{2} \ln \left| \frac{x^2}{1 - \mathrm{Ma}_\infty^2} + y^2 \right|$$

satisfies the linearized compressible potential Equation (19.2).

The character of the flow represented by this potential strongly depends on the Mach number. In the subsonic case, the absolute value bars are not required since their contents are necessarily nonnegative and the corresponding velocity perturbation is easily obtained by partial differentiation, in accordance with Equation (2.28):

$$u = \frac{\partial \phi}{\partial x} = \frac{2x/(1 - \mathrm{Ma}_\infty^2)}{\frac{x^2}{1 - \mathrm{Ma}_\infty^2} + y^2}$$

$$v = \frac{\partial \phi}{\partial y} = \frac{2y}{\frac{x^2}{1 - \mathrm{Ma}_\infty^2} + y^2}.$$

Therefore the slope of the perturbation streamlines is

$$\frac{v}{u} = (1 - \mathrm{Ma}_\infty^2)\frac{y}{x}$$

$$\tan \beta = (1 - \mathrm{Ma}_\infty^2) \tan \theta,$$

where $\beta = \arctan \frac{v}{u}$ is the polar angle of the velocity (as in Section 2.6.2) and $\theta \equiv \arctan \frac{y}{x}$ is the usual polar coordinate from Equation (2.20b). When $\mathrm{Ma}_\infty = 0$, $\beta = \theta$ the streamlines are rays from the point source, as in Figure 3.3. As Ma_∞ increases, $\tan \beta < \tan \theta$, the velocity vectors point below the rays, and the streamlines curve towards the x-axis. As Ma_∞ tends upwards towards unity $\tan \beta \to 0$, so β tends towards zero or π and the streamlines are asymptotically horizontal.

In the supersonic case, $\mathrm{Ma}_\infty > 1$, the denominator in the first term is negative and the absolute value bars have a very important effect: the gradient of the solution is discontinuous across

$$\frac{x^2}{1 - \mathrm{Ma}_\infty^2} + y^2 = 0$$

$$\frac{y^2}{x^2} = \frac{1}{\mathrm{Ma}_\infty^2 - 1}$$

$$\tan^2 \theta = \tan^2 \mu'.$$

This new angle μ' defined by

$$\mu' \equiv \arctan \frac{1}{\sqrt{\mathrm{Ma}_\infty^2 - 1}} \equiv \arcsin \frac{1}{\mathrm{Ma}_\infty} \qquad (19.5)$$

for $\mathrm{Ma}_\infty > 1$ is called the *Mach angle*.

19.5.2   The Speed of Sound

The significance of the Mach angle is not restricted to this special solution; indeed, the second part of Equation (19.5) permits a sense to be given to the idea of the 'speed of sound' introduced in Section 18.1, since it can be rearranged to give:

$$a_\infty \equiv \frac{q_\infty}{\mathrm{Ma}_\infty} \equiv q_\infty \sin \mu'.$$

Consider, as in Figure 19.1, a particle of fluid travelling from left to right with the steady uniform stream along the x-axis through the point source at the origin (remember that the velocity disturbances generated by the disturbance potential ϕ here are supposed small compared to the velocity of the free-stream). A time t later it is at $x = q_\infty t$ on the x-axis and its perpendicular distance from the two rays $\tan \theta = \pm \tan \mu'$ along which the perturbation solution is discontinuous is

$$q_\infty \sin \mu' \equiv \frac{q_\infty}{\mathrm{Ma}_\infty} \equiv a_\infty,$$

the 'speed of sound'.

The mathematics of the discontinuity of the solution may seem somewhat mysterious— mostly because they are so different to those applicable to subsonic flows. There is a good framework for dealing with supersonic flows, but it is completely different to the one for

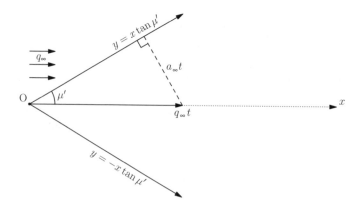

Figure 19.1   A two-dimensional linearized disturbance in a uniform stream, obtained from the incompressible point source (at O) by the Prandtl–Glauert transformation, at a Mach number greater than one, $\mathrm{Ma}_\infty = \mathrm{cosec}\, \mu'$

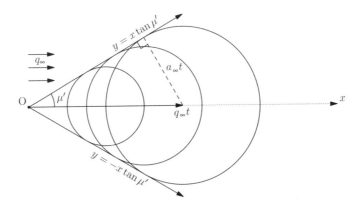

Figure 19.2 Another representation of Figure 19.1, showing three of the circles centred on fluid particles which have passed with the free-stream through the disturbing point, and with radius equal to the product of the speed of sound and the time elapsed since the passage. The Mach rays form the envelope to this family of circles

incompressible flows and so it is not developed here. Instead, experimental observations come to our aid. Wing sections with sharp leading edges placed in supersonic wind tunnels (as the point source here could be considered a model of) also display these 'Mach rays' emanating from the leading edge at the Mach angle from the flow direction downstream. Upstream of the rays, the oncoming stream is just as if there were no disturbance downstream of it; only inside the sector bounded by the rays is the influence of the disturbance felt. It is as if influence of the disturbance travelled outward from the point on the free-stream passing through the disturbing point in circles of radius $a_\infty t$, and to which the Mach rays form an envelope; a sequence of such circles is sketched in Figure 19.2.

19.6 Application of the Prandtl–Glauert Rule

Although, as shown in Section 19.5.1 it is possible to find solutions of the linearized continuity Equation (19.2), a simpler approach is to transform the geometry of interest via the longitudinal dilation of Section 19.5, solve a Laplace equation for the distorted disturbance potential by any one of the techniques already developed for incompressible aerodynamics, and then re-transform the results to the finite subsonic Mach number of interest.

19.6.1 Transforming the Geometry

The first step is to transform the geometry by stretching all the x-coordinates by the factor $\mu = \left(1 - \mathrm{Ma}_\infty^2\right)^{-1/2}$. For $0 < \mathrm{Ma}_\infty < 1$, this factor exceeds one and is therefore a proper stretching rather than a compression.

The next question is how the thickness and camber of the geometry should be transformed in the y-direction, since it is already being treated as thin in the linearized theory of compressibility. The essential property of the geometry in inviscid flow is impermeability; i.e. the surface should be composed of streamlines, or, the component of velocity normal to the surface should vanish. This boundary condition was used in various ways in Chapters 6, 7, 8, 11, and 14, but not in terms of a velocity potential, so this is briefly derived here. In two dimensions, say the slope of the surface at the point under consideration is $\frac{dy}{dx} \equiv \tan \lambda$, then the unit outward normal will be $\hat{n} = -\mathbf{i} \sin \lambda + \mathbf{j} \cos \lambda$ and so the normal component of velocity is

$$\hat{n} \cdot \{q_\infty (\mathbf{i} \cos \alpha + \mathbf{j} \sin \alpha) + \nabla \phi\}$$

$$= (-\mathbf{i} \sin \lambda + \mathbf{j} \cos \lambda) \cdot \{q_\infty (\mathbf{i} \cos \alpha + \mathbf{j} \sin \alpha) + \nabla \phi\}$$

$$= -q_\infty \sin \lambda \cos \alpha - \frac{\partial \phi}{\partial x} \sin \lambda + q_\infty \cos \lambda \sin \alpha + \frac{\partial \phi}{\partial y} \cos \lambda$$

$$\sim q_\infty (\alpha - \lambda) + \frac{\partial \phi}{\partial y}$$

and so its vanishing gives the condition

$$\frac{\partial \phi}{\partial y} = q_\infty (\alpha - \lambda),$$

for the linearized compressible potential which implies

$$\frac{\partial \Phi}{\partial Y} = q_\infty (\alpha - \lambda),$$

for the transformed Prandtl–Glauert potential. These two boundary conditions are formally identical so the corresponding profiles in the two spaces will be geometrically similar. Note though that to keep λ the same, the y-coordinates of the profile (or camber line) have to be dilated as the x-coordinates were; i.e. by $\left(1 - \mathrm{Ma}_\infty^2\right)^{-1/2}$.

Since the Laplace equation is linear and isotropic, its solution about an impermeable scaled object in a free-stream at incidence is the same, subject to the same scaling; i.e. the potentials are geometrically similar too. There is a subtle point here in that the Prandtl–Glauert transformation is not isotropic, the points on the two surfaces do not correspond under the transformation; however, the whole Prandtl–Glauert theory is part of the broader idea of thin aerofoil and wing theory, so the thicknesses or distances from the chord are neglected and in that sense the nominally corresponding points do correspond.

19.6.2   Computing Aerodynamical Forces

Aerodynamical forces can be computed directly from pressures or indirectly from circulations and the Kutta–Joukowsky thereom.

Circulation

In solving the Laplace equation for the potential with the same free-stream and geometrically similar but dilated boundary, the potential field will come out to be geometrically similar but also scaled. The scalings of some quantities are easily inferred from dimensional analysis. Circulation is defined as a path integral of velocity and so scales in proportion with both velocity and length; this may be verified by inspection of the exact solution in Equation 5.6

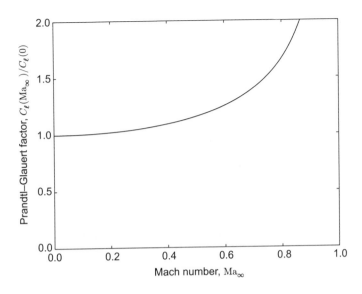

Figure 19.3 The Prandtl–Glauert Equation (19.6) for the effect of compressibility on lift

for the thin flat plate with Kutta–Joukowsky condition at the trailing edge, $\Gamma = \pi c q_\infty \sin \alpha$ – the circulation is proportional to the product of the velocity and the length, $q_\infty c$. Similarly observe that in our numerical methods the vortex strengths have come out proportional to the same product of free-stream velocity and length-scale; see e.g.

- the matrix–vector Equation (7.3) of the lumped vortex method;
- Section 8.1.6 for the CUSSSP program; or
- Equation (14.1) for the calculation of the lift coefficient in the vortex lattice method.

In the Prandtl–Glauert transformation under consideration, the velocity is unchanged but the length is scaled by $\left(1 - \mathrm{Ma}_\infty^2\right)^{-1/2}$. Therefore, by the Kutta–Joukowsky theorem, the lift coefficient will be increased by the same amount:

$$C_\ell(\mathrm{Ma}_\infty) = \frac{C_\ell(0)}{\sqrt{1 - \mathrm{Ma}_\infty^2}}. \qquad (19.6)$$

This simple result, known as the Prandtl–Glauert rule and plotted in Figure 19.3, says that the first-order effect of compressibility as speeds increase from zero Mach number while still remaining strictly subsonic is that the lift coefficient increases monotonically and without bound as the Mach number tends towards one.

Pressure Coefficient

The pressure coefficient, as originally given by Equation (2.11), was quadratic in the ve-locity; therefore for use with a velocity field obtained from a linearized approximation like

Equation (19.2), it should be linearized too. From the differential Bernoulli Equation (2.13),

$$\delta p = -\rho \delta \left(\frac{q^2}{2} \right)$$

and then using the basic decomposition of the flow into free-stream and disturbance potential by Equation (19.1):

$$q^2 = |q_\infty(\mathbf{i} \cos \alpha + \mathbf{j} \sin \alpha) + \nabla \phi|^2$$

$$= q_\infty^2 + 2q_\infty \left(\frac{\partial \phi}{\partial x} \cos \alpha + \frac{\partial \phi}{\partial y} \sin \alpha \right) + |\nabla \phi|^2$$

$$\sim q_\infty^2 + 2q_\infty \frac{\partial \phi}{\partial x}$$

for small disturbances and so, since the first term is constant, to leading order the difference is like

$$\delta \left(\frac{q^2}{2} \right) \sim \delta \left(q_\infty \frac{\partial \phi}{\partial x} \right),$$

and the Bernoulli equation becomes

$$\delta p \sim -\rho q_\infty \delta \left(\frac{\partial \phi}{\partial x} \right).$$

The appropriately approximated pressure coefficient is

$$C_p \equiv \frac{p - p_\infty}{\frac{1}{2} \rho q_\infty^2}$$

$$= \frac{-\rho q_\infty \frac{\partial \phi}{\partial x}}{\frac{1}{2} \rho q_\infty^2}$$

$$= \frac{-2}{q_\infty} \frac{\partial \phi}{\partial x}.$$

Alternatively, in terms of the Prandtl–Glauert transformed potential Φ of Equation (19.3) which satisfies the Laplace Equation (19.4) in the dilated (X, Y, Z) coordinates,

$$C_p = \frac{-2}{q_\infty \sqrt{1 - \text{Ma}_\infty^2}} \frac{\partial \Phi}{\partial X}. \tag{19.7}$$

Thus the pressure coefficient also scales with Mach number according to the Prandtl–Glauert Equation (19.6).

19.6.3 The Prandlt–Glauert Rule in Two Dimensions

That Equation (19.7) for the pressure coefficient contains the same scaling factor as the Equation (19.6) for the lift coefficient, derived from dimensional analysis of circulation in the Prandtl–Glauert transformation, is as it should be, since it should be equally possible to compute the aerodynamic forces from pressures as well as from circulations, and the answers obtained should be the same.

From this uniform scaling of the pressure coefficient around the profile, it may be deduced that the theoretical drag coefficient increases in the same proportion; being zero in the incompressible limit it remains zero within the scope of the Prandtl–Glauert theory.

Experimentally, when streamlined two-dimensional aerodynamics is approximated, e.g. by extending the wing span right across the wind tunnel into the walls, the measured drag does indeed remain low as the Mach number is increased, and the measure lift coefficient does increase in line with Equation (19.6), but the lift coefficient does not increase to infinity as the speed of sound is approach. Quite suddenly the lift drops or even reverses and the drag jumps. The reason is that the flow over a lifting surface is always accelerated somewhere, usually on the upper surface to induce suction, i.e. negative pressure coefficient. As this negative pressure coefficient in this region of suction will scale in magnitude according to the Prandtl–Glauert Equation (19.6), it becomes large and negative which means the velocity there is increased in proportion and eventually locally reaches the speed of sound. This is accompanied by the generation of shock waves, and a 'compressibility burble' (Stack 1933, 1935) related to the discontinuities sketched in Figures 19.1 and 19.2. These discontinuities violate the assumptions on which the Blasius's theorem (Equation 3.12) was based; the shock waves do significantly rearrange the pressure field and cause *wave drag*.

The free-stream Mach number at which the local Mach number somewhere on the aerofoil reaches unity is called the critical Mach number for the given profile and incidence. Although this is a precisely defined quantity, the experimental curves are not so severe, at least for thin aerofoils.

For example, Figure 19.4 shows data extracted from figure 8 of NACA Report No. 463 (Stack 1933) on the measured lift coefficient and Mach number for a wing with Clark Y section of thickness ratio 0.06. This test piece was said to have a zero-lift incidence of about $\alpha_0 = -2.5°$ and was placed for this sequence of measurements at an incidence of $\alpha = 2°$. For comparison with the experimental data, Figure 19.4 also contains the theoretical curve based on thin aerofoil theory Equation 6.11 for the zero-Mach number lift coefficient and Equation 19.6 for the effect of Mach number, i.e.

$$C_L = \frac{2\pi(\alpha - \alpha_0)}{\sqrt{1 - \text{Ma}_\infty^2}}. \tag{19.8}$$

Note here in comparing the experimental measurements with theory that the effects of span are not quite clear and so the use of the theoretical two-dimensional lift–incidence slope of 2π from Equation 6.11 is only a first guess. A slightly lower value might account for this and reduce the overall vertical offset. The main point to take from Figure 19.4 is that the lift coefficient does indeed increase with Mach number at low Mach numbers, with a similar slope and curvature to that predicted by the Prandtl–Glauert Equation (19.6), but that this only persists up to some Mach number below unity (here about 0.75) above which the measured lift falls sharply rather than increasing without bound as in the theory. The discrepancy due to the unknown incompressible lift–incidence coefficient can be largely removed by dividing the

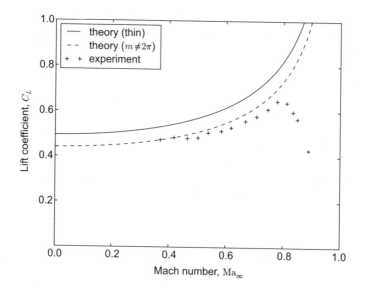

Figure 19.4 Data extracted from Figure 8 of Stack (1933) for the lift coefficient of a 6% thick Clark Y wing section in a compressible wind tunnel. The curves are Equations (19.8) and (19.9)

Prandtl–Glauert Equation (19.6) for the rest of the data by its value for the first member of the sequence, which gives

$$C_L = C_{L,1} \sqrt{\frac{1 - \mathrm{Ma}^2_{\infty,1}}{1 - \mathrm{Ma}^2_{\infty}}}, \tag{19.9}$$

which also appears in Figure 19.4.

19.6.4   The Critical Mach Number

Since the pressure coefficient distribution can be estimated from the incompressible flow solution and the Prandtl–Glauert Equation (19.7), it is possible to estimate this critical Mach number for any given profile and incidence. These do correlate reasonably well with the experimental Mach numbers at which the sudden loss of lift and jump in drag is observed, but this is not pursued here.

19.7   Sweep

When the vortex lattice method was developed in Chapter 14, it was important to give it the generality to handle swept wings (as in Figures 14.1–14.3), but within the context of incompressible flow, there is no reason to use swept wings. Indeed, wings designed for flight at low Mach numbers are generally kept straight. However, the few preliminary ideas about compressible flow introduced in this chapter are enough to suggest the reasons for sweep.

If the flow is supersonic so that the nose of the aircraft sheds a bow wave, or Mach cone, the axially symmetric three-dimensional analog of the two-dimensional Mach rays of Figure 19.1, the wings can be pulled back inside the cone by sweeping them at an angle greater than the Mach angle, defined by Equation (19.5). However, not enough else has been said about supersonic aerodynamics to make it worthwhile developing this idea any further here; see the references in the Further Reading section at the end of this chapter.

In the subsonic regime, the intention is the same: to keep the local speed everywhere less than the speed of sound; i.e. to increase the critical Mach number. In the subsonic regime though there is no Mach angle, since it is evident that the definition in Equation (19.5) does not return a real angle unless the Mach number is greater than one. Instead the idea is that in relating the three-dimensional wing to the two-dimensional Prandtl–Glauert theory developed in Section 19.6, it is only the component of the free-stream velocity normal to the leading edge that counts; i.e. $q_\infty \cos \Lambda$, if Λ is the angle of leading edge sweep, as in Section 14.1. Since the free-stream speed of sound is unaffected by geometry, it remains constant and the apparent local free-stream Mach number then is not $\mathrm{Ma}_\infty = q_\infty/a_\infty$ but $\mathrm{Ma}_\infty \cos \Lambda$. This approach can be taken further, leading to the incorporation of factors of $\cos \Lambda$ into the Prandtl–Glauert transformation (Jones 1947; Jones and Cohen 1960), but this is not pursued here.

19.8   Exercises

1. (a) Show that $\phi(x, y, z) = \ln \sqrt{x^2 + y^2 + z^2}$ is a three-dimensional point source solution of the incompressible continuity Equation (9.10).
   (b) What is its flow rate?
   (c) Find the corresponding disturbance potential solution of the linearized compressible continuity Equation (19.2) by the Prandtl–Glauert transformation.
   (d) Show that for $\mathrm{Ma}_\infty > 1$, the gradient of the disturbance potential solution is discontinuous across a cone making the Mach angle $\mu' = \arcsin \mathrm{Ma}_\infty^{-1}$ with the downstream x-axis. This is called a *Mach cone*.

2. Find the solution of Equation (19.2) corresponding to the two-dimensional incompressible doublet of Section 3.1.7.

3. In Exercise 1.7.5b, the free-stream Mach number was calculated for one of the wind tunnel experiments reported by Silverstein (1935). What Prandtl–Glauert correction factor would this imply for the lift coefficients?

19.9   Further Reading

The Prandtl–Glauert small-disturbance theory of linearized compressible flow is discussed by Liepmann and Roshko (1957), Jones and Cohen (1960), Belotserkovskii (1967), Milne-Thomson (1973), Ashley and Landahl (1985), Kuethe and Chow (1998), Katz and Plotkin (2001), Bertin (2002), Moran (2003), and Anderson (2007). It has been much used in practice (DeYoung and Harper 1948; Gray and Schenk 1953; Margason and Lamar 1971).

Many more examples of two- and three-dimensional fundamental solutions of the linearized potential Equation (19.2) are given by Bousquet (1990).

For proper treatments of supersonic flow in terms of Prandtl–Meyer expansions, shock waves, and the method of characteristics, see Liepmann and Roshko (1957), Carafoli *et al.* (1969), Milne-Thomson (1973), Kuethe and Chow (1998), Bertin (2002), and Anderson (2007).

Further experimental measurements of lift and drag showing agreement with the Prandtl–Gluaert rule at small Mach numbers and a sudden departure more or less in the vicinity of the critical Mach number are repeated and discussed by Jones and Cohen (1960), Kuethe and Chow (1998), Bertin (2002), and Moran (2003).

References

Anderson, J.D. (2007) *Fundamentals of Aerodynamics,* 4th edn. New York: McGraw-Hill.

Ashley, H. and Landahl, M. (1985) *Aerodynamics of Wings and Bodies.* New York: Dover.

Belotserkovskii, S.M. (1967) *The Theory of Thin Wings in Subsonic Flow.* New York: Plenum.

Bertin, J.J. (2002) *Aerodynamics for Engineers,* 4th edn. New Jersey: Prentice Hall.

Bousquet, J. (1990) *Aérodynamique: Méthode des singularités.* Toulouse: Cépaduès.

Carafoli, E., Mateescu, D. and Nastase, A. (1969) *Wing Theory in Supersonic Flow.* Oxford: Pergamon.

DeYoung, J. and Harper, C.W. (1948) Theoretical symmetric span loading at subsonic speeds for wings having arbitrary plan form. Report 921, NACA.

Gray, W.L. and Schenk, K.M. (1953) A method for calculating the subsonic steady-state loading on an airplane with a wing of arbitrary plan form and stiffness. Technical Note 3030, NACA.

Jones, R.T. (1947) Wing plan forms for high-speed flight. Technical Report 863, NACA.

Jones, R.T. and Cohen, D. (1960) *High Speed Wing Theory* vol. 6 of *Princeton Aeronautical Paperbacks.* New Jersey: Princeton University Press.

Katz, J. and Plotkin, A. (2001) *Low-Speed Aerodynamics,* 2nd edn. Cambridge: Cambridge University Press.

Kuethe, A.M. and Chow, C.Y. (1998) *Foundations of Aerodynamics,* 5th edn. New York: John Wiley & Sons, Ltd.

Liepmann, H.W. and Roshko, A. (1957) *Elements of Gasdynamics.* New York: John Wiley & Sons, Ltd.

Margason, R.J. and Lamar, J.E. (1971) Vortex-lattice FORTRAN program for estimating subsonic aerodynamic characteristics of complex planforms. Technical Note TN D-6142, NASA.

Milne-Thomson, L.M. (1973) *Theoretical Aerodynamics,* 4th edn. New York: Dover.

Moran, J. (2003) *An Introduction to Theoretical and Computational Aerodynamics.* New York: Dover.

Silverstein, A. (1935) Scale effect on Clark Y airfoil characteristics from N.A.C.A. full-scale wind-tunnel tests. Report 502, NACA.

Stack, J. (1933) The N.A.C.A. high-speed wind tunnel and tests of six propeller sections. Report 463, NACA.

Stack, J. (1935) The compressibility burble. Technical Note 543, NACA.

Appendix A

Notes on Octave Programming

A.1 Introduction

Programs written in interpreted programming languages such as Octave are generally slower to run than those written in compiled programming languages such as Fortran, since the compiler can rewrite the instructions beforehand to make them simpler and more suited to the computer's architecture whereas the interpreter can only try doing that a line at a time, during execution.

This disadvantage is compensated for by programs in interpreted languages being generally easier and quicker to write, and if we value our time as programmers more highly than that of the computer's in executing, this compensation might be decisive. Further, the run-time for many small programs, such as all of those considered in this book, is quite short so that the difference between interpreted and compiled implementations wouldn't be noticed. Indeed, in assessing speed for such programs which are only intended to be run once, or at most a few times, one must include the time of compilation as well as of execution; for small programs this will tilt the balance in favour of interpreters.

Nevertheless, there is a single key to making interpreted programs run faster and it is so important in scientific programming that it has been consistently exploited throughout this text. It is to ensure that most of the work is done inside the precompiled subroutines which are called by operators built in to the language. For example, it would not be difficult, and it might even be deemed instructive, to write an Octave function for solving a system of linear equations using Gaussian elimination; however, Octave possesses the backslash operator for this purpose and on seeing it Octave will pass all the work to an appropriate compiled and optimized binary library function, which will have originally been written in a language like Fortran or C++. This step will then run as fast in Octave as it would in the compiled language, and if this is the dominant computational step, we need only look to ensuring that it is not eclipsed by overheads in other parts of the program.

A.2 Vectorization

The main things to look for in this regard are explicit iterations at the interpreted level over the elements of arrays. For example, preparing a graph like Figure 1.5 requires evaluating

---

Theory of Lift: Introductory Computational Aerodynamics in MATLAB®/Octave, First Edition. G. D. McBain.
© 2012 John Wiley & Sons, Ltd. Published 2012 by John Wiley & Sons, Ltd.

Listing A.1   Explicit iteration to compute the speed of sound at each of a collection of temperatures: speed_of_sound_for.m.

```
n = 1e6;
t = linspace (-60, 100, n);
tic
for i = 1:n
  a(i) = speed_of_sound (t(i) + 273.15);
end
toc
```

the scalar Equation (1.4) for enough values of temperature to enable a smooth curve to be interpolated. This is a very common task.

A.2.1   Iterating Explicitly

This could be done in a for-loop as shown in Listing A.1, but (for the admittedly large number n = 1e6 of abscissae) this takes 34 seconds on my laptop.

A.2.2   Preallocating Memory

Part of the time here is taken by having to repeatedly allocate memory to store the values of the speed of sound. This can be removed by prefilling the result array with dummy values, e.g. as generated by the built-in function **zeros** by adding the line before the for-loop.

```
a = zeros (size (t));
```

With this improvement, the program takes slightly under thirty seconds.

A.2.3   Vectorizing Function Calls

Avoiding repeated memory allocation for a single array saved over a tenth of the run-time, which is some improvement, but much more can still be achieved. The key is already visible in the first two versions of the program: notice that the built-in functions **linspace** and **zeros** have returned complete arrays rather than single numbers. What if speed_of_sound could return the whole array? Then it would only need to be called once. The beauty of matrix-languages like Octave is that that is exactly what they are designed to do. We can indeed simply pass the whole array of abscissae to the function and get back the whole array of ordinates from a single call; see Listing A.2.

In this case there is no need to preallocate since the result vector is not resized.

The new program is not only equivalent in terms of answers but also arguably much more clearly and naturally expressed; moreover, it takes only 34 milliseconds, a thousand times less than the original program.

Listing A.2   Vectorizing Listing A.1.

```
n = 1e6;
t = linspace (-60, 100, n);
tic
a = speed_of_sound (t + 273.15);
toc
```

Five seconds is long enough to make an interactive program begin to seem unresponsive; thirty seconds will likely induce thoughts that it is broken. It is true that one million points are not required to draw Figure 1.5, but a million degrees of freedom are by no means large in contemporary computational aerodynamics, and since elementwise functions of scalars are common and natural operations in the same field, this technique is worth mastering, and worth mastering early. There are many more like it, some of which are exploited in this book and explained below.

A.2.4   Many Functions Act Elementwise on Arrays

The first thing to know is that most of the functions built in to Octave act on arrays and not just scalars. Mathematical functions that really only make sense for scalars, such as **sqrt** or **exp** or **sin**, are typically applied elementwise to arrays. For example, while **sqrt** (-1) will return 0 + 1i, as expected, **sqrt** (-1:1) will return the array [1i, 0, 1]. This was exploited above in ListingA.2, which was only able to return an array of outputs for an array of inputs because **sqrt** can.

Actually there is another way that scalar functions have been generalized to act on matrices, using the Taylor series representation of the function (Moler and Van Loan 1978); however, where such matrix functions are implemented in Octave and there is a corresponding elementwise version, the matrix version is given a different name; e.g. **expm**, **logm**, and **sqrtm**. While these matrix functions do have uses, they are not required for any of the problems in this book.

A.2.5   Functions Primarily Defined for Arrays

Of course there are also many functions which are primarily defined for arrays or matrices as a whole; e.g. **max**, **det**, or **flipud**. These generally treat a scalar argument as a singleton; i.e. as an array containing a single element, a vector of length one, or a one-by-one matrix. In the three examples mentioned, this leads to rather trivial results since the maximum element of an array of length one is just the sole element, the determinant of a one-by-one matrix is again the sole entry, and a vector of length one looks the same after flipping as it did before.

A.2.6   Elementwise Arithmetic with Single Numbers

Often in writing a function which is essentially scalar but which, in accordance with Section A.2.3, we would like to be able to act elementwise on array inputs, it is necessary to vectorize arithmetic. This works naturally for scalar multiplication or division; i.e. the multiplication or division of every element of an array by the same real or complex number.

An operation which resembles scalar multiplication (but without the fundamental mathematical significance) is the addition of the same number to each element of an array. This arose in Listing A.2 in converting the temperatures from degrees Celsius to kelvins: 273.15 K was added to each entry. The syntax for this (or the corresponding subtraction of a common subtrahend) is simple; in particular note that the syntax is identical whether the addend (or minuend) is a matrix or a single number.

Raising an array to a power which is a number or a number to a power which is an array is different, because these operations can have a special meaning for square matrices (analogous to that of **expm**). To specify elementwise exponentiation, one precedes the operator with a dot; e.g. **pi** .^ [1, 2] or [1, 2] .^ **pi**.

The same dot-prefix is required when dividing a single number by an array; e.g. **pi** ./ [1, 2].

A.2.7   Elementwise Arithmetic between Arrays

Since vector and matrix addition and subtraction are defined elementwise in mathematics, there is no difficulty with these operations.

Matrix multiplication and division are defined specially though, with matrix division being used to solve linear systems of equations as exploited in Listings 1.4, 7.4, 8.6, 12.1, and 14.5, for example. Elementwise multiplication or division of two arrays with identical shapes is indicated by preceding the operator with a dot, as in A .* B or A ./ B. Elementwise multiplication was used in cusssp.m to correct the direction of the velocity at the midpoint of a panel due to the panel's own strength. Elementwise division was used in Listing 1.3 to calculate the density from the ideal gas law for multiple altitudes each with their own pressure and temperature.

Raising each element of an array to a different power can be done with .^, as in Listing 1.3.

A.2.8   Vector and Matrix Multiplication

In Octave, the operator * denotes multiplication, and more specifically matrix multiplication, so that if the symbol is proceded by a row-vector and followed by a column-vector, the answer will be a scalar, being the sum of the products of the corresponding elements. This was used in Listing 6.1.

Multiplying a column by a row gives a rectangular matrix, which can be useful too.

A.3   Generating Arrays

Arrays of any shape and size with all elements equal may be generated using the built-in function ones and then multiplying by the required value. The special case **zeros** is also provided.

A.3.1   Creating Tables with **bsxfun**

One way of generating a rectangular matrix was noted in Section A.2.8: as the product of column-vector and a row-vector. The elements of the product matrix are all the possible pairwise products of the elements of the factors. A generalization of this operation to other pairwise operations is provided in Octave by **bsxfun**; see e.g. Listings 7.3 and 8.5.

A.4 Indexing

A.4.1 Indexing by Logical Masks

Sometimes the same operation is done to every element of an array, as in the speed-of-sound example discussed above, in which case indexing can and should be left to the internal workings of the interpreter or compiler; however, special cases often intrude, for which no single overarching operation suffices. An example of this was met early in the book, in Listing 1.3: the thermal variation of the atmosphere is defined piecewise, and so it was necessary to classify the input altitude according to whether they were in the troposphere, (lower) stratosphere, etc. Similarly, the four-digit NACA camber lines have piecewise definitions and so chordwise stations had to be classified in Listing 6.3 as fore or aft. To access just the elements of an array satisfying some condition, one forms a logical array which has true or false elements in the corresponding positions and then indexes the original array with it using parentheses; examples may be spotted in the two cited listings.

The required logical 'mask' can be formed beforehand and stored in a variable, like troposphere and strat in Listing 1.3, or even just formed temporarily inside the reference-parentheses of the array using an array logical operation like greater-than, as in Listing 8.1. If the mask is only required once, this is both more compact and clearer, since the definition is present right where it is used.

A.4.2 Indexing Numerically

Other times, the locations of the elements in the array are known to begin with and it isn't necessary to compute a logical array to find them. For example, in Listing 8.6 which implements the two-dimensional panel method, there were n panels defined by $n + 1$ nodes (the last being a repeat of the first), so the starting points of the panels were all those points but the last; this subarray can be extracted, by passing the numerical index range 1:n to the array, in parentheses, as in zp(1:n).

That requires n to be known, and in that example it was computed and stored using the **length** built-in function by n = **length** (dz).

If the required numerical range involves the end of the array, as in this example, there is even a special numerical indexing keyword **end** (which automatically knows how long the relevant array is) with z(1:(**end**-1)). This was used in Listing 7.4; however, since the value n was required several times in Listing 8.6, it was not used there.

A.5 Just-in-Time Compilation

A different approach to the problem of vectorization is that the interpreter could be taught to spot patterns like the above simple for-loop and automatically replace them with the same fixes that have been implemented manually here. This would mimic what compilers for languages like Fortran did (before, influenced by interpreted array languages like Lisp and APL, Fortran 90 introduced array operations), and so is called 'just-in-time' compilation. This is an ongoing area of research and goes back to the earliest Lisp interpreters (McCarthy 1960).

Just-in-time compilation involves compromise, since the more patterns the interpreter tries to spot, the longer it takes analysing the code before processing it. Spotting the patterns is not trivial, since if the functions called inside the for-loop include functions defined by the user, they might depend on external or persistent data which they in turn could modify, so while each call might look the same, it might not be the same. Since an interpreter should not change the effect of code, it must check that each iterated call is equivalent.

In practice, just-in-time compilers tend to accelerate very impressively simple test examples like the above but may not do so well on real code; hence this is one reason why programmers should avoid developing a reliance on a just-in-time compiler.

The main advantage of array programming should be though that it permits expression in the computer language of the idea in the mathematics and the physics, which very often is in terms of collections of values conceived collectively and not as individuals. Thus, thinking in the computer language becomes closer to thinking in the terms of the intrinsic theory of the phenomenon. The features of an array programming language which permit the development of such idioms should therefore be exploited and not avoided, whereas it is primarily the latter avoidance that just-in-time compilation facilitates, encouraging adherence to old-fashioned and obsolete programming paradigms.

A.6 Further Reading

The canonical reference for Octave is Eaton *et al.* (2008); another popular and excellent guide is Quarteroni *et al.* (2010).

Most works dealing with either Octave or MATLAB® e.g. Trefethen (2000) or Higham and Higham (2005), are usable more or less interchangeably with the other program.

References

Eaton, J.W., Bateman, D. and Hauberg, S. (2008) *GNU Octave: A High-Level Interactive Language for Numerical Computations* 3rd edn. Network Theory Ltd, UK.

Higham, D.J. and Higham, N.J. (2005) *MATLAB Guide*, 2nd edn. Philadelphia, PA: Society for Industrial and Applied Mathematics.

McCarthy, J. (1960) Recursive functions of symbolic expressions and their computation by machine, Part I. *CACM* **3**.

Moler, C. and Van Loan, C. (1978) Nineteen dubious ways to compute the exponential of a matrix. *SIAM Rev.* **20**(4).

Quarteroni, A., Saleri, F. and Gervasio, P. (2010) *Scientific Computing with MATLAB and Octave*. Texts in Computational Science and Engineering, 3rd edn. Berlin: Springer.

Trefethen, L.N. (2000) *Spectral Methods in Matlab*. Philadelphia, PA: Society for Industrial and Applied Mathematics.

Glossary

This glossary is for terms which are defined in one part of the book but used in others; terms used only in the vicinity of their definition are not glossed.

adiabatic without heat transfer; flows in which a fluid parcel is deformed rapidly are nearly adiabatic, since the transfer of heat takes time; the adiabatic approximation permits the treatment of compressible flow without a partial differential equation for temperature

adverse (of pressure) increasing in the direction of flow (Section 17.1.1); an adverse pressure gradient retards the flow and if strong enough may lead to *separation*; opposite to *favourable*

aerodynamic centre point along the chord of an aerofoil about which the *pitching* moment is independent of incidence (Section 5.3.1)

aerodynamic force part of the force exerted by air on a body immersed in it which is associated with relative motion, i.e. excluding buoyancy or mean compression (Section 1.1)

aerodynamic pressure sum of the true pressure and the potentials of any conservative external forces such as gravity (Section 2.4.4); using aerodynamic pressure means that the conservative external forces can be ignored

aerofoil *wing section* or *profile*; called *airfoil* in American English

analytic possessing a well-defined derivative; wherever a function of a complex variable is analytic, it corresponds to a plane ideal velocity field (Section 2.6); a function of a complex variable is analytic if it satisfies the *Cauchy–Riemann Equations* (2.22)

anhedral negative *dihedral*; like the wings of most moths at rest, as opposed to butterflies

anticommutative like *commutative* but involving a change of sign; e.g. as in subtraction and the *vector product* (Section 9.3.2)

argument the argument of a complex number is the arctangent of the ratio of the imaginary and real parts; also called *phase*; the argument of a complex coordinate of a point in the plane is its angle from the positive x-axis (Section 2.6)

Theory of Lift: Introductory Computational Aerodynamics in MATLAB®/Octave, First Edition. G. D. McBain.
© 2012 John Wiley & Sons, Ltd. Published 2012 by John Wiley & Sons, Ltd.

the word *argument* is also used to refer to the input to a mathematical or computational function; these senses are not connected

aspect ratio geometric parameter of the planform of a wing, roughly corresponding to the ratio of *span* to *chord* (Section 1.2.2)

barotropic of a fluid in which the pressure is a function of the density; this is a generalization of *incompressibility* which still has many important properties, such as persistance of *circulation* and a well-defined speed of sound; air is barotropic in particular in *adiabatic* processes, which includes subsonic and supersonic flow over wings, to a good approximation

Bernoulli's equation a relation between the pressure and the magnitude of velocity in a parcel of *irrotational* fluid, or along a streamline in a flow of ideal fluid satisfying Euler's equation even if the vorticity is nonzero

Blasius's theorem a relation holding in plane ideal flow between the aerodynamic force on an aerofoil and a complex integral of the square of the complex velocity around it, Equation (3.12)

bound vortex segment of a *horseshoe vortex* conceptually fixed inside a wing

boundary layer momentum equation simplified form of the Navier–Stokes momentum equation applying approximately to the layer along a nonslip boundary of an almost inviscid fluid

Buckingham's Π theorem that a function of n physical quantities involving d independent physical dimensions can be expressed as a function of $n - d$ dimensionless quantities; one of the main results in dimensional analysis (Section 1.5)

camber deviation of the *camber line* of a wing section from its chord

camber line of a wing section, a line or curve halfway between the upper and lower surfaces, curved if section is cambered; the *chord* is sometimes (e.g. for the NACA 4- and 5-digit sections) defined as the line segment joining the ends of the mean; see Figure 1.2

Cauchy's First Law of continuum mechanics that the product of the density and the acceleration of an infinitesimal piece of continuous matter is equal to the difference between the externally imposed forces and the divergence of the internal stress; the continuum-mechanical equivalent of Newton's Second Law of motion

Cauchy–Riemann equations relations equivalent to the *analyticity* of a complex function of a complex variable

centre of pressure the intersection with the chord or *camber line* of the line of action of the resultant of the aerodynamic force (Section 5.3.1)

chord reference line segment characterizing an *aerofoil*, e.g. by joining the ends of the *camber line*, though several slightly different definitions exist (Section 1.2)

circulation the integral of the tangential component of velocity around an circuit, traced clockwise in two dimensions (Section 2.5.3)

collocation procedure for determining an unknown quantity by enforcing a condition (which holds more broadly) at a definite point (Ashley and Landahl 1985); used in finite representations of continuous systems such as the lumped vortex method of Chapter 7 and the panel method of Chapter 8, and not only in spatial discretizations but also for truncated series such as the *lifting line* theory as shown in Figure 12.1; collocation points are also known as 'pivotal' (Falkner 1946) and 'control' points (Van Dorn and DeYoung 1947; Margason and Lamar 1971; Hess 1972; Anderson 2007)

commutative said of a binary operation which is not affected by switching the operands; e.g. for numbers, addition and multiplication are commutative but subtraction (which is *anticommutative*) and division aren't; multiplication is generally not commutative for matrices though addition is

complex coordinate complex number uniquely representing a point in the Cartesian plane (Section 2.6)

complex potential an *analytic* complex function of the *complex coordinate* representing the *potential* and *stream function* of a plane ideal flow; an antiderivative of the *complex velocity*

complex velocity complex number uniquely representing a two-dimensional velocity; the complex velocity is *analytic* if and only if the velocity field is *divergence*-free and *irrotational*

compressible of a substance which responds to an increase of pressure with an increase of density; the relative rate of decrease of density along a particle trajectory is proportional to the *divergence*; effects of compressibility on flow remain negligible if the *Mach number* is small; opposite to *incompressible*

continuity the conservation of mass, which for an *incompressible* fluid means a *divergence*-free *velocity field*

cross product *vector product* of two three-dimensional vectors (Section 9.3.2)

curl vector formed with components from the limiting ratios of the circulations to areas of three infinitesimal plane loops normal to the Cartesian coordinate axes in three dimensions (Section 9.3.3)

dihedral the angle by which each wing of a pair is raised from the *longitudinal–spanwise* plane (Section 1.2.2); like the wings of a butterfly at rest, as opposed to those of most moths; called *anhedral* if negative

deviatoric of a second-order tensor (such as stress), having had an *isotropic* tensor subtracted so as to leave zero *trace*; the deviatoric stress is basis of the rheology of the linearly *viscous* fluid (Section 15.2.2)

difference, forward amount by which a term of a sequence is surpassed by the one following it (Section 6.4); the *backward* difference, which is not required here but is very useful in constructing numerical schemes for unsteady aerodynamics (Moran 2003), is the amount by which a term exceeds the one before it

dimension *dimension* has three related but distinct meanings in aerodynamics

1. the irreducible non-numerical aspect of a physical quantity; quantities of different dimension are incommensurate; it is to this sense that *dimensional analysis* (Section 1.5) or *theory*, or *dimensionless* or *nondimensional* refer
2. the number of independent such dimensions of length that a domain possesses, and thus the number of coordinates required to define a point in it; e.g. a plane is two-dimensional but space is three-dimensional
3. the quantitative extent of a object in one of these dimensions; e.g. the chord of a wing section or the span of a pair of wings; this usage is less formal and not found in this book

dimensionless *dimensionless* never means of zero extent, or relating to a point rather than a line, plane, or space, but rather that the physical statement in question has been expressed in such a way as to be independent of the units used to attach numbers to the quantities in each dimension; physical quantities are generally rendered dimensionless by choosing as unit not an arbitrary standard quantity like metre or second but a quantity of that dimension which is relevant to and characteristic of the problem (Section 1.5)

displacement thickness amount a nonslip surface would have to be thickened into the fluid to explain the reduction in flow-rate it causes due to boundary layer friction; see Equation (16.10)

divergence the net efflux per unit volume of a vector field from an infinitesimal volume; the divergence of the velocity field of an incompressible fluid should be zero everywhere, which is Equation (2.3)

divergence theorem equivalence between the net efflux of a vector field though a closed surface and the integral of the *divergence* over the enclosed volume; see Equation 9.7

dot product *scalar product* of two vectors

doublet an ideal flow *singularity* having zero net outflow and circulation but which entrains fluid in one direction; can be conceived as the coalescing limit of a source–sink pair or of a counterrotating *vortex* pair; see Figure 3.4

downwash vertical component of velocity experienced by a wing of finite span, induced by *trailing vortices*

drag component of *aerodynamic force* parallel to the free-stream (Section 1.1.3); composed of *induced drag* and *profile drag*

drag coefficient drag rendered *dimensionless* by the product of the free-stream *dynamic pressure* and a characteristic area; e.g. the planform area of a wing

drag polar the plane curve with x- and y-coordinates traced out by the lift and *drag coefficients* as the angle of *incidence* is varied

dynamic pressure kinetic energy per unit volume, which according to Bernoulli's Equation (2.14) can be interchanged with pressure; i.e. the sum of the pressure and the dynamic pressure is constant along a streamline for a *perfect fluid* (Section 1.5)

eccentric angle a transformed coordinate running along the chord in *thin aerofoil theory* in two dimensions (Equation 5.3b), or along the *lifting line* in three (Equation 11.11)

effective incidence angle of incidence corrected for the addition of the *downwash* to the apparent free-stream (Section 11.3.1)

elliptic lift loading spanwise variation of lift per unit span which is optimal, according to *lifting line* theory, in that it leads to the least *induced drag* (Figure 11.5)

equation of state relation between the pressure and density of a substance, possibly involving the temperature, and for mixtures the composition ratios; the equation of state supplements the equations of *continuity* and momentum balance which feature density and pressure and velocity and so have more unknowns than equations; the two most important equations of state in aerodynamics are *incompressibility* in which the density is constant (while pressure remains free to be determined by *continuity* and the momentum balance) and the ideal gas equation, which does involve the temperature and therefore requires another equation to close the system, the energy or heat equation

Euler equations coupled nonlinear partial differential equations governing the evolution of the velocity and pressure fields, expressing in the *spatial description* continuity, Equation (2.3), and a force balance, Equation (2.6); compressible fluids further require a relation between pressure and density; for an incompressible fluid, these equations are complete and require only initial and boundary conditions for full specification of a problem

favourable (of pressure) decreasing in the direction of flow (Section 17.1.1); opposite to *adverse*

gradient vector indicating the steepest increase of a scalar function of position (Section 9.3.3); *irrotational* vector fields can be defined as the gradient of a *potential*

horseshoe vortex infinite vortex line consisting of a segment, typically along the *lifting line*, extended indefinitely downstream from both ends (Section 11.2.4)

ideal flow a *velocity field* which is both *divergence*-free and *irrotational* (Section 2.5.4)

incidence angle between the chord of an aerofoil and the free-stream velocity it is immersed in; also called *attack*

incompressible having a density which is constant, in particular not being influenced by variations in aerodynamic pressure; *perfect fluids* are incompressible; air may be treated as incompressible if the *Mach number* is small; opposite to *compressible*

induced drag *drag* resulting from the tilting backwards of lift by the tilting downwards of the free-stream by *downwash* (Section 11.3.2)

induced incidence correction to the angle of incidence due to *downwash*

inner product *scalar product* (Section 9.3.2)

influence coefficient 'effect' of some kind at one place of a unit 'cause' of another kind at a second place; e.g. the velocity induced at a test point by a unit vortex somewhere else, as in Section 7.5

irrotational for a velocity field, having zero vorticity; for a general vector field, having zero *curl*. Irrotational vector fields can be expressed as the gradient of a scalar *potential*. Parcels of irrotational ideal fluid remain irrotational and obey Bernoulli's equation.

isobaric at constant pressure

isochoric at constant volume

isothermal at constant temperature (Section 18.3.4); cf. *adiabatic*

isotropic independent of direction; pressure is an isotropic stress (Section 1.4.2); a *perfect fluid* supports only isotropic stresses (Section 2.1)

Joukowsky's transformation conformal mapping taking the exterior of the unit circle to the entire complex plane slit by a line segment, and other nearby circles to shapes somewhat reminiscent of aerofoils (Section 4.3)

Kármán integral relation form of the *momentum integral equation* in terms of the *displacement* and *momentum thicknesses* (Chapter 16)

Kutta–Joukowsky condition condition that the velocity leave smoothly from the sharp trailing edge of an aerofoil (Section 5.1.2); sometimes known as the 'Kutta condition' (Kuethe and Chow 1998; Anderson 2007) or 'Joukowski's hypothesis' (Glauert 1926; Milne-Thomson 1973).

Kutta–Joukowsky theorem that an aerofoil in plane ideal flow experiences no drag and a lift proportional to the circulation around it; Equations (3.15a) and (3.15b)

lapse rate negative vertical component of temperature gradient in the atmosphere (Section 1.4.4)

lifting line simple model of a wing as a segment of vortex line, leading to quantitative estimates of the dependence of induced drag on aspect ratio (Chapters 11–12)

longitudinal in general, running the direction of the long axis of a body; for a wing section, in the direction of the chord, and of the root-chord for a wing, from leading to trailing edge; typically chosen as the x-axis herein (Section 1.1.2)

Mach number ratio of the speed to the speed of sound; Mach numbers less than and greater than one are called subsonic and supersonic, respectively; in the small Mach number limit, aerodynamic pressure variations do not lead to significant density variations and so the air may be treated as *incompressible*

material description of motion description of flow using tags identifying definite pieces of fluid as the independent variables (Chapter 2); an alternative to the *spatial description*; the laws of motion apply to matter rather than space and so are generally more naturally expressed in material terms; sometimes called the Lagrangean description (misleadingly, accordingly to Truesdell and Rajagopal 2000)

maximum camber greatest perpendicular distance of the *camber line* from the *chord* (Section 1.2.1)

mean line *camber line*

modulus square root of the sum of the squares of the real and imaginary parts of a complex number; also called magnitude and absolute value (Section 2.6); the distance from the origin of the point which a complex number represents; the modulus of the complex velocity is the speed

momentum integral equation result of integrating the momentum equation along a path through the boundary layer from the *nonslip* surface to the ideal free-stream

momentum thickness amount a *nonslip* surface would have to be thickened into the fluid to explain the reduction in momentum-transfer it causes due to boundary layer friction; see Equation (16.11)

nonslip of the interface between a *viscous* fluid and a solid surface: not admitting any discontinuity in tangential component of velocity

one-seventh profile popular approximation for the transverse profile of tangential velocity in a turbulent boundary layer (Section 16.3.3)

perfect fluid a continuous homogeneous substance incapable of supporting shear stress; an inviscid *incompressible* substance; see Section 2.1

pitch rotation about the spanwise axis, taken positive to raise the leading edge, i.e. negative rotation about the z-axis (Section 5.3)

polar form description of a position in the plane in terms of distance from a fixed point and angle (anticlockwise) from a fixed ray emanating from that point; description of velocity in the plane in terms of magnitude and direction; description of a complex number by *modulus* and *argument* (Section 2.6)

potential in general a field from which another field can be derived in such a way that the derived field will automatically satisfy some property; in particular, a scalar field of which the *gradient* is the (necessarily *irrotational*) velocity (Section 3.6.1); in plane ideal flow, the potential is the real part of the *complex potential* (Section 2.7); also, a *conservative* force field has a potential (Section 2.4.4)

pressure coefficient the *aerodynamic pressure*, relative to the far-field, and nondimensionalized by the far-field dynamic pressure; see Equation (2.11)

profile usually *wing section* (Section 1.2); but also the variation of some quantity along a line, e.g. as in the *one-seventh profile*

profile drag that part of the *drag* attributable to the *profile* (i.e. two-dimensional *wing section*) and therefore not attributable to finiteness of the span, i.e. the drag minus the *induced drag*; profile drag is itself the combination of form drag (due to *aerodynamic pressure*, and which vanishes in unseparated ideal flow) and *skin friction*

Reynolds number dimensionless number indicating the relative importance of viscosity in an aerodynamical flow; for complete dynamical similitude, the Reynolds number of the model should match that of the application, although often simply having the Reynolds number large in both cases suffices; while a high Reynolds number corresponds to small viscosity, there will always be a viscous boundary layer along any *nonslip* surface immersed in a viscous flow, so the rules for ignoring viscosity are rather involved

rheology study of the mechanical behaviour of matter, the main goal being to relate the stress to the flow or deformation via a *constitutive law*; in choosing a rheological model, consideration needs to be given to the type of flow as well as simply the composition of the fluid; important rheological models in aerodynamics include the *perfect fluid* and the *incompressible linearly viscous fluid* (Section 15.1)

right-handed system triad of mutually orthogonal directions ordered as are east, north, and up, or backwards, up, and left; any cyclic permutation of a right-handed system is also right-handed; a right-handed rotation is like that of the Earth about its axis (taken as running from the south to the north pole), or that of hands of a clock about its axis directed into its face (Section 1.1.2); ropes are conventionally described as 'S' or 'Z' depending on which upper case letter's central diagonal lines up with the strands of a vertical length—a Z-twist is right-handed

roll rotation about *longitudinal* or *x*-axis (Section 11.4.2)

scalar multiplication multiplication of each element of a vector or matrix by the same number; not to be confused with *scalar product*

scalar product sum of the products of corresponding elements of two vectors of equal length (Section 9.3.2); not to be confused with *scalar multiplication*

scale analysis the rational approach to neglecting terms in an equation based on order-of-magnitude arguments, as in boundary layer theory (Chapter 16)

separation failure of a flow to tangentially follow the surface of an aerofoil or other obstacle. Separation implies a curve on the surface from which a sheet of limiting separating streamlines emanates. The viscous shear stress vanishes at this curve.

shape factor ratio of *displacement thickness* to *momentum thickness* in a viscous boundary layer (Section 16.3.1)

similarity representation in terms of a single coordinate of the variation of a dependent variable in two directions, or one and time (Section 15.6.1)

singularity point (or curve or surface) at which an otherwise smooth function is ill-defined, discontinuous, or not differentiable; by extension, point (etc.) at which velocity field which elsewhere satisfies the governing equations involves a source of mass or *circulation*; important examples in aerodynamics include *sources*, *sinks*, *vortices*, *vortex sheets*, and *doublets*

sink the opposite of a *source* (Section 3.5.4)

skin-friction *viscous* shear stress on a *nonslip* surface

skin-friction coefficient *skin-friction* normalized by the *dynamic pressure*

slip a *perfect fluid* is unable to sustain shear stress and therefore can slip over a solid surface; a *viscous* fluid can't

small-angle approximation representation of the trigonometric functions by the first terms of their *Taylor expansions* about zero

source point, line, or other concentrated region from which fluid issues; typically fictitious and therefore needing to be placed outside the physical domain of flow

span distance between *wing-tips*

spanwise from right to left, in particular between a pair of wing-tips

spatial description of motion description of flow using a system of coordinates positioned relative to space, e.g. fixed relative to the aircraft or the ground, as the independent variables (Chapter 2); an alternative to the *material description*; if the laws of motion can be expressed spatially, they are often more easily solved in that form; sometimes called the Eulerian description (misleadingly, according to Truesdell and Rajagopal 2000)

stagger longitudinal offset of aerofoils, e.g. in a biplane (Section 7.6); also called *décalage*

stagnation condition (e.g. at a point or along a curve, particularly but not necessarily on a surface) of vanishing velocity; if Bernoulli's differential Equation (2.13) holds, this will be a maximum point of the pressure

stall state in which the flow over a lifting surface massively separates, typically because the angle of incidence has become too high, resulting in a severe loss of lift and increase in drag

stationary of a real function of a real variable, at a point, neither increasing nor decreasing, and therefore having zero derivative if differentiable; of an analytic function of a complex variable, as in Equation (4.10), having zero derivative

steady a flow is *steady* in the *spatial description* if the velocity at each point remains constant; moving particles may still experience acceleration in a steady flow

Stokes's theorem that the integral over a surface of the normal component of a curl of a vector field is equal to the integral over its boundary of the tangential component of the vector field (Equation 9.9); Stokes's theorem relates the local differential property of vorticity to the integral property of circulation.

stream function volumetric flow-rate per unit span across a curve between the point in question and a fixed reference point whenever this flow-rate is independent of the curve joining the two points, which holds whenever the velocity is *divergence*-free through the region; the imaginary part of the *complex potential* of a plane ideal flow

strength (of a *singularity*) an integral measure of effect; e.g. the strength of a point *source* is the net rate of fluid issuing from it (Section 3.5.4); for a line or surface source it would be the net rate per unit length or area, respectively; an integral measure might be required if the usual local flow properties such as velocity are unbounded at or along the singularity

substantial derivative substantial is a synonym for *material*; the substantial derivative is the rate of change experienced by a (possibly moving) particle; see Equation (2.4)

substitution vortex early name for the lumped vortex of Chapter 7

Taylor series approximation of a function in the vicinity of a point in terms only of its value and those of its derivatives at the point (Section 9.3.3)

thin aerofoil theory approximation of a wing section by its camber line, ignoring thickness, and assuming the *Kutta–Joukowsky condition* (Chapter 6)

trace sum of the diagonal coefficients of a second order tensor; the *deviatoric* part of a tensor has zero trace; the trace of the deviatoric stress tensor (Equation 15.3) of an incompressible linearly viscous fluid is twice the viscosity times the *divergence*

trailing vortices semi-infinite vortex lines, extending indefinitely downstream from the *bound vortex*; each *horseshoe vortex* contains a pair of parallel but counter-rotating trailing vortices

twist spanwise variation of the angle of the chord

vector product a vector perpendicular to both of two given vectors and forming a *right-handed system* with the first and the projection of the second onto the first; the magnitude is proportional to the area of the parallelogram defined by the two given vectors (Section 9.3.2); the vector product is *anticommutative*

velocity field velocity is fundamentally defined as the rate of change of position of a particle, but a velocity field is the *spatial description*, referring the velocity to the point in space that the particle is passing through (Chapter 2); alternatively, to begin from a spatial description, velocity is the limiting ratio of the volumetric flow-rate through a surface to its area as the area shrinks to zero

viscosity tending to resist relative internal motion; a linearly viscous fluid has a contribution to the stress which is proportional to the rate of strain; the proportionality constant (or constants, though only one is required for *incompressible* linearly viscous fluids) is the coefficient of *viscosity*

vortex a *singular* region of concentrated vorticity or circulation in a flow, being point-like or tube-like in two and three dimensions, respectively; see Figure 3.6 for a plane ideal vortex

vortex sheet *singular* concentration over a surface of vorticity in an otherwise *irrotational* flow; surface inside a flow-field across which the tangential components of velocity are discontinuous (Section 5.1.7)

vorticity *curl* of velocity; local differential quantity corresponding, via Stokes's theorem, to *circulation*; see Equation (2.16) for vorticity in plane flow

wake region of flow downstream of a wing or other obstacle. In idealized theories of aerodynamics, the wake is sometimes pictured as a *vortex sheet*, consisting of *trailing vortices*.

wing section intersection of a wing with a plane parallel to the plane of left–right symmetry (Section 1.2); also called *aerofoil* (or *airfoil*) and *profile*

wing-tip point on the wing furthest from its left–right plane of symmetry, usually one of a left–right pair (Section 1.2.2)

wing-tip vortices pair of bundles of *trailing vortices* emanating from the ends of the *bound vortex* associated with a wing of finite span; curling up of the *vortex sheet* constituting a *wake*

zero-lift incidence attitude in which a wing or wing section generates no lift; the thin-aerofoil theoretic formula is Equation (6.10); the excess of the angle of incidence of this reference value is sometimes called the *absolute incidence* (Kuethe and Chow 1998)

References

Anderson, J.D. (2007) *Fundamentals of Aerodynamics*, 4th edn. New York: McGraw-Hill.

Ashley, H. and Landahl, M. (1985) *Aerodynamics of Wings and Bodies*. New York: Dover.

Falkner, V.M. (1946) The calculation of aerodynamic loading on surfaces of any shape. Reports and Memoranda, 1910. London: Aeronautical Research Committee.

Glauert, H. (1926) *The Elements of Aerofoil and Airscrew Theory*. Cambridge: Cambridge University Press.

Hess, J.L. (1972) Calculation of potential flow about arbitrary three-dimensional lifting bodies. Final Technical Report MDC J5679-01. Long Beach, CA: Douglas Aircraft Company.

Kuethe, A.M. and Chow, C.Y. (1998) *Foundations of Aerodynamics*, 5th edn. New York: John Wiley & Sons, Ltd.

Margason, R.J. and Lamar, J.E. (1971) Vortex-lattice FORTRAN program for estimating subsonic aerodynamic characteristics of complex planforms. Technical Note TN D-6142, NASA.

Milne-Thomson, L.M. (1973) *Theoretical Aerodynamics*, 4th edn. New York: Dover.

Moran, J. (2003) *An Introduction to Theoretical and Computational Aerodynamics*. New York: Dover.

Truesdell, C. and Rajagopal, K.R. (2000) *An Introduction to the Mechanics of Fluids*. Basel: Birkhäuser.

Van Dorn, N.H. and DeYoung, J. (1947) A comparison of three theoretical methods of calculating span load distribution on swept wings. Technical Note TN 1476, NACA.

Nomenclature

A aerodynamic force, Section 1.1

A area, with its normal direction, as a vector; for a closed surface the normal is usually taken as outward by convention

A area, e.g. the cross-sectional area of a stream tube in Equation (10.2)

A with positive integer subscript, one of the coefficients in Glauert's expansion for the chordwise distribution of circulation in thin aerofoil theory in Equation (6.6) or for the spanwise loading in lifting line theory in Equation (11.13)

$\mathcal{R}$ aspect ratio, Section 1.2.2 and Equation (11.16)

a speed of sound, Equation (18.4)

b span, Section 1.2

C with a subscript L or D, a nondimensionalized aerodynamic force coefficient, corresponding to one of its components; with a subscript ℓ or d, the two-dimensional analogue of the same, in a plane perpendicular to the span

c chord, Section 1.2

c_v *isochoric* specific heat, Section 18.3.1

c_p *isobaric* specific heat, Equation (18.8)

D drag force, Section 1.1

d two-dimensional drag force (per unit span), Section 3.7

$\Im$ imaginary part of a complex number, Section 2.6

$\mathbf{i}$ unit vector in x-direction, Section 1.1

i an integer index, e.g. for the ith element of a column vector; for a matrix more often used to indicate the row than the column

i imaginary unit, defined by $i^2 = -1$

Theory of Lift: Introductory Computational Aerodynamics in MATLAB®/Octave, First Edition. G. D. McBain.
© 2012 John Wiley & Sons, Ltd. Published 2012 by John Wiley & Sons, Ltd.

j unit vector in y-direction, Section 1.1

j an integer index, e.g. for the jth element of a column vector; for a matrix more often used to indicate the column than the row, e.g. in the lifting line matrix Equation (12.4), β_{ij} refers to the coefficient in the ith row and jth column

k unit vector in z-direction, Section 9.2.1

L lift force, Section 1.1

ℓ two-dimensional lift force (per unit span), Section 3.7

$\hat{\ell}$ unit vector defining the direction of a rectilinear vortex line, Equation 10.4

Ma Mach number, Equation 1.12

m maximum camber of a wing section, expressed as a fraction of chord, Figure 1.2

m slope of lift–incidence curve, Equation (11.22); thin aerofoil theory predicts $m = 2\pi$ in Equation (6.2.1)

p pressure

Q heat, only ever appearing as δQ, in the First Law of Thermodynamics, Equation 18.6

q relative chordwise position of maximum camber of an wing section, Figure 1.2

q speed, Chapter 2

$\Re$ real part of a complex number, Section 2.6

R gas constant, Equation (18.5)

r distance from the origin, magnitude of a complex number

r one more than the number of collocation points used to solve the lifting line problem in Equation (12.4)

Re *Reynolds number*, Equation (1.11)

T temperature, Section 1.4

t wing section thickness as a fraction of chord, Figure 1.2

u x-component of velocity

u specific internal energy, Section 18.3

V velocity of an aircraft, Section 1.1

v y-component of velocity

W complex potential, Equations 2.26

w complex velocity, Section 2.6

w z-component of velocity in three dimensions

x longitudinal coordinate, positive along the chord in the direction from the leading to the trailing edge, Section 1.1

y upward coordinate, at a right angle to x, but still in the plane of symmetry, Section 1.1

z spanwise coordinate, at right angles to x and y and increasing from right to left, Section 1.1

Greek letters

α geometric angle of incidence, Section 1.3

β angle between (two-dimensional) velocity and positive x-axis, Section 2.6.2

Γ lapse rate in the troposphere, Section 1.4.4

Γ circulation, Equation (2.15)

γ ratio of specific heats of a gas, Section 1.4.1

γ (two-dimensional) vortex sheet strength density, Section 5.2.2

δ small change operator: δx is a small change in x

δ boundary layer thickness, Chapter 16

δ^* boundary layer displacement thickness, Equation 16.10

ϵ a small quantity, Section 3.4.2

ζ two-dimensional vorticity (Equation 2.16), or in three dimensions the z-component of vorticity (Equation (9.13))

ζ a second complex variable, after z, Equation 4.3

η imaginary part of complex variable ζ, Equation (4.3)

η similarity coordinate, Equation (15.10)

Θ divergence, Equation (2.3)

θ polar angle, argument of complex number, Equation (2.20b)

θ momentum thickness of a viscous boundary layer, Equation (16.11)

κ compressibility, Equation (18.2)

λ angle of a panel-segment, Chapters 7, 8

λ the parameter representing pressure gradient in Thwaites's method for laminar boundary layers, Equation (17.3)

λ taper ratio, Chapters 12 and 14

μ coefficient of dynamic viscosity, Section 1.4.2

μ longitudinal dilation factor for the linearized compressible potential Equation (19.2)

ν coefficient of kinematic viscosity, defined by Equation (1.6) and used in the Navier–Stokes Equation (15.4)

ξ real part of complex variable ζ, Equation (4.3)

π ratio of circular circumference to diameter (3.141592653589793...)

ρ density

σ stress, Section 15.1

τ deviatoric (Section 15.2.2) or shear (Section 16.2) stress

$\hat{\tau}$ tangent vector to a curve, Equation 2.15

Φ energy potential of a conservative external force, Section 2.4.4

Φ dilated velocity potential for linearized compressible flow, Section 19.5

ϕ velocity potential; in plane ideal flow, real part of complex velocity potential, Equation (2.27)

χ the argument of the auxiliary complex variable ζ (e.g. Section 4.2.1); hence thin aerofoil theory's eccentric angle coordinate of Equation (5.3b)

ψ stream function; in plane ideal flow, the imaginary part of the complex velocity potential (2.27)

ω influence coefficient, e.g. in a lumped vortex (Section 7.5) or panel (Section 8.1.3) method

$\boldsymbol{\omega}$ vorticity vector, Section 9.5.3

Super- and Subscripts

0   at rest (stagnation)

0   the zeroth term in a expansion such as a power series

∞ in the free-stream, at an infinite distance from the aerofoil

r radial component, defined by Equation (2.25) and used in Equation (3.2a), for example

x component in the x-direction (Section 1.1)

y component in the y-direction (Section 1.1)

α geometric angle of incidence (Section 1.3)

β angle of velocity (Section 2.6.2)

δ with reference to boundary layer thickness (Chapter 16)

ζ auxiliary complex coordinate, Equation (4.3), related to z by a conformal mapping

θ azimuthal component, defined by Equation (2.25) and used in Equation (3.2a), for example

θ in Section 16.3.3, with reference to the momentum thickness of Equation (16.11)

Index

Abbott, I. H., *see also* NACA Report No. 824
  lift, measured, 185
  nonelliptic loading, 182
  perfect fluid, definition of, 25
abs, 49, 55, 57, 71, 73, 134, 188, 210
adiabatic
  compression, 9, 271
  process, 268, 293–4
  speed of sound, 269
aerofoil, 6, 293–4
  Clark Y, 6, 18, 139, 283
  discretization, 129
  Joukowsky, 75, 298
  NACA, 6, 22
    2412, 6, 105, 122, 130, 135
    five-digit, 101, 109, 129, 294
    four-digit, 101, 104, 109, 129–130, 139, 294
  reversed, 81
air
  gas constant, 9, 266
  humidity, 21
  ideal gas, as, 9
  molar mass, 266
  specific heat ratio, 268
  viscosity, 11
analytic, 37–8, 40–41, 47, 62, 293–5, 301,
    see also Cauchy–Riemann equations
angle
  attack, *see* incidence
  eccentric, 80, 91, 95, 105–106, 125, 175, 197,
    297, 308
  small, approximation, 82, 88–9, 122, 194, 300
angle, 65, 70, 121, 131, 134
anhedral, 8, 293, 295, *see also* dihedral

anticlockwise, 26, 71, 82
  aerofoil discretization, 129
  plane polar angle, 299
anticommutative, 146, 293, 295, 302
approximation, *see* angle: small; boundary layer;
    compressibility: linearized; profile:
    boundary layer; scale analysis; Taylor
    Series; thin aerofoil theory
arg, *see* **angle**
argument
  complex, 36, 40, 70, 293, 299, 307–308, *see*
    also **angle**
  function, 206, 219, 294
  symmetry, 80, 101, 138, 164, 176–7, 202
aspect ratio, 7, 143, 182, 188, 191, 294
atmosphere, 11, 298
  International Standard, 12, 20–22, 270
atmosphere, 13, 21, 291
AVL, 221

barotropic, 264–6, 294
Bernoulli
  equation, 33, 294, 301
    boundary layer, 239, 241, 251–2
    compressibility, 263, 265, 269, 282
    three-dimensional, 154
    trailing edge, 170
    vortex core, in, 173
Bertin, J. J.
  lifting line calculation, 188
  thin aerofoil calculation, 105
  vortex lattice method, 218, 221
Biot–Savart law, 161, 163, 173
biplane, 123, 301

Theory of Lift: Introductory Computational Aerodynamics in MATLAB®/Octave, First Edition. G. D. McBain.
© 2012 John Wiley & Sons, Ltd. Published 2012 by John Wiley & Sons, Ltd.

span, 5–8, 14–15, 18, 143, 169, 294, 301
spanwise, 144, 170–5, 185–90, 295, 301
spatial description, 25, 297–8, 301–302
specific heat
  isobaric, 9, 267
  isochoric, 9, 266–7
  ratio, 9, 268, 271
speed_of_sound, 10, 13, 288
squeeze, 217–19
Stack, J., 20, 283
stagger, 123, 301
stagnation, 65, 77–8, 80–83, 138, 140, 158,
    257–8, 268, 301
  temperature, 270
stall, 81, 174, 225, 301
state
  equation of, 297
    rheological, 227, *see also* constitutive law
    thermal, 9, 263–6, *see also* gas: ideal
Stokes, *see also* Navier–Stokes equation
  First Problem, 235, *see also* plate: suddenly
    sliding
  theorem, 148, 301
    vortex tubes, 159
    vorticity and circulation, 152–4, 169, 302
stratosphere, 12–13, 21, 270, 291
stream filament, 158
stream tube, 158
streamline, 157
  Bernoulli's equation, 35
  plotting, 42, 50
stream function, 42, 57–62, 295, 301
strength
  lumped vortex, 112
  panel, 127
  singularity, 301
  source, 61
  stream filament, 158
  vortex, 61, 85, 159, 162, 281
    horseshoe, 193, 204, 212, 218
stress, 3, 10, 26, 225–8, 300, 308
  shearing, 25
Stringfellow, John, 3, 22
subsonic, 265, 285, 294, 298
substantial derivative, 27–8, 44, 149–51, 301
suction
  above wing, 32, 266, 283
  leading edge, 82, 84, 91
sum, 59, 122, 134–7, 217–19
supersonic, 264, 266, 274, 277, 285, 294, 298

Sutherland law, 11, 22
sweep, 7–8, 209–21, 284–5
symmetry
  axial, 285
  spanwise, 4, 7, 176–7, 189, 196, 302
    vortex lattice method, in, 215–18
  tensor, 227
system
  coordinate, 4, 25, 42, 225, 275
  linear, 185–6, 199, 216
    solution, 16–17, 115, 187
  right-handed, 4, 145, 148, 162, 300, 302
  thermodynamic, 267
  units, *see* units

taper, 7, 182, 188, 209–16
Taylor
  series, 9, 28, 147, 289, 300–301
temperature, 12, 298, 306, *see also* lapse rate
  absolute, 10–11, *see also* atmosphere
  advected quantity, as an example of an, 27
  partial differential equation, 293
  specific heat ratio, effect on, 268
  speed of sound, effect on, 9–10
  viscosity, effect on, 11, 22
tensor, 225, 295
thermodynamics, 9, 266
thickness
  boundary layer, 233, 245
    displacement, 243–5, 296, 298, 300
    momentum, 243–5, 298–9, 300, 307
  momentum, 298–9, 307
  wing, 129–30, 139, 221, 306
    ignored, 87, 93, 169, 302
thin, 107–108, 290
thin aerofoil theory, 87–90, 93, 99, 297, 301
  lumped vortex method comparison, 122
three-quarter chord, 112, 120–25, 200, 202
Thwaites
  momentum integral equation, 247, 255, 259
  pressure gradient parameter, 254
toeplitz, 198–9, 206
topology (streamline), 158
TORNADO, 221
trace
  camber line, 210
  loop-integral, 26–7, 294
    three-dimensional, 150
  streamline, 157
  tensor, 227, 295, 302

transpose (matrix), 108, 211
triad, right-handed, 162, 300, *see also* system: right-handed
tropopause, 12
troposphere, 12–13, 291, 307
turbulence
  boundary layer, 244, 247, 299
  transition, 260
twist, 8, 302
  aerodynamic, 210
  geometric, 209–10

UIUC
  Airfoil Coordinates Database, 129, 221
units
  Imperial, 16
  SI, 10–11, 16, 270

vector
  column, 108, 214, 290, 305
  component, 39
  Euler equation, 29
  hydrostatic force, 30
  identity, 145–8
  momentum, rate of change, 28
  normal, 148, 215
  notation, 3, 145–8
  row, 108, 131, 209, 290
  unit, 25, 44, 54, 64, 145, 160, 170, 203, 305–306
    normal, 26
    tangent, 34
  velocity, 25
vectorization, 287–91
velocity
  complex, 38, 79, 295
    aerodynamic force, 62–4
    drawing with Octave, 49
    integer power, 48, 50, 64
    rotated, 53–4
  field, 25
Venturi, 265
viscosity, 10–11, 26, 225, 299, 302
  kinematic, 11–12, 228, 239, 248, 307
`viscosity`, 11, 13
`vlm_demo_bertin`, 218
`vlm_demo_bertin2`, 219
`vlm_demo_campbell`, 216, 290

Von Doenhoff, A. E., *see* Abbott, I. H.
vortex
  bound, 170, 193, 213, 294
  counter-rotating, 296, 302
  horseshoe, 170, 294, 297
    Octave, 165, 197
  lattice method, 212, 218
  line, 159, 164
    segment, 161, 165, 213–15, 298
  plane ideal, 54
  sheet, 86, 93, 300, 302–303, 307
    thin aerofoil as, 82, 85, 87, 93, 98, 113
    wake as, 170
  substitution, *see* vortex, lumped
  trailing, 170, 173, 302
    numerical, 165
  tube, 159–60
  wing-tip, 173, 302
`vortex_horseshoe`, 204–205, 217–18
`vortex_segment`, 203–204
`vortex_semiinf`, 204
vorticity, 34, 40–43, 48, 59, 62, 152, 294, 302, 308

wake, 170, 173, 302
Ward, K. E., *see* NACA Report No. 460
water
  compressibility, 270
  vapour, 21
  viscosity, 20, 229
`'Waypoints'` (**quadgk**), 65
wind tunnel, 3, 22, 105, 144, 279
wing
  delta, 7
  planform, 7, 182, 296
    elliptic, 182
    rectangular, 187, 191
  section, 5–7, 14, 31–2, 293, 299
wing-tip, 7, 143, 176, 196, 301–302
  vortex, *see* vortex: wing-tip
work (thermodynamic), 9, 266–7

yaw
  horseshoe vortex, of a, 205, 216, 221

zero-lift incidence, *see* incidence: zero-lift
zeros, 131, 191, 198–9, 205, 288, 290